ALSO BY CONRAD RANZAN

Guide to the Construction of the Natural Universe

The Nature of Gravitational Collapse (*How the photon, the particle of light, is responsible for mass, gravity, and superneutron stars*)

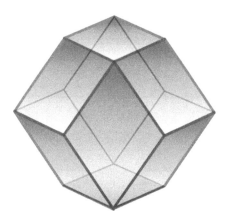

Unifiers are people whose driving passion is to find general principles which will explain everything. They are happy if they can leave the universe looking a little simpler than they found it. Diversifiers are people whose passion is to explore the details. They are in love with the heterogeneity of nature ...
–Freeman J. Dyson, *Infinite in All Directions*

This book was written by a unifier.

Laws of Physics Twentieth-Century Scientists Overlooked

Conrad Ranzan

DSSU Research
5145 Second Avenue, Niagara Falls, Canada L2E 4J8

Laws of Physics Twentieth-Century Scientists Overlooked

Copyright © 2022 by Conrad Ranzan,

All rights reserved, including the right of reproduction in whole or in part in any form. Permission to quote passages is granted in connection with research or reviews written or prepared for inclusion in any print publication, electronic post, or audio-visual broadcast; provided that the source passages are properly accredited.

For additional copies of this book visit **www.CellularUniverse.org** where retailer/distributor information is posted.

Cover image: Beams of truth penetrating the infinite Cosmos, emanating from a stylized Mainspring object (a Terminal neutron star). The truths facilitate cognition of our Universe. The Mainspring objects sustain its perpetual steady-state existence.

Ranzan, Conrad, 1945-
    Laws of Physics Twentieth-Century Scientists Overlooked / by Conrad Ranzan

    Includes references and index.
    ISBN #978-0-9938823-1-9   (Softcover)
    1. Cosmology.   2. Science—Physics   3. Science—Philosophy

    BISAC Codes: SCI055000   SCIENCE / Physics / General
                       SCI015000   SCIENCE / Space Science / Cosmology
                       SCI024000   SCIENCE / Energy

    Copyright © 2022 by Conrad Ranzan. All rights reserved.

    www.CellularUniverse.org
    DSSU Research, 5145 Second Avenue, Niagara Falls, Canada L2E 4J8

    Title: Laws of Physics Twentieth-Century Scientists Overlooked

Momentous ideas

We have a natural tendency to romanticize breakthrough innovations, imagining momentous ideas transcending their surroundings, a gifted mind somehow seeing over the detritus of old ideas and ossified tradition. But ideas are works of bricolage; they're built out of that detritus. We take the ideas we've inherited or that we've stumbled across, and we jigger them together into some new shape.

– Steven Johnson, *Where Good Ideas Come From, The Natural History of Innovation* (2010)

Yes indeed, some of us take old discarded concepts and bring them together into some new and totally unexpected shape. –C.R.

Laws of Physics Twentieth-Century Scientists Overlooked

CONTENTS

Prologue 7
Something that they all missed.. 7
Overviews and highlights of the various overlooked Laws 8
The question of validity... 13
The obvious question... 13
Terminology 20
Words of warning 21

1. Cosmic Redshift and the Velocity Differential Propagation of Light... 23
What is the Cosmic Redshift? 23
The search for the cause of the Cosmic Redshift 24
Overlooked aspect of light propagation 30
Proof based on gravity as a force-effect 32
Proof based on a dynamic space-medium 34
Some relevant aspects.. 38
Evidence... 41
What do the experts believe?.. 43
Discussion of validity... 44
Revisiting the unavoidable question.. 45
Momentous misinterpretation 48

First Interlude 51
Reviewers Confronted with a New Exciting Theory................................ 51
Proof using Cartesian coordinate system 55
Velocity differential effect holds the key to an amazing positional effect. 57

2. Energy Generation via Velocity Differential Blueshift 59
Gravity well and aether flow profile 61
Process of Blueshift Accrual –the proof.. 63
Proof based on gravity as a force effect 66
Environment most favorable for Blueshifting-mode propagation............ 68
The Terminal neutron star and the Amplification process...................... 70
Rotating Terminal neutron star 73
The energy Amplification process as the driver of astrophysical Jets........ 77
Energy Amplification process as the source of ultra energy particles...... 78
The energy Amplification process provides a bonus feature 80
Summary... 81

Second Interlude 83
Scholarly and erudite 83
Acceptance turns into rejection! 84
Strange reasons for rejecting the Law of Energy Regeneration/Amplification 85

3. Noninteraction Mass-to-Energy Conversion 91
Introduction 91
Conventional motion versus "stationary" motion 92
The energy triangle 95
Energy triangle applied to Scenario 1 97
Energy triangle applied to Scenario 2 100
Noninteraction mass-to-energy conversion 101
Mass-to-energy conversion mechanism 105
Collisional mass-to-energy conversion 111
Some relevant aspects of mass-to-energy conversion 114
Implications 115

Third Interlude 119
The Unhinged Theoretical Physicist 119

4. Mass Extinction/Vanishment 131
Introduction 131
Extreme gravitational collapse 132
Aether Deprivation Annihilation 140
Mass Extinction in its supreme manifestation 143
Density challenge, black holes, rotation, Higgs 145
Summary comments 147

Fourth Interlude 151
The Key to Decoding the Universe 151
Selected issues raised during peer review 154
Stephen Hawking on black holes 155

5. Centrifugal Effect Curtailment and Negation 157
The basic centrifugal effect 157
How special relativity can restrict rotation 159
Centrifugal effect curtailment 162
Total cancellation of centrifugal effect 164
Summary discussion 172
Implications 175

Fifth Interlude 179
Historical perspectives on the nature of rotation 179
Strange world of science journals 181
Invoking the centrifugal effect to support a mythical cosmology 184

6. Energy Conservation of the Universe 187
Two questions 187

Energy defined at the most fundamental level 189
Energy classification 191
How the photon unifies all fundamental Negative Energy 195
Aether unifies Positive and Negative Energy 199
Cosmic-scale energy conservation 201
Summary and implications 205

Sixth Interlude 209

Mytho-Science "Wisdom" 209
Reactions to the original research paper from several journals 209
Mytho-science non-conservation views 212

7. Law of Cosmic Cellular Structure 215

First Determiner of cosmic-scale structure: the empirical proposition, EXPANSION 216
Second Determiner: matter regeneration 219
Third Determiner: matter gain is balanced by matter loss 224
Fourth Determiner: gravitation based on aether 226
Prediction of a natural cosmic pattern 227
Structure prediction matches observational evidence 230
Thoughts on cells non-expanding, order eternal, aether uniqueness, reason triumphant 240

Seventh Interlude 247

Experts, Obstructers, and Fraudsters 247
Orthodox experts say orthodox cosmology does not match reality 247
Journals not interested in evidence-validated cosmology 250
Obstructers 256

8. Aether Theory of Unified Gravity 259

Background 260
Cause of mass property and its connection to Primary gravity 263
Primary convergent effect 266
Secondary convergent effect 268
Divergent gravity, the antigravity effect 274
Vorticular effects 281
Gravity waves 292
Summary of stresses, the unifying factor, and theory subsummation 295

Epilogue 301

Extraordinary Connectedness of the Grand Design 301

Appendices 305

Appendix A: Basic aether flow velocity equation 305
Appendix B: Finding the radius of the Terminal structure 307
Appendix C: Quantifying the self-dissipation of aether (the cause of the secondary gravitation effect) 308
Appendix D: Central mass expressed in terms of its companion's orbital radius and period 310

Glossary 312

Index 328

> *When the parts of nature are considered according to their design and a discovered plan, there emerge certain properties in it which are otherwise overlooked and which remain concealed when observation is scattered without guidance over all sorts of objects.*
>
> –Immanuel Kant, *Universal Natural History and Theory of the Heavens* (1755)

In other words, Nature's grand design to be properly perceived requires the interpretive guidance of a well-constructed theory. –C.R.

Rewards of science

Science is the art of our time. Science has several rewards, but the greatest is that it is the most interesting, difficult, pitiless, exhilarating, and beautiful pursuit that we have yet found.
–Horace Freeland Judson, *The Great Betrayal, Fraud in Science* (2004)

Surprises in science

In science, surprises come very often. In each decade over the past few centuries there have been several unexpected discoveries that forced everyone to change their ideas about the way nature behaves.
–Sheldon L. Glashow, *Interactions*

It seems like an odd and special property of the universe that it does seem to like the simple possibilities. ...
–Astronomer Edwin Turner (1995)

When a light brightens and illuminates a corner of a room, it adds to the general illumination of the entire room. Over and over again, scientific discoveries have provided answers to problems that had no apparent connection with the phenomena that gave rise to the discovery.
–Isaac Asimov, *Atom, Journey Across the Subatomic Cosmos*

Tradition in science

Ignoring great insights is a venerable tradition in cosmology and indeed in science more generally.
–Max Tegmark, *Our Mathematical Universe ...*

... the prevailing scientific beliefs of any era have always been confidently professed, widely accepted, actively supported and staunchly defended —including those now known to be false.
–Mark McCutcheon, *The Final Theory, Rethinking Our Scientific Legacy*

Advancement in science

Every great advance in natural knowledge has involved the absolute rejection of authority.
–Thomas Henry Huxley (British scientist and humanist, defender of Darwinism, 1825-1895)

*Throughout history, opposing views, vigorous debate and openness to new ideas have been the bedrock of scientific progress. Any **major advance in science** has been arrived at by practitioners vigorously questioning "official" narratives and following a different path in the pursuit of truth.*
–Declaration of Canadian Physicians for Science and Truth. Issued in 2021 by a diverse group of Canadian Physicians in response to threats and censorship. (www.CanadianPhysicians.org)

Hope in science

Steven Weinberg expressed a hopeful attitude about the unification of the forces of nature and our efforts to decipher the ultimate laws of the universe using new theories. "I don't know if the human race is intelligent enough to decipher the laws of physics," *he said*, "but I hope that it is."
–per interview, Amir D. Aczel, *Present at the Creation*

Incompleteness in science

[F]or all its magnificent achievements, modern science remains a flickering candle surrounded by inky mystery. With refreshing candor, Roger Penrose forcefully underscores the extent of our ignorance about the origin —and indeed the very nature— of the physical laws that govern the operation of our magnificent universe. What the enterprise of science has compiled to date, in Penrose's view, is a decidedly **incomplete** guide to the laws of nature. –James Gardner, *Intelligent Universe*

Goal in science

"My goal is simple," Stephen Hawking once said. "It is a complete understanding of the universe, why it is as it is and why it exists at all." ... ***That***, *in a nutshell, is the scientific manifesto.* –Gilead Amit, *New Scientist* (8 August 2015)

[I]t is clear ... our road to understanding the nature of the real world is still a long way from its goal. Perhaps this goal will never be reached, or perhaps there will eventually emerge some ultimate theory, in terms of which what we call "reality" can in principle be understood. If so, the nature of that theory must differ enormously from what we have seen in physical theories so far.
–Roger Penrose, *The Road to Reality: A Complete Guide to the Laws of the Universe*

✠ ✠ ✠

Prologue

> *"It is quite likely that the 21st century will reveal even more wonderful insights than those that we have been blessed with in the 20th. But for this to happen, we shall need powerful new ideas, which will take us in directions significantly different from those currently being pursued. Perhaps what we mainly need is some subtle change in perspective*
> *—something that we all have missed."*
> –Roger Penrose, *The Road to Reality: A Complete Guide to the Laws of the Universe*; from the concluding passage of the book

Something that they all missed

This book is about the processes —the laws of nature— that sustain the existence of the Universe and define its nature and manifest its phenomena. It is primarily a journey into cosmology. But let me immediately declare, it is not a speculative exploration through some jungle of wild hypotheses and outrageous extrapolations. No exploding universe, no higher dimensions, no multiverses, no cosmos replications, no runaway inflation. Be assured, all the laws and ideas presented in this book have been peer reviewed by practicing physicists. It may be broadly stated that they reported no flaws in the logic of the arguments, no errors in mathematical proofs, and no violation of the basic laws of physics.

The book is a presentation of a *validated* theory of The Universe. And giving it a unique and lively thematic focus is this: The book is about what has long been suspected (not only by Roger Penrose but more than a few others): **It is about something that they all missed**.

The "something" they missed was crucial to the way the Universe works. As a supremely-important factor, it underpins several overlooked laws of physics.

Consequently, what cosmologists and astrophysicists need is not something subtle, not "some subtle change in perspective" as Roger Penrose stated in another one of his books *The Road to Reality*; but rather, a radical reassessment beginning with a recognition and proper appreciation of the various laws of physics that were actually missed.

The focus of considerable researched over the last two decades, these Natural Principles that rule the Universe have all been recently published, after having undergone a peer review process, and now form part of the scientific literature. Each Law is summarized below. Each is accompanied by highlights that underscore significant features and implications —all things that will be detailed in the chapters to follow (chapters correspondingly titled).

Overviews and highlights of the various overlooked Laws

1. Cosmic Redshift and the Velocity Differential Propagation of Light

The velocity-differential mechanism of cosmic redshift (discovery in 2013) is one of the consequences of inhomogeneous aether flow: The combination of (i) the fact that aether is the conducting medium of light and (ii) the fact that aether is not static but is involved in a dynamic flow, in accordance with the aether theory of gravity, leads directly to a new mechanism of *cosmic redshift*. This law is supported by the proof that contraction of aether *can* cause spectral redshifting. What this means is that lightwaves stretch not only in expanding "space," as has long been known, *but they also stretch in inhomogeneously contracting "space."* [1]

Highlights:

• A surprisingly simple law governing the prolonged interaction between electromagnetic radiation and gravity gradients.

• Light acquires a redshift while crossing a gravity well; it does so while entering the well and also while emerging from the well. This basic phenomenon, completely missed by 20th-century scientists, is supported by a velocity-differential proof AND a force-differential proof. One proof is based on *aether-flow differential*, the other is based on a force-like differential (technically called a gravity gradient).

• The historic oversight of scientists to recognize this cause of redshift led to a momentous misinterpretation of Edwin Hubble's redshift observations. The redshifted light from distant galaxies was misinterpreted as the evidence for the recession of those galaxies. Their oversight of a straightforward principle connecting light and gravity had revolutionary consequences —it led to a bizarre new vision of the universe. The misinterpretation, of the cosmic redshift, of Hubble's galaxy measurements, gave birth to the exploding universe paradigm!

• This Law explains exactly why the big bang hypothesis has no scientific foundation.

• The implications for cosmology, needless to say, are profound.

---

[1] C. Ranzan, *Cosmic Redshift in the Nonexpanding Cellular Universe: Velocity-Differential Theory of Cosmic Redshift*, American Journal of Astronomy & Astrophysics Vol.**2**, No.5, 2014, pp.47-60. (Doi: http://dx.doi.org/10.11648/j.ajaa.20140205.11)

2. Energy Generation via Velocity Differential Blueshift

We have here a truly revolutionary mechanism —Nature's *fundamental energy amplification process*. According to this law, the **Universe's source energy** is identified as the Blueshifting of photons and neutrinos compelled to propagate in-place. This law specifies the origin (the source process) of Nature's limitless font of energy. The process occurs only on the surface of what are descriptively termed *end-state neutron stars* —Nature's most extreme contiguous mass objects.

Highlights:

• As a bonus, it explains the mechanism that powers astrophysical jets (those beams of ejecta associated with extreme mass concentrations). Remarkably, aether is the driver in the extraction of energy from the inner side of the critical boundary of gravitationally collapsed stars known as Terminal stars or, equivalently, as *end-state neutron stars*.

• The process represents the ultimate source of ALL the radiation energy of the Universe. (This specifically refers to electromagnetic radiation and neutrinos.)

• The process is both conceptually unambiguous and scientifically unprecedented; its presentation, therefore, can benefit both the casual science enthusiast and the specialist.

• An amazing aspect of this process (of potentially limitless amplification of energy) is that the energy gain is not derived from the composition of aether itself; but rather from the characteristic motion that aether undergoes.

• Remarkably, astrophysical jets are powered by the extraction of energy from the *inner* side of what is popularly called the "event horizon" of gravitationally collapsed stars. Such energy transfer occurs even in the absence of rotation.

• The mechanism provides the resolution to the long-standing mystery of the source of ultra-high energy particles, as well as extreme bursts of emission.

• An unexpected but powerful implication associated with of the Principle of propagational blueshifting of radiation is that *an energy-particle creation process is not a necessary feature of the Universe!*

3. Noninteraction Mass-to-Energy Conversion

According to this principle of nature, mass is able to undergo a complete 100-percent conversion to radiant energy without involving any sort of particle-antiparticle annihilation! ... There are two manifestations of this: In the one case, the conversion involves no interaction whatsoever. In the other, the conversion does involve collision, but again, with a total transformation of mass (and its kinetic energy) to radiant energy —this latter being trapped radiant energy. ... The Law applies during and after extreme gravitational collapse. That is, it plays a key role in the formation and subsequent functioning of *Terminal stars* and *end-state neutron stars*. This principle governs the very mechanism by which gravitationally collapsed bodies first produce and subsequently sustain their unique energy surfaces.

Highlights:

• The mechanism, although new to physics, actually requires little more than basic physics and a dynamic space medium.

• This new pathway for the conversion of mass into energy leads to the resolution of long-standing paradoxes associated with the traditional view of stellar black holes.

• This Principle, along with one other, functions in a way that strictly regulates the size stability of TOTALLY collapsed stars. The other is what the 4th overlooked Law is all about. Together, the two Laws prevent the formation of singularity-type black holes.

• Represents much of the energy input for the mechanism that actually drives the astrophysical and cosmic jets associated with *Terminal stars* and *end-state neutron stars*.

4. Mass Extinction by Aether Deprivation

This Law is as straightforward as it can be: Mass, when it is deprived of aether, vanishes —it literally ceases to exist. It holds the extraordinary explanation of how Nature annihilates mass; but not in the sense of ordinary destruction, rather, this is annihilation in the irreversible terminal sense.

Highlights:

• The Principle addresses the issue of excess matter. Given that a contiguous mass of less than 3.4 solar masses is insufficient to form a lightspeed boundary (analogous to what surrounds a black hole of general relativity theory); and given that this same mass is just sufficient to form such a barrier; the question then is *What happens if the collapsing body is greater than 3.4 solar masses? Significantly greater?* And, say, none of the material is outwardly expelled. … Twentieth-century scientists never did find the answer.

• Nature, it turns out, has a maximum density state and there exists a mechanism to prevent the density from going higher. A simple rule, a basic process, assures that it is never exceeded —regardless of the quantity of additional mass.

• The extinction of mass by the process of *aether deprivation* logically follows from the fact that the existence of mass is sustained by the continuous absorption of aether. Cut off the supply of aether and the mass vanishes.

• This is the key process by which TOTALLY collapsed stars are able to maintain strict size stability.

• Although the star's excess mass actually disappears, remarkably, there is a way to stay within the important law of matter conservation as applied on the cosmic scale!

5. Centrifugal Effect Negation

The fundamental law governing circular motion. This Law is most relevant to compact objects of extreme gravity. It governs the attenuation and even complete cancellation of the *centrifugal effect* associated with rotational motion. It explains how rotational motion *with respect to the universal space medium* (aether) actually determines the potency of the centrifugal effect.

Highlights:

• The Law deals with the outwardly directed pseudo-force primarily as it affects compact objects of extreme gravity. The chapter devoted to this topic very clearly shows how and why the Effect can actually become completely impotent.

• As a fundamental law governing circular motion, it also plays an important role in maintaining the structural cohesion of spiral galaxies *without the need for so-called dark matter.*

• The proof can be most easily demonstrated.

• The implications are startling and profound.

6. The Law of Energy Conservation of the Universe

The ***cosmic-scale conservation of energy*** is maintained by the balancing tendency of certain relevant processes of two distinct realms —*the physical and the sub-physical.* There exists a self-regulating balance within the physical realm; and there exists a self-regulating balance within the nonphysical realm. The amazing aspect —the underlying synthesizing factor— is that they totally depend on each other.

Highlights:

• Overturns the mistaken conclusion of 20th-century scientists that the conservation-of-energy principle does not apply to the Universe as a whole.

• This conservation mechanism made it possible to construct the first true scientific theory of the Universe —one that complies with the paramount law of nature.

• Yet no radical physics was required, rather, the conservation-of-energy mechanism emerged as a logical extension of the highly-successful *unified theory of gravity* based on a very specific type of aether.

• This Law holds the secret of how cosmic-scale energy conservation is maintained within the physical realm *and* within the sub-physical realm; and how physical existence is entirely dependent on the aether medium.

• Arrives at the profound conclusion that we live in a two-particle Universe.

7. Law of Cosmic Cellular Structure

This one involves a basic proposition (*expansion*), accepted by virtually all theorists, plus a combination of several other overlooked Laws, including (i)

"Cosmic Redshift and the velocity differential propagation of light," (ii) "Energy generation via velocity differential blueshift," and (iii) "Mass extinction by aether deprivation." Together they manifest the dynamic system that sustains cosmic structure —specifically, the universe's *dynamic steady-state cellular structure*.

Highlights:
- A combination of simple processes (the various laws) predict very specific cosmic structure patterns.
- This system of processes, as a mechanism, drives the Universe. This system of processes, as a testable theory, reveals an astonishingly accurate match between model-specific prediction and physical evidence —a wonderful correspondence between expectation and the actual observational cosmic-scale structure.
- From the nature of the processes, it logically follows that cosmic structure is sustained by a *perpetual mechanism* —a timeless steady state system.
- The profound conclusion is that the patterns of the Universe's cosmic-scale structure are not phenomenological, as had long been believed, but are inherent. Cosmic structure exists by virtue of a perpetual self-sustained mechanism. ... We live in a steady state universe.

8. Law of Mass Property Acquisition & the Aether Theory of Gravity

The simple Law of Property-of-Mass Acquisition actually leads to a remarkable unification of gravity.

The mass attribute and gravity unification are based on that "something that they all missed." Both mechanisms are based on a very specific kind of dynamic aether —the aether that is foundational to the cosmology presented herein. The **Aether Theory of Gravity** that will be detailed is a remarkably elegant unification of the five manifestations of gravity. Although it is presented as the last of the overlooked laws, this aether-gravity theory serves as the foundation for all the others, and also as the framework tying the collection together into a comprehensive and validated cosmology.

Nature uses the universal space medium to produce the gravity effect. What had eluded scientists was the exact combination of properties possessed by the gravity-producing space medium.

The Gravity Unification Law simply specifies the essential characteristics of the aether, then spells out the logical consequences —the observable effects.

Highlights:
- The Mass property acquisition mechanism is far superior to the Higgs hypothesis of the academics.
- A perfectly natural mechanism, a revolutionary unification of gravity

theory, a synthesis of the primary cause and the four manifestations of gravity. All that was needed was the one key Factor that is common to gravity's convergent, divergent, vortex, and wave effects.
- The first-ever theory of gravity to pass *the ultimate test of validity* by predicting the key identifying patterns of our Universe —by predicting what astronomers have observed.
- It changes our understanding of how rotation, particularly grand-scale galactic rotation, actually *increases* the potency of gravitational attraction.
- In fact, the potency of gravitational attraction, in the case of high-mass compact bodies, can become truly extreme —bringing about a reduction in the centrifugal pseudo-force. In the maximally extreme situation there is a total cancellation of the centrifugal effect.
- Leads to a remarkable new development in the theoretical nature of gravity waves. This Law clearly explains why the gravity waves being detected are not energy waves. What astronomers in this century have detected were not gravitational *energy* waves but rather subquantum-level gravity waves.

There is a dual purpose in examining the above overlooked elements of the natural world. One is to expose the adverse consequences of the failure of 20th-century science to uncover the essential characteristics of the universal ethereal medium and to incorporate the medium into mainstream physics. The most glaring consequences are the ongoing crises in Physics and Cosmology. The main purpose, however, is to present, under correspondingly-titled chapters, a substantiation and confirmation of *DSSU Cosmology* —historically the first true steady-state universe.

The question of validity

Granted, the successfully passage through the peer-review process gives a certain assurance that the material is original and has no obvious errors, contradictions, and omissions. But how do we know the Overlooked Laws are not just some cleverly constructed conjectures and hypotheses? What certitude is there for the validity of these Laws?

Within the framework of a naturally cellular universe, the Laws are glaringly self-evident. All are supported by sound easy-to-express proofs. All make testable predictions. All share a degree of interconnectedness. Practically all are supported by incontrovertible evidence. None suffer from contradictory evidence.

The obvious question

Given this situation of the Laws governing the Universe being so seemingly obvious, and readily provable, and well-supported by observational evidence, how in the world did so many great minds of the 20th century miss them? It borders on the inconceivable! I've been asking myself this sort of question since the year 2001.

An initial response to the question would be to blame the confusion over the historical Michelson-Morley experiment conducted in 1887. In that year, Albert Michelson and Edward Morley, working in their lab at the Case Institute of Technology in Cleveland, measured the velocity of the aether wind to be between 5 and 7.5 kilometers per second. But, because the speed was considerably less than what had been expected, their finding became popularly reported as *a null result.*[2]

The confusion resulted in the rejection of a mechanical aether and the eventual adoption of a mathematical —or geometrized— space medium. But it's not quite that simple.

The aether certainly had lost its appeal and status. However, as the 20th century progressed, all the characteristics that had been rightly or wrongly attributed to the aether and rejected became incorporated into a new space medium. Generically, it was called the *vacuum.* Einstein called it the *spacetime continuum* and gave it the ability to somehow carry or convey the energy of gravity.

At first the vacuum had but one property, namely, that of wave propagation. "But with the development of quantum electrodynamics," science historian Sir Edmund Whittaker tells us, "the vacuum has come to be regarded as the seat of the 'zero-point' fluctuations of electric charge and current, and of a 'polarization' corresponding to a dielectric constant different from unity." Whittaker adds, "It seems absurd to retain the name 'vacuum' for an entity so rich in physical properties, and the historical word 'aether' may fitly be retained."[3] By the end of the century, this heavily overburdened and supposedly "vacuous" medium became known as *the quantum foam.*

Evidently the theorists had thoroughly confused themselves.

• They claimed, without any hint of irony, that light does not require a medium to effect its own propagation. But if space is not nothingness, then light *must* be propagating through some sort of medium. (This only complicated matters for the deniers. They still had to devise a way for light to travel *through* the space medium, the vacuum or the quantum foam, *without*

---

[2] Michelson and Morley had measured the shifts in the interference pattern produced by two perpendicular light beams. Those fringe-shifts were between 0.004 and 0.008 of a full wavelength and corresponded to an aether wind having velocity of 5 to 7.5 kilometers per second. See A.A. Michelson and E.W. Morley, *On the Relative Motion of the Earth and the Luminiferous Ether*, The American Journal of Science, Vol.**34**, No.203, pp.333-345 (1887) p341. (https://history.aip.org/exhibits/gap/Michelson/Michelson.html)

[3] I. Bernard Cohen, *Scientific American* May 1952, p80.

light being conducted by it!)

- The young Einstein, influenced by the alleged "null result" of the Michelson-Morley experiment, declared aether to be a redundant concept. The mature and more experienced Einstein declared that, yes, aether does exist. It must exist. *"According to the general theory of relativity space without aether is unthinkable; for in such space there not only would be no propagation of light, but also no possibility of existence for standards of space and time (measuring-rods and clocks)."*[4] Was he serious? or was he confused? Einstein did not pursue the aether issue further and virtually no one paid his assertion any attention.

- The Nobel organization did its part in adding to the confusion. Albert Michelson, the man who was the first to detect the aether flow, was awarded the 1907 Nobel Prize for Physics. The citation stated it was "for his optical precision instruments and the spectroscopic and metrological investigations carried out with their aid"; but it made no mention of the aether-flow discovery. The Nobel Awards Committee honored him essentially for his high-precision measurements. Ironically, the Scientific Community *dishonored* him by ignoring his meticulous measurements revealing the aether flow.

Professor Dayton Miller was likewise dishonored. His significantly more precise measurements, collected during the 1920s, of the aether flow were ignored by most scientists and actively discredited by some. Confusing indeed.

- It was generally agreed that the aether couldn't be detected. Nevertheless, the conviction of nondetectability did not prevent experimental physicists from actually detecting the space medium. In fact it was detected at least six times during the 20th century —with a Michelson-type interferometer no less.[5]

- What added to the confusion was the fact that when the Michelson-Morley experiment was repeated under ordinary air pressure conditions (*gas mode* in scientific terminology), the results were consistently positive —a fringe shift was detected over a 90 degree rotation of the apparatus; but when conducted in the absence of any air or gas (when in *vacuum mode*), the results were always null. In other words, a gas-mode test proves the existence of aether; while a vacuum-mode test supports the *nonexistence* of aether.

- Back to the problem of light propagation: According to the eminent physicist Julian Schwinger, the vacuum is a quantum field consisting of

---

[4] A. Einstein, *Sidelights on Relativity*, Lecture delivered on May 5th, 1920, at the University of Leyden. Translated by G.B. Jeffery and W. Perrett (Methuen & Co. Ltd., 36 Essex Street, London, 1922); p23.

[5] R.T. Cahill. (2004) *Absolute Motion and Gravitational Effects*, Apeiron, Vol.11, No. 1 (January 2004). Posted at Apeiron website: http://redshift.vif.com/journal_archives.htm

quantum harmonic oscillators at each and every point of the vacuum-occupied space.[6] If this is so, and this oscillator-saturated medium does not propagate light (since light supposedly does not require a conducting medium), then what do the photons of light do? —zig-zag between the oscillators? or do they plough their way through pushing them out of the way?! But this would mean that some form of conduction by a medium is occurring after all. It sounds rather contradictory —or just plain confusing.

• The aether side in the debate merely claimed, most reasonably, that the aether was luminiferous and detectable. The aether deniers, on the other hand, claimed the *vacuum* possessed incredible amounts of energy and was even capable of spontaneously producing pairs of particles (particle-antiparticle twins). In fact the vacuum-energy load was mindboggling and represented one of the most embarrassing predictions, but was downplayed as the *vacuum energy problem*. Actually, it stands as the biggest discrepancy error between prediction and observation in the history of science. Aether theorists had given their space medium only two very reasonable properties versus the highly speculative properties for the vacuum acting as the quantum foam.

• Evidence. Think about how confused one must be to reject the medium *backed by evidence* and embrace the medium *lacking evidence*. ... On the one hand we have (i) the definitive detection of aether (including its direction of flow) by Dayton Miller[7] during the 1920s, and (ii) light evidently propagates through it. On the other hand there is no detection of the vast energy said to imbue the vacuum and no detection of particle-antiparticle pairs popping into existence.

• The name game. As it turned out, as the century progressed, the experts paid lip service to an abstract spacetime continuum while concocting a replacement type of aether —but, of course, never ever calling it "aether." The name game continues even as this book is being written.

In short, *aether* was out and the *vacuum / quantum foam* was in.

But it was mostly smoke-and-mirrors craftiness. Aether, because it allegedly had not been and could not be detected, was denied scientific status; while the vacuum was loaded up with various dubious properties.

There was another important factor in explaining the oversight. Scientists really were not specifically looking for new-perspective laws of physics.

Consider the sentiments of physicist Albert A. Michelson (1852-1931)

---

[6] Robert: R. Oerter, *The Theory of Almost Everything* (Pi Press, New York, 2006) p127-133.

[7] Dayton C. Miller, *The Ether-Drift Experiment and the Determination of the Absolute Motion of the Earth*, Reviews of Modern Physics Vol.**5**, pp.203-242 (July 1933). (http://dx.doi.org/10.1103/RevModPhys.5.203)

whose speech (at the University of Chicago) back in 1894 is often cited: "While it is never safe to say that the future of Physical Science has no marvels even more astonishing than those of the past, it seems probable that most of the grand underlying principles have been firmly established and that further advances are to be sought chiefly in the rigorous application of these principles to all the phenomena which come under our notice."[8]

Michelson was awarded the 1907 Nobel Prize for physics. It acted as a bestowal of legitimacy to his earlier "words of wisdom." The grand underlying principles have been firmly established, so don't bother looking for new ones.

Moreover, the prevailing attitude was that new discoveries would emerge not so much from new ideas but, rather, from greater quantification accuracy. The 1907 Nobel Presentation speech in honor of Albert Michelson emphasized that **achieving ever more precise measurements "is the very root, the essential condition, of our penetration deeper into the laws of physics — our only way to new discoveries."**

As an aside: Interestingly, there was no mention in the awards presentation and speeches, contrary to popular belief, about Michelson having demonstrated the non-existence of the luminiferous aether. Take this as a reflection of the fact that there was, at the time, a continuing strong belief in aether's existence by some segments of the scientific community.

Next, consider the words of one of the most brilliant physicists of the 20th century. "The age in which we live is the age in which we are discovering the fundamental laws of nature, and that day will never come again. It is very exciting, it is marvelous, but this excitement will have to go."[9] That quote from one of Richard Feynman's now famous Cornell University lectures reflects the popular feeling in the 1960s that physics, especially at the fundamental level, was coming to an end.

Near the end of the century, John Horgan, a popular writer for *Scientific American*, published *The End of Science: facing the Limits of Knowledge in the Twilight of the Scientific Age*. Horgan's message was that *all* the sciences are in their completion stages. He cleverly couched an apocalyptic prophecy in seemingly reasonable and assuring tones. From scores of interviews and other contacts with scientists of all stripes, he put together a patchwork of quotations, chapter by chapter: "The End of Philosophy," "The End of Cosmology," "The End of Evolutionary Biology," and so on, and of course, "The End of Physics." A stream of luminaries, including Sheldon Glashow,

---

[8] Albert A. Michelson, *Dedication of Ryerson Physical Laboratory*, quoted in Annual Register 1896, p.159.

[9] Horace F. Judson, *The Great Betrayal, Fraud in Science* (Harcourt Inc., Orlando, 2004); p445.

Hans Bethe, Steven Weinberg, John Archibald Wheeler, expressed various degrees of doubt over whether anything is left to do.[10]

Thus, by the end of the 2nd millennium, they honestly believed that new discoveries in physics were unlikely, all the important laws of physics had already been found.

Then there is what has been called the *Einstellung effect*.

Scientists for the most part see what they are taught to see —what they expect to see. Psychologists call it the *Einstellung effect*, the brain's tendency to stick with the most familiar solution to a problem and stubbornly ignore alternatives.[11] It dissuades a person from looking for potentially superior solutions or models.

> **"The more important fundamental laws and facts of physical science have all been discovered,** and these are now so firmly established that the possibility of their ever being supplanted in consequence of new discoveries is exceedingly remote. ... Our future discoveries must be looked for in the sixth place of decimals."
> –Albert Michelson (from an address at the University of Chicago in 1894)

The Effect is the basis for many cognitive biases. English philosopher, scientist and essayist Francis Bacon explained the bias in his 1620 book *Novum Organum:* "The human understanding when it has once adopted an opinion ... draws all things else to support and agree with it. And though there be a greater number and weight of instances to be found on the other side, yet these it either neglects or despises, or else by some distinction sets aside and rejects. ... Men ... mark the events where they are fulfilled, but where they fail, though this happen much oftener, neglect and pass them by. But with far more subtlety does this mischief insinuate itself into philosophy and the sciences, in which the first conclusion colours and brings into conformity with itself all that comes after."[12]

The Effect is also known as the *confirmation bias*. It has been demonstrated, in controlled experiments, that even when people attempt to test theories in an objective way, they tend to seek evidence that confirms their preformed ideas and to ignore anything that contradicts them.[13]

Once the "leading experts" and the grandees of the scientific establishment adopted the idea that aether was redundant and that the universe was expanding, the confirmation bias took hold. Aether was relabeled or just swept

---

[10] Horace F. Judson, *The Great Betrayal, Fraud in Science* (Harcourt Inc., Orlando, 2004); p405.
[11] *Scientific American* March 2014, p76.
[12] As in *Scientific American* March 2014, p77.
[13] Ibid.

under the rug; and *apparent cosmic expansion* was interpreted as the continuing effect of a cataclysmic big bang. Everything, all evidence, became preferentially interpreted in the light the big bang hypothesis. However, the validated Laws of the Universe have no relevance, no connection, to such a scenario. Broadly speaking the new Laws were not needed; thus, they were not sought.

The compilation of reasons behind the failure would not be complete without briefly discussing Authoritarian science.

Accredited experts, at times, completely miss the mark. As the Editors of Scientific American (December 2002) pointed out, "Inevitably, scientists will sometimes be just plain wrong —they make mistakes. Interpretation of evidence leaves room for error. Moreover, scientists aren't saints. They can be swayed by careerism, by money, by ego. Biases and prejudices can blind them. As individuals they are no more or less flawed than those from any other walk of life."

Physicist Max Tegmark expresses the same theme; but he goes into what I believe is the heart of the problem: "No matter how emphatically we scientists claim to be rational seekers of truth, we're as prone as anyone to human foibles such as prejudice, peer pressure and herd mentality. Overcoming those shortcomings clearly takes more than just talent for calculating."[14]

Did you catch that? ... Scientists can and do succumb to *peer pressure and herd mentality*.

And these can have devastating effects on science and the search for truth.

To better understand this, one must realize that the education system is a form of classical conditioning in the Pavlovian mode. "We [post-graduate students] were its dogs," relates Hilton Ratcliffe, a South African-born physicist, mathematician, and astronomer, "and both the tricks we were to perform and the rewards we would consequently receive were made abundantly clear to us." And recognize this: "There is no doubt that science is governed. There are rules, and the rules are enforced."[15]

Professor Ratcliffe continues: "Scientists are above all human beings, with all the foibles and limitations that their species generally expresses. Scientists are neither superhuman nor divinely privileged. Scientists, let me tell you right now, are simply plodding bricklayers in the wall of knowledge." Meanwhile there are those who rule and are venerated. "The kings of knowledge are all-powerful in the realm they administer, and it has surely corrupted them."

The kings of knowledge decided aether does not exist, light needs no medium, general-relativity theory commands the universe, and the universe is

---

[14] Max Tegmark, *Our Mathematical Universe* ... (Alfred A. Knopf, New York, 2014); p50.

[15] Hilton Ratcliffe's website: www.hiltonratcliffe.com (accessed 2015-8-13).

obediently expanding.

The Pavlovian scientists responded in keeping with their training, performed their tricks, ignored unapproved ideas, suppressed any serious critical thinking, and paid a heavy price. *In consequence of subservience to the corrupted philosopher kings, the brilliant minds of their century missed the essential laws that rule the Universe.*

Terminology

I want to emphasize that the technical terms used in this book are purposely kept to a minimum. New terms, of which there are precious few, have self-evident meanings. There are instances where familiar terms take on meanings somewhat modified from the conventional interpretation. All terms —old, new, modified, and multiple-meaning terms— are clearly defined in the Glossary at the end of the book.

With one exception, there are no acronyms. The only acronym used is "DSSU" which stands for **Dynamic Steady State Universe** and refers to the 21st-century cosmology that is founded on validated laws and gives unprecedented agreement with the observational evidence.

More specifically, the Dynamic Steady State Universe (DSSU) is the cosmology theory, based on a dynamic aether space-medium, in which aether continuously expands and contracts *regionally and equally* thereby sustaining a cosmic-scale cellular structure. It models the real world on the premise that all things are processes. Historically, it is the first *true* Steady State universe —steady-state nonexpanding, steady-state cellular, steady-state infinite, steady-state perpetual.

Mathematics is kept to a minimum. The math that does appear is quite basic. Anything more advanced is relegated to the Appendix and is not essential for understanding the particular concept being presented.

In accordance with common practice, I have used "Universe" when referring to the world we live in, and "universe" when referring to a world model or speaking in the generic sense. The distinction also applies to "Cosmos" versus "cosmos."

Clarification as to the meaning of "space" in the context of astrophysics and cosmology: Conventionally, *space* is a general term for *the vacuum, the quantum foam, the cosmic fabric*, etc. It is an ambiguous term for the background medium of the universe. However, as used in this book *space* has the most self-evident meaning possible —it simple means an *empty volume*.

The term "space" as used herein is the 3-dimensional background and is completely permeated by the universe's essence medium (which, as will be explained, is a ubiquitous, non-mass, non-energy, discretized aether). Space is a nothingness volume; it has no properties; none whatsoever. Its only function is to serve as an empty container of three spatial dimensions.

Outer space does not change; it still means any region removed from astronomical bodies.

Words of warning

Throughout the 20th century, the perceptive and prescient writings of Aldous Huxley, Albert Jay Nock, George Orwell, Kurt Vonnegut, Ayn Rand and many, many others exposed the insidious guises of collectivism / totalitarianism and warned about its inevitable praxis of spreading tyranny. But too few paid attention, too few heeded the warnings, too few understood the seriousness.

As this book is being written (2021), the Western Nations of the world are continuing with their decades-long suicidal policies of egalitarianism while continuing to submit to the Globalists' agenda of the enslavement of the nations and peoples of the planet. Moreover, there are dire warnings that "multiracial societies as they currently exist in Western nations are doomed to strife, conflict, and ultimate collapse."[16]

Tragically, the cancers of collectivism and egalitarianism are spreading. We are witnessing an utter corruption of freedom and liberty. Truth is being censored on an unprecedented scale.

Civilization is fading.

In these dark times, the recognition of truth is our only hope. The pursuit of truth is the cornerstone of civilization … and it is so much more.

Truth-seeking is the essence of human dignity.

–Germar Rudolf (imprisoned in Germany for his historical research)

What follows is essentially a pursuit of truth —the true nature of the Universe.

✠ ✠ ✠

---

[16] Arthur Kemp, *The War Against Whites, The Racial Psychology Behind the Anti-White Hatred Sweeping the West* (Ostara Publications, ostarapublications.com, 2020); p269.

Law of Velocity Differential Propagation of Light

1. Cosmic Redshift and the Velocity Differential Propagation of Light

The *cosmic redshift* is a phenomenon of the *law of velocity differential propagation of light*

What is the Cosmic Redshift?

Cosmic redshift refers to an aspect of the received light (or any electromagnetic radiation) from distant astronomical sources, which typically are the stars within distant galaxies. Such light invariably has undergone an increase in its wavelength. And *that* change gives us the connection with the color red. In a rainbow-like spectrum of visible light, the red light has the longer wavelength; hence, when any light, visible or otherwise, has undergone wavelength elongation, the change is called a *red*shift.

In short: "Cosmic" stands for cosmic distance; "red" stands for weaker, lower-energy, light, when compared to its original state; "redshift" stands for the process, whatever it may be, that induces the weakening of the light.

Incidentally, the actual *spectral-shift* measurement consists of first projecting onto a screen, or digitally scanning, two spectra; one spectrum is the light received from the distant source, the other spectrum is from a lab-produced light with similar characteristics; then it is a matter of comparing the position of corresponding *emission lines* or *absorption lines* present in the spectra. The difference is the shift recorded for subsequent calculations.

What made all this so important was that the more distant the light source, the greater was the spectral shift. Here was a key element, if properly interpreted, for gaining a deep understanding of the universe.

Let us look briefly at the historical background.

The early part of the 20th century witnessed the discovery of what was then called the *astronomical redshift* but, in time, became known as the cosmic redshift. Found was a completely unexpected phenomenon that altered the light spectrum of distant galaxies. The light had somehow been stretched and the degree of stretch increased with the distance of the source. During the 1910s and 1920s, the work of several astronomers, notably, Vesto M. Slipher (1875-1969), Carl W. Wirtz (1876-1939), Knut Lundmark (1889-1958), Milton Humason (1891-1972) and, the man who is generally given most of the credit, Edwin P. Hubble (1889-1953), firmly established the existence of a direct relationship between a galaxy's measured redshift and the galaxy's distance from Earth.

The relationship was quite unambiguous. The greater the measured wavelength elongation, the greater was the estimated distance to the source galaxy. Although only an empirical relationship, it was elevated to the status of a physical law called the *redshift-distance law* [1].

But by altering the empirical relationship slightly (by multiplying both sides by the speed of light c), the phenomenon could alternatively be expressed as a *velocity-distance law*. That is, because the redshift could be interpreted as a measure of velocity —the apparently receding speed of the light source— astronomers also like to use what is known as the *velocity-distance law* [2].

The search for the cause of the Cosmic Redshift

The correlation was as remarkable as it was unexpected. How could it be explained? What mechanism was altering the light spectrum? Was the light originally altered at the source, or was it being changed during its long cosmic journey over many millions of years? Maybe it was both.

The simple and familiar interpretation initially proposed was the straightforward *Doppler effect*. Galaxies were just displaying their recessional motion through the vacuum of space. And because under this mechanism the redshift was interpreted as a measure of actual motion, it became known as the *velocity-distance law* (for explaining the cosmic redshift). It did not take long

---

[1] **Redshift-distance law:** (redshift index) = $(1/c)$ × (empirical constant) × Distance
$$z = (HD)/c.$$
[2] **Velocity-distance law:** (recession velocity) = (empirical constant) × Distance
$$cz = H \times D,$$
where z is the measured redshift index, c is the speed of light, H is the Hubble constant, D is the distance of the light source at the time of reception (the "here and now" time).

for investigators to realize this can't be right. If all distant galaxies were truly fleeing away from our location in the universe, it would mean our position must be at the center of the visible universe and, hence, a violation of the Copernican principle. A special location, a universe center-point, is verboten. Moreover, there was absolutely no way to explain what stupendous force could possibly propel whole galaxies to the speeds suggested by the redshift factor (eventually, data was acquired that suggested Doppler recession speeds close to that of light itself). And so, the basic Doppler interpretation was quickly abandoned.

Not abandoned, however, was the belief that the *apparently* receding galaxies were *actually* receding. It so happened, there was another conceptual way to impart recessional velocities to galaxies. Intervening space was said to be slowly continuously expanding, continuously increasing the distance to the galaxies, as those galaxies are carried by the vacuum farther and farther away. Einstein's geometric medium, *spacetime*, even allowed for this expansion (although it also allowed for contraction). Galaxies were not speeding away *through* the vacuum; they were speeding away *with* the vacuum. The cause of redshift was then self-evident; while the light travels through the expanding vacuum the lightwaves expand as well. Lightwaves and gaps undergo elongation during the cosmic journey. The consensus among theoretical astronomers settled comfortably on the belief that the cosmic redshift was really an *expansion redshift*. And so the velocity-distance law, as something having a connection to reality, was replaced by the new *expansion-redshift law* [3]. The "best" part of this interpretation was that it made explicit the most exciting scenario imaginable for the evolution of the universe. Yes indeed.

As might be expected, there were problems. Universal expansion was not at all compatible with growing observational evidence of grand systematic structure. Furthermore, when conceptually extrapolated into universe-wide expansion, one encounters violations of philosophical principles. It all leads to a very messy cosmology. In fact, the experts eventually, many decades later, called it the "Preposterous universe." But, at the time, in the enthusiasm inspired by the opportunity to explore a dynamic new hypothesis, the problems were either unrecognized or simply ignored.

But not by everyone.

Consider the remarkably perspicacious warnings of the foremost redshift expert himself, the eminent Edwin P. Hubble: "[L]ight *may* lose energy during its journey through space, but if so, we do not yet know how the energy loss

---

[3] **Expansion-redshift law:** is expressed by the Friedmann-Lemaître expansion-redshift equation, $z = (R_0/R) - 1$, where R is the value of the scaling factor at the time of emission of the light at the source and R_0 is the value at the time of reception (the "now" time). The scaling factor changes in proportion to the radius of the expanding universe.

can be explained."[4] Hubble is referring to the connection between energy and redshift —the fact that light being redshifted is entirely equivalent to it losing energy. In other words, he is lamenting the fact that "we [Hubble and his contemporaries] do not yet know how the *cosmic redshift* can be explained."

Moreover, he expressed serious doubts about the universal expansion concept: "... this [is a] dubious world, the expanding universe of relativistic cosmology, ..."[5]

Hubble spoke of the dilemma facing astronomers confronted with the two interpretations available at the time and the need for a better theory. The choice between Doppler and Expansion was "a dilemma, and the resolution must await improved observations or *improved theory* or both."[6]

> "If the nebulae [i.e., galaxies] are not rapidly receding, redshifts are probably introduced between the nebulae and the observer; they [must then] represent *some unknown reaction* between the light and the medium through which it travels."[7]

Hubble concluded his 1937 book, The Observational Approach to Cosmology, with these words:

> "But the essential clue, the interpretation of redshifts, must still be unraveled. The former sense of certainty has faded and the clue stands forth as a problem for investigation." ... "We seem to face, as once before in the days of Copernicus, a choice between a small, finite universe, and a universe indefinitely large plus **a new principle of nature.**"

There you have it. It was his conviction that cosmic redshifts "represent some unrecognized principle of nature."

Decades later, beginning in the 1960s, the German astronomer Halton Christian Arp (1927-2013) began expressing strong doubts about the *hypothesis of receding galaxies*. He had amassed evidence of galactic redshifts that could not have been caused by recession velocities. Conventional-thinking astronomers implied that his controversial anomalous observations of distant galaxies "violated the known laws of physics" and must therefore be wrong. Galactic redshift, they asserted, was the measure of radially outward motion. The "recession" interpretation had, thanks to the Pavlovian response mechanism (as mentioned in the Prologue), been elevated to an unimpeachable

---

[4] E. P. Hubble, *The Observational Approach to Cosmology* (Oxford University Press, Oxford, 1937)

[5] Ibid., Preface, p1.

[6] Ibid., Preface. Emphasis added.

[7] Ibid., Chap. 3, Possible Worlds. Emphasis added.

law. Responding to the skepticism and criticism from fellow astronomers, Halton Arp stated that their attitude was akin to saying "At this moment in history we know all the important aspects of nature we shall ever know."[8] Arp accused his critics of presuming that the then-known laws of physics were the only laws we will ever know; there are no other laws to be discovered. A scientist, he argued, reasoning deductively only from known laws *will never discover anything new!*

The controversy continued throughout the 20th century. The phenomenon was called the cosmological redshift but its cause was repeatedly questioned. Universal expansion was just too radical an idea. Other interpretations were sought, advanced, and debated. For the most part, they fall into two categories, the *gravitational shift* and the various *tired light* proposals.

> Astronomer Halton Arp warned, a scientist reasoning deductively only from known laws *will never discover anything new!*

Let us first examine the gravitational spectral shift.

Maybe, somehow, gravity caused a weakening of light. For this alternative explanation, some researchers turned to Einstein's general relativity theory, according to which there exists a time dilation effect within any gravity well[9] and this could manifest as a change in the wavelength of light. The effect is called a gravitational spectral shift, sometimes called an *Einstein Shift*; and what it does is weaken the light climbing out of a gravity field and strengthen the light entering a gravity field. So, in one direction of propagation there is an energy loss (a redshift) and in the other direction there is an energy gain (a blueshift). Right away one can see a serious problem in trying to apply this theory to the universe. The universe, obviously, is full of gravity fields/wells. Whenever light descends into a gravity well it will be subjected to the Einstein Blueshift; and whenever light ascends a gravity well it will be subjected to the Einstein Redshift. Clearly the two tendencies applied over cosmic distances will, more or less, cancel out. Any net shift would be negligible.

It should be pointed out here that the cosmic redshifting mechanism can stretch a lightwave to several times its original wavelength (it simply depends on the total propagation distance). In fact there is virtually no theoretical limit.

The Einstein's gravitational spectral shift has the advantage of being a proven effect —at least as a localized effect and confined for just one side of a gravity well. It was detected for the Sun's well and for the Earth's

---

[8] H. C. Arp, *Quasars, Redshifts and Controversies* (Cambridge University Press, U.K., 1988)
[9] Every gravitating object sits at the bottom of its own gravity well or "gravity sink." The Sun generates and resides in a deep well; the Earth does the same in a comparatively shallow well.

comparatively weaker gravity depression. It was confirmed, by James W. Brault in 1954, for spectral lines emanating from the Sun.[10] The laboratory proof came in the early 1960s with the work of R. V. Pound, G. A. Rebka Jr., and J. L. Snider; and involved the remarkably precise Mössbauer effect.

There was also Fritz Zwicky's Gravitational Drag model from the 1920s and 1930s. But, whatever the reason, it never became a serious contender as an explanation for the cosmic redshift.

Swiss-born Fritz Zwicky was the sole astronomer at California Institute of Technology. He was tall, irascible, volatile, had been trained as a solid state physicist, and was brilliant. Brilliant, but he had so many ideas it was almost impossible for other astronomers to distinguish the good from the off-the-wall. One of his "major discoveries" was that 90 percent of the matter in the universe seemed to be invisible.[11]

Not surprisingly, he also contributed to the tired light concept.

The catch-all category in the interpretation-of-redshift debate is *tired* or *fatigued* light. It was probably Fritz Zwicky (1898-1974) who was the first person to propose the tired-light idea. The interpretation is that light from distant galaxies might somehow become fatigued on its long journey to us, in some way expending energy during its travels. The loss of energy is evident in the stretching of the wavelength. Although there was considerable speculation by accredited experts (George Gamow, for instance) intrigued by the tired-light idea as they sought explanations by altering the laws of Nature and adjusting the constants of Physics, a convincing cause for the energy loss remained elusive. In some versions, the light during its extended journey through space is required to interact with something along its path —encounter some perturbation, disturbance, or interaction that in one way or another robs the photons of some of their energy. The longer the duration of the journey, the greater is the energy loss. The problem is that any type of interplay will inevitably cause a deviation in the flight path of the light. Even if the deviations are ever so slight, the image of the emitting object would acquire a degree of fuzziness. However, to the detriment of the tiredness idea, there is no such evidence; on the contrary, high redshift objects appear as clear and sharp as low redshift objects. As astrophysicist Edward Wright has stated, "There is no known interaction that can degrade a photon's energy without also changing its momentum, which leads to a blurring of distant objects which is not observed."[12]

---

[10] J. W. Brault, *Gravitational Redshift of Solar Lines*, in Bull. Am. Astron. Soc. Vol.8, 28 (1963). (Based on Brault's PhD Thesis: https://ui.adsabs.harvard.edu/abs/1962PhDT........57B/abstract)

[11] Dennis Overbye, *Lonely Hearts of the Cosmos* (Little, Brown and Co.; Boston, 1999); p18.

[12] E. L. Wright, *Errors in Tired Light Cosmology*, http://www.astro.ucla.edu/~wright/tiredlit.htm

Table 1-1. Five categorical ways of explaining the observed cosmic redshift.

COSMIC REDSHIFT overview		
POSSIBLE CAUSE	MODUS OPERANDI	PROBLEMS
Basic DOPPLER Effect:	• Galaxies are receding. • Galaxies are moving away THROUGH the vacuum. (Hence, recession velocity is limited to c.) • KEY POINT: The redshift occurs ONLY at the time of emission (at the original source galaxy).	• Violates Copernican rule. • Lacks a driving force.
Expanding Vacuum (or space medium):	• Galaxies are receding due to ongoing cosmic expansion of the vacuum. • Galaxies are moving away WITH the vacuum. • KEY POINT: The redshift occurs DURING the time of the light's cosmic journey.	• Violates observational evidence of systematic structure. • When extrapolated, violates philosophical principles as it leads to a "preposterous" cosmology.
Gravitational (Einstein Shift):	An effect predicted by General Relativity (and proved by the R. L. Mössbauer experiment)	• Predicts near equal degrees of redshift and blueshift; thus, resulting in near complete cancellation.
Tired Light: (and related exotic causes)	Light during its extended journey through space encounters some perturbation, disturbance, or interaction that in one way or another robs the photons of energy.	• Predicts fuzziness, blurred images, but is not observed. • Fatal flaw: Can't explain the time-stretch between pulses.
Overlooked Mechanism:	A fundamental interaction between electromagnetic radiation and gravity	No problems whatsoever

But there is an even worse problem. Even if the energy-loss mechanism can be made to work, there is a critical feature that simply cannot be explained. There is no way to explain the increased delay between weakened pulses; the increased time intervals that actually occur between redshifted light pulses. No explanation for the elongation of the "gaps" between photons! It's a fatal flaw.

During the second half of the 20th century, astrophysicists such as Geoffrey. Burbidge and Halton Arp, while struggling to explain extreme redshifts associated with quasars, tried to exploit the weakened-light idea but were simply unable to accommodate the essential time-stretch feature. The photons may become tired, but not those in-between gaps.

Lastly, there is the "unrecognized principle of nature" of Hubble's prescient opinion. Having examined the traditional ideas for the cause of the cosmic redshift —Doppler, expanding vacuum, gravitational/Einstein shift, tired light— we now turn to the Overlooked Mechanism.

For quick-reference convenience, a summary of the five possible interpretations is presented in **Table 1-1**. Profiled are four familiar interpretations plus the mechanism that was overlooked —the "unrecognized principle of nature" of Hubble's long-ago forewarning.

Overlooked aspect of light propagation

What the theorists overlooked was the law governing the prolonged interaction between electromagnetic radiation and gravity gradients. Missed was an elementary rule for in-flight activity occurring between lightwaves and the ever-changing gravitational environment through which the light is propagating. One might also say they forgot to take into account a basic fact of Quantum theory: Photons are not located at a point, rather, they are spread out.

New law of physics: The overlooked law can be simply stated as the *principle of velocity differential propagation* (the Principle). It affects all electromagnetic radiation, quanta of light, and the in-between gaps, and the photons that constitute neutrinos[13].

Its most easily understood manifestation may be described this way: Light waves and neutrinos traversing the external portion (external to the mass body or mass aggregation) of a gravity well will intrinsically lose energy. The next part you will definitely need to read twice: They will lose energy during the inbound propagation AND during the outbound propagation. In other words, light undergoes redshifting and neutrinos lose energy —*throughout the entire journey*.

The effect accumulates, without limit, over multiple gravity wells; and it is the integral of all the wavelength stretching occurring during the cosmic journey that becomes observable as the cosmic redshift. Regarding the calculations of the effect, those must be done in the frame of background Newtonian space —"space" in the sense of *an empty container*.

Corollary: Lightwaves, light pulses, and neutrinos propagating radially in the *interior* of a gravitating body will *gain* energy during both the inbound direction AND the outbound direction. In other words, light undergoes blueshifting and neutrinos gain energy. This corollary effect, although not important for the present chapter, is presented here for the sake of completeness. It is, however, extremely important for total gravitational collapse (traditionally called black-hole physics).

---

[13] A neutrino is a twinning (a superposition) of two phase-shifted photons (resulting in the effective cancellation of their normal electromagnetic effects).

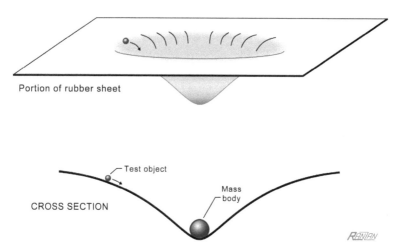

Figure 1-1. Schematic of a rubber sheet "gravity well"; shown in perspective (top) and cross section (bottom). The distortion of the elastic sheet increases with proximity to the central mass. Test objects accelerate as they roll or slide into the well —analogous to the acceleration of freefalling objects towards a gravitating body, such as a planet.

As counterintuitive as the Principle may seem, the reasoning behind it is surprisingly self-evident.

Three proofs are available. One is based on gravity being treated as a familiar force-like effect. It will be presented first. The second proof is based on the gravitation theory that actually incorporates the cause of the gravity effect. It employs a dynamic space-medium. Surprisingly straightforward, the mathematical part of the proof uses just basic algebraic addition and subtraction. This second proof will be given immediately after the first.

The third proof is just a more math-intensive version —using basic integral calculus. It too is constructed within the framework of a dynamic space-medium. For the details see the published article ... *The Velocity Differential Propagation of Light* (Physics Essays Vol.**33**, No.2, 2020) [14].

The proof based on gravity as a force/effect has two requirements: The first is simply that light quanta must possess wavelengths; photons must have a degree of spread. The second is that the force/effect we call gravity must vary inversely with distance from the center of mass.

Nature, in fact, does conform to these demands. Lightwaves and photons do have a longitudinal dimension. And gravity's force/effect does vary according to the inverse-squared relationship to distance —the simple Newtonian formulation.

---

[14] Available at www.CellularUniverse.org

In order to demonstrate the proof, we need a gravity well.

The popular way of illustrating a gravity well is to use a rubber sheet analogy. A mass resting on the sheet distorts the shape; the depression so formed is supposed to represent the *well*. It's a poor analogy, granted, but it is easy to visualize and the cross-section is quite useful. See **Figure 1-1**.

We can use the cross-section profile to make a simple graph. We take the profile and stick it into an *x-y* grid (**Figure 1-2**). The horizontal axis is for the radial distance from the center; the vertical axis is for the strength of gravity as measured by *gravitational acceleration*. A precise scale is not necessary; a visual comparison of relative strength is sufficiently adequate.

Notice that the graph (imitating the curve of the profile) has been placed in the negative region of the vertical axis (acceleration axis). This corresponds to the fact that the acceleration of gravity is a vector and the direction is always negative, always in the direction of the center of mass.

The schematic gravity well (**Figure 1-2**) —in which the curve measures the relative strength of the gravitational force— provides the *force gradient* necessary for the proof.

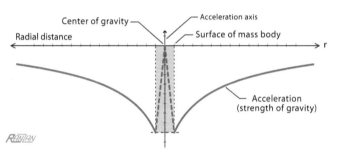

Figure 1-2. Graph of a typical Newtonian gravity well (with key features labelled). The curve traces the relative strength of the gravitational force at the radial location represented by the horizontal axis.

Proof based on gravity as a force-effect

The "force" argument depends only on self-evident factors: Namely, light quanta are extended entities, that is, they possess wavelengths; an understanding that a photon can change its dimension, its "extension," unlike a mass particle; and further, that gravity "pulls" on photons (and neutrinos). Gravity's ability to influence and accelerate light has long been known from the proven phenomenon of gravitational lensing. And what has been said here of photons also applies to lightwaves and pulses.

It is a fact of physics that the influence of gravity applies to electromagnetic radiation. It can cause a change in the direction of propagation and the spacing between lightpulses and the wavelength of light itself.

When a lightpulse descends into a gravity well, the leading end is nearer to the source of the gravity, thus, making the gravitational effect acting on the pulse's leading end greater than the effect on the trailing end. See **Figure 1-3**. This difference, or tiny differential, in the acceleration exists throughout the inbound journey. It follows that, in the frame of the pulse itself (and with respect to the background Euclidean space), there will occur a progressive separation between the two ends. An elongation of the wavelength will accrue. In other words, the lightpulse will be redshifted!

Then, when the pulse ascends the gravity well, it is the tail end that is closer to the center of gravity. The gravitational "pull" on the back end is ever so slightly more intense than is the pull acting on the leading end. There exists a gravitational acceleration differential as evident in **Figure 1-3**. The trailing pulse-end "feels" a stronger backwards pull throughout the outbound propagation. Once more, it follows that there will be an intrinsic separation manifesting as wavelength elongation. Again there will be a redshift.

The argument applies, just as well, to a train of lightpulses; but it also applies to the spacing between mass objects (aligned along a radial axis and undergoing inertial freefall). Objects falling in tandem from a very great height will, as a matter of fact, experience an increase in their vertical separation. Such a scenario exhibits a basic effect due to a gravitational potential differential and serves as an analogy of a gravity differential *redshift*. Yet, the conventional view predicts a gravity differential *blueshift!*

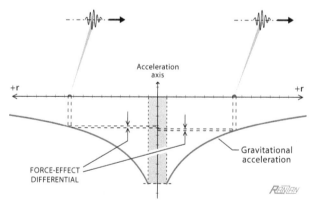

Figure 1-3. *Lightpulse transiting a gravity well.* During the descent, the gravitational acceleration acting on the leading end of the pulse is slightly more intense than the acceleration on the trailing end. This differential in the acceleration (equivalently a force-effect) manifests as an intrinsic elongation. During the ascent, the situation is reversed; the gravitational acceleration acting on the fore end is slightly *less intense* than what is experienced by the back end. In other words, the trailing end is being "dragged back" more than is the front end. Consequently, there is again an intrinsic stretch. (Gravitational acceleration curve: $a = -GM/r^2$) (Note the use of a cylindrical coordinate system.)

Clearly, there is a deep principle here that has been overlooked.

The foregoing proof treated gravity solely as a force without any reference to its fundamental cause. The following argument focuses on the actual causative process. It uses the dynamic aether that permeates the universe.

Proof based on a dynamic space-medium

The proof, under this context, has two requirements: The first is, as before, that light quanta possess wavelengths. Photons are spread out. The second is the existence of a dynamic space-medium through which light propagates. Instead of calling it the vacuum or the quantum foam, we will call it *aether*. Specifically, this refers to DSSU aether. Like Einstein's aether it is *nonmaterial* and *dynamic*, but unlike his aether it is not a continuum. Rather, it consists of discrete units; and it is kinetic. This property of aether having motion is a crucial element in the causal mechanism of gravitation, as will be explained in detail in Chapter 8.

In order to demonstrate the proof, we need a gravity sink —not a "well", not a depository, but a *sink*. This is not an analogy. Gravitating mass, in fact any mass whatsoever, is literally a sink into which aether flows.

The universal space medium indeed flows. While Einstein's aether "flows" in an abstract geometro-dynamic sense, DSSU aether flows in the very real fluid-dynamic sense (in conjunction with a self-dissipative process). Mass acts as a sink into which aether flows. In the absence of rotation and other disturbances, the flow is symmetrical as shown in the upper part of **Figure 1-4.** Notice that the inflow vectors increase in length with proximity to the spherical body. This means the aether flow is accelerating; and it is this acceleration (the rate of *change* of the velocity) that is the actual cause of the gravity effect. But for the purpose of demonstrating the velocity-differential redshift all we need is the velocity itself; and the rate of this flow, in accordance with the derivation shown in Appendix A, is:

$$\text{Aether flow velocity:} \quad v = -\sqrt{2GM/r}, \quad (1)$$

where G is the empirical gravitational constant and r is the radial distance (from the center of mass M) to any external point of the gravity sink. The equation represents a spherically symmetrical inflow field, and gives the speed of *inflowing aether* at any radial location specified by r. Incidentally, it also expresses the upper limit velocity of freefalling objects.

For any particular mass body, the expression simplifies to $v = -\frac{\text{CONSTANT}}{\sqrt{\text{RADIUS}}}$.

And this we use to generate the curves of the gravity sink shown in the lower part of **Figure 1-4.** As before, numerically-scaled axes are not necessary. The analysis will be a comparative one —simply based on relative values.

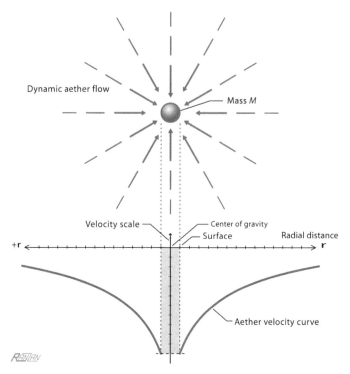

Figure 1-4. Schematic of the gravity sink (top) shows the flow of aether into the central nonrotating mass. The flow is an accelerating stream that nominally models Newtonian gravity. The graph (bottom) representing the *velocity* of the flow of the space medium. The flow is inward, hence, negative and is inversely proportional to the square root of the radius.

A few things to keep in mind: (i) The gravitating body is deemed to be at rest within the *space medium*. (ii) Consequently, the aether flow is simply proportional to the square root of the radial distance. (iii) The aether flow is with respect to the center of gravity, but it can also be thought of as being with respect to the background Euclidean space. And (iv) all the gravity-sink graphs presented here use a radial coordinate system. That is, the radius axes on the left side and on the right side are both POSITIVE. And the sign convention is: Any motion away from the origin (the center of gravity) is considered positive and motion toward the origin is deemed negative. The reason for using this method is to keep things as intuitive as possible. Needless to say, using a regular Cartesian coordinate system will give the same results.

Lightpulse during inbound journey

The representative gravity sink from **Figure 1-4** is reproduced in **Figure 1-5**; there, we examine a photon or lightpulse propagating along an inward path. By simple inspection of the diagram, it should be apparent that the front end of the pulse is moving inbound faster than the back end. It is a straightforward matter to show that the two pulse ends are moving apart.

At the instant that the lightpulse is located at the radial position indicated as r_1 and r_2, the two ends of the lightpulse will have velocities $-c+v_1$ and $-c+v_2$ respectively.

That is, the velocity of the pulse-end *lower down* in the gravity well is $-c+v_1$; while the velocity of the pulse-end *higher up* the well is $-c+v_2$. Next, subtract the two velocities: From the one *higher* up the gravity well, subtract the one *lower* in the well. An expression for the end-to-end relative velocity, then, follows.

(Relative velocity between ends of lightpulse)
= (vel. of higher end) − (vel. of lower end)
$$= (-c + v_2) - (-c + v_1)$$
$$= (v_2 - v_1) > 0, \qquad (2)$$

where v_2 and v_1, are the radial velocities of the aether flow. Both, of course, are negative; but, as plainly evident in the diagram, v_2 is higher on the velocity scale than v_1. Therefore, the expression must be positive. Hence, there is a velocity of separation between the two ends of the pulse.

Needless to say, the lightpulse's wavelength, and therefore also the velocity differential, is greatly exaggerated in the drawing. However, in a normal gravitational situation, the actual differential and corresponding velocity of separation, which over time manifests as a redshift, is *extremely tiny*.

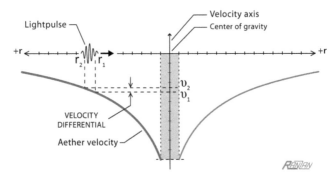

Figure 1-5. Lightpulse entering a gravity sink "experiences" a flow differential between its front and back ends. In other words, because the medium (aether) through which the pulse is traveling has a speed gradient between the two ends of the pulse, the pulse ends are being conducted with two slightly different speeds. As a consequence, the pulse tends to elongate; the pulse undergoes redshifting. (Note, the use of a cylindrical coordinate system, making the radius axis positive in both directions.)

Lightpulse during outbound journey

Next, consider the lightpulse propagating through the ascending half of the gravity sink. Once again, it is found that the front end of the photon is moving faster (in the direction of propagation) than is the back end. See **Figure 1-6**. The proof is a simple matter of showing that the two ends are moving apart; which means showing that the two ends have a positive relative velocity.

At the instant that the lightpulse is located at the radial position indicated as r_3 and r_4, the two ends of the lightpulse will have velocities $+c+v_3$ and $+c+v_4$ respectively.

As before, subtract the two velocities: From the one *higher* up the gravity sink, subtract the one *lower* down.

(Relative velocity between ends of lightpulse)
$$= \text{(vel. of higher end)} - \text{(vel. of lower end)}$$
$$= (+c + v_3) - (+c + v_4)$$
$$= (v_3 - v_4) > 0, \tag{3}$$

where variables v_3 and v_4 represent the radial velocities of the aether flow. Since v_3 is higher on the velocity scale (**Figure 1-6**) than v_4, the expression must be positive. Hence, there is a velocity of separation between the two ends of the pulse.

This confirms there does exist a positive end-to-end relative velocity —a situation of wavelength elongation and redshifting.

Summarizing the journey: A lightpulse enters the gravity sink, bypasses the central mass, proceeds to exit the well, and continues on its way to eventually enter into another gravity sink, and so on. All the while, and this is the essential point, the pulse elongates —it accumulates a redshift. And over cosmic distance, as the pulse passes through galactic sinks and galaxy-cluster sinks, the effect of differential propagation becomes perceptible as a *cosmic*

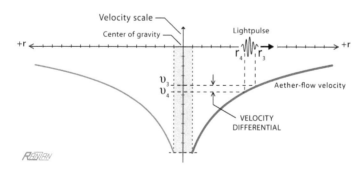

Figure 1-6. Lightpulse ascending a gravity sink "experiences" a flow differential between its front and back ends (whose instantaneous positions from the center of gravity are r_3 and r_4 respectively). As a consequence, the pulse undergoes elongation and acquires a spectral redshift —a *velocity differential redshift*. (Pulse is greatly exaggerated in size.)

redshift.

The velocity difference of the ends of the lightpulse is the consequence of two concurrent factors: The velocity gradient of the conducting medium (aether); and the constancy of the pulse's speed (lightspeed c) with respect to that medium —specifically and emphatically *with respect to the medium*. So, when the medium's own velocity is not exactly the same at the front and back ends, the difference is imparted to the pulse.

Some relevant aspects

Total redshift across gravity well/sink

Let us return to using the more common term *gravity well*. But always keep in mind that gravity is produced by a kinetic-and-dynamic aether —not by some abstract curvature of spacetime and not by some magical force carriers.

The total velocity differential effect across a gravity well is not simply the sum of the redshifts of the in and out paths. Realize that the redshifting process occurring during the outbound path applies to the original wavelength PLUS the stretched portion (its so-called delta-λ) acquired during the inbound path. In other words, redshifting is a compounding process.

Two simple definitions:
- **Redshift** is the stretched portion, the decimal fractional change in the wavelength, and is symbolized by the unitless index z.
- **Redshift factor** is the unitless term $(1+z)$. It just includes the unit factor for the original wave.

Back to the compounding process. Assuming an uninterrupted total light-path, the overall effect is calculated by multiplying the redshift factors as follows:

(Total redshift factor) = (Inbound redshift factor) × (Outbound redshift factor)

$(1+z_{total}) = (1+z_{inbound})(1+z_{outbound})$

For multiple gravity wells the expression would look like this:

$(1+z_{total}) = (1+z_{well\ 1})(1+z_{well\ 2})(1+z_{well\ 3}) \ldots (1+z_{well\ n})$

The more gravity wells that the light traverses during its cosmic journey the more important this compounding effect becomes. Just like an unpaid monetary debt grows with the relentless compounding of interest on the growing outstanding balance —the original principle plus seemingly endless interest accruals. The amount owing can, over time, far exceed the original loan.

Over multiple wells, compounding is the great redshift booster. The wave can be stretched by over 1000 times its original length. In fact, extreme redshifts in the range of 1000 to 2000 are routinely detected by radio

ONE• Cosmic Redshift and the Velocity Differential Propagation of Light 39

astronomers. They call this ultra-redshifted starlight the *cosmic microwave background radiation*.

What if the light is mirror-reflected from the surface of a gravitating body?

If the light is reflected, nothing would change; redshift acquisition would still take place during both legs of the path.

Observability aspect

The Principle of velocity differential propagation makes the following prediction: Light that enters a gravity well and reaches the bottom, will have acquired a certain amount of redshift. We proved this to be the case.

However, an observer at the bottom, that is, on the surface of the gravitating body, will observe the light as having undergone a blueshift. The observer at the surface does not detect the redshift of the Principle!

The question is *Why not*?

It's simply that surface observers are *not* inertial spectators. Their seemingly stationary light-measuring apparatus is actually "experiencing" acceleration —an acceleration that is built into the force holding up the observatory and everything in it.

What the observer is actually detecting is the wavelength of relativity theory, the gravitational shift (which, for incoming light, is a blueshift).

The intrinsic shifts are not directly observable from inside the gravity well. The truth of this applies to just a single gravity well (but does not apply over multiple wells). The underlying reason is that any observer inside the well is always, and everywhere, under the influence of accelerated motion with respect to the inflowing space medium (aether). For instance, the "stationary" observer positioned on the Earth's surface is subject to an upward acceleration of 9.8 meters per second per second. And as part of the same mechanism, the observer is subject to a constant relative-to-aether motion of 11.2 kilometers per second (if one ignores the background aether flow through which the Earth's gravity well moves, or just thinks in terms of an isolated earth-like object). What this means is that Earth-surface detectors, by virtue of location, are undergoing radially upward "motion"; consequently, incoming light waves and pulses are subject to an underlying Doppler effect. Also, measuring instruments are subject to a clock-slowing factor.

How large is the resulting *Doppler* blueshift effect? (It hardly needs stating but this Doppler shift is quite unrecognized within conventional gravity theory.) In the case of the earth-like example, the associated surface speed of the inflow (per equation from Appendix A) is 11.2 kilometers per second. Effectively, measuring instruments are in motion vertically into the aether at this same speed. This "measured" Doppler shift turns out to be −0.000,03733 (a blueshift as indicated by the negative sign). Meanwhile, the velocity differential redshift is +0.000,03733 for the same incoming light. This means the two effects cancel each other to within 4 significant digits.

The main reason, then, that the intrinsic redshift is not observed is attributed to the canceling effect of the Doppler blueshift. (For the earth-like example of at-the-surface measurements, the velocity-differential redshift of +0.000,03733 cancels the Doppler blueshift of –0.000,03733.)

The proper way, or most effective way, to measure the velocity differential redshift is to take reading across the width of the gravity well or some reasonable portion thereof. A light source needs to be on one side and the detector on the other.

Nevertheless, when the light passes through multiple gravity wells, including those of cosmic scale, the redshift imprint of each is accumulated and compounded; the opportunity for detecting the shift is then most favorable. In that case, the two shifts introduced by the receiving gravity well, the Doppler blueshift and the gravitational blueshift, become relatively negligible; and so, the ground-based spectrometers are able to measure the much larger velocity differential shift and identify it as the *cosmic redshift*.

Putting this in perspective for the earth-like gravity well: The gravitational blueshift is a miniscule $z_{grav} = -6.965 \times 10^{-10}$; and the Doppler shift is $z_{Dop} = -0.000,03733$. When compared to the spectral shifts of galaxies that astronomers commonly deal with, both of these contaminants are inconsequential.

Earth astronomers, however, do have to consider another Doppler effect. They are careful to make compensating corrections for Earth's Doppler motion caused by its orbit about the Sun. They then refer to the so-corrected redshift as being heliocentric. The idea is to remove the known significant contaminants.

Why the neutrino is subject to the Principle

Neutrinos are subject to the velocity differential effect by virtue of being a composite of linearly propagating photons.

A neutrino is simply a pair of photons locked together in such a way that they internalize their electromagnetic fields (and effects) causing the neutrino to lose, almost entirely, its ability to interact with the rest of the electromagnetic world. All the while, the locked pair propagates as a unit at the speed of light.[15] In the words of neutrino experts E. Kearns, T. Kajita, and Y. Totsuka, "Described quantum-mechanically, the neutrino is apparently

---

[15] C. Ranzan, *Natural Mechanism for the Generation and Emission of Extreme Energy Particles*, Physics Essays Vol.**31**, No.3, pp.358-376 (2018). Figure 6.
(Doi: http://dx.doi.org/10.4006/0836-1398-31.3.358)

a superposition of two wave packets of different mass."[16] But ignore the claim of neutrinos possessing mass; it has never actually been confirmed.

The take-away point is that a neutrino travels at the speed of light and possesses a wavelength and, consequently, is subject to the velocity differential effect just like ordinary quanta of light. And so it is, neutrinos will lose energy when traversing a gravity well or gravity domain.

Evidence

There is good experimental evidence of the velocity differential redshift available from signals propagating across the solar gravity well —evidence that wavelengths and the intervals (the gaps) between pulses are affected by the Principle.

Evidence from 1967 of additional redshift when star passes behind the Sun: Every year in the month of June the star Taurus A (a radio source also known as Aldebaran) aligns very close to the Sun. During such annual approach, it is possible to measure the change in the star's light spectrum, notably its 21-centimeter absorption line, as the radiation passes through the Sun's gravity well. Any shift in this absorption line to a lower frequency (or longer wavelength) would indicate that a redshift had been imparted.

In 1968, the journal *Science* published the results of such an experiment. It was reported that the 21-centimeter signal coming from Taurus A suffered a redshift of 150 hertz while grazing the Sun at a distance of 5 solar radii on 15 June 1967. The authors noted the shift was much larger than what is predicted by the general theory of relativity. This redshift cannot be explained by the Einstein shift which predicts a change of only ±0.16 hertz.[17]

So here is a situation in which the redshift was measured in terms of a change in the light's frequency —and that change was found to be a significant 150 cycles per second more than what had been expected.

Evidence from 1974 relating to the *Pioneer-6* anomaly: It was reported in the journal Astronomy and Astrophysics that the 2292 megahertz signal from *Pioneer-6* was found to be subjected to a redshift when it passed behind the Sun, that is, when the Sun lies near the signal path. The authors pointed out *there is no satisfactory quantitative explanation of these phenomena.*[18]

---

[16] E. Kearns, T. Kajita and Y. Totsuka, *Detecting Massive Neutrinos*, Scientific American Special Edition (2003), Vol.**13**, No.1, p72.

[17] D.S. Sadeh, S.H. Knowles, and B.S. Yaplee, *The Taurus A Experiment*, Science, **159**, 307 (1968).

[18] P. Merat, J.C. Pecker, and J.P. Vigier, *Possible Interpretation of an Anomalous Redshift Observed on the 2292 MHz Line Emitted by Pioneer-6 in the Close Vicinity of the Solar Limb*,

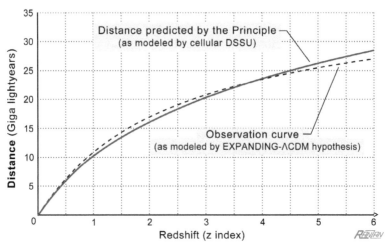

Figure 1-7. Cosmic redshift is interpreted in accordance with the *Principle of velocity differential propagation* (solid curve). The dashed curve represents the cosmic-redshift-versus-cosmic-distance relationship based on astronomical observations conducted over many decades using sophisticated techniques. The DSSU "prediction" curve agrees remarkably well with the "observation" curve. Note that the Observation curve is modeled by the expanding-ΛCDM hypothesis; reasonable agreement to actual observations is obtained by adjusting the latter's various parameters, none of which has a connection to physical reality. (ΛCDM stands for Lambda Cold Dark Matter.)
Specs: The DSSU model (solid curve) uses gravity wells 350×10^6 lightyears in diameter. Each has a calculated redshift of 0.0242.
Specs: The expanding-ΛCDM model (dashed curve representing the "now" distance) uses $H_0 = 70.0$ km/s/Mps, $\Omega_M = 0.30$, $\Omega_\Lambda = 0.70$; and the plotting was done with Edward Wright's *Cosmology Calculator*, www.astro.ucla.edu/~wright/CosmoCalc.html

Evidence from 1976 relating to the Viking mission to Mars: Several experiments have been performed with planetary probes; one of the most precise was with the Viking landers on Mars. It was found that when Mars is on the far side of the Sun, signals from the Vikings must pass through the deepest portion of the Sun's gravity well resulting in observed delays of about 100 microseconds. It is as if Mars had jumped some 30 kilometers out of its orbit! [19] According to the overlooked Principle, this represents a redshift, or fractional elongation, of 0.0810×10^{-6}. And if the source signal had had the

---

Astronomy and Astrophysics Vol.**30**, pp.167-174 (1974).
(http://adsabs.harvard.edu/full/1974A%26A....30..167M)

[19] G. O. Abell, *Exploration of the Universe* 4th Ed. (Saunders College Publishing, New York, NY, 1982); p579.

same radio frequency as Taurus A, the delay would correspond to a frequency decrease of about 116 hertz. This supports the earlier results of the 1967 *Taurus A Experiment*, in which the frequency shift was 150 hertz acquired over a larger portion of the Sun's well.

Remember, the *Einstein shift* of general relativity —because it practically cancels out— cannot explain this.

Then there is the evidence relating to the cosmic-scale gravity wells. When the Principle is applied to the gravity domains of rich galaxy clusters, when the line of sight passes through cluster after cluster and still more clusters, one finds the most amazing agreement with astronomical distance measurements. **Figure 1-7** shows, on the one hand, what is predicted by the DSSU with its *nonexpanding* cosmic gravity wells, and on the other, the observation-based distance as a function of redshift. The graphic comparison is truly remarkable; all the more so because the DSSU is strictly a nonexpanding universe.

A more in-depth analysis of the DSSU redshift-distance relationship (and additional graphical comparisons) is presented at **www.CellularUniverse.org**. Go to the Directory webpage (www.CellularUnivese.org/Directory.htm) where the following articles are posted:
    —*DSSU the Nonexpanding Universe: Structure, Redshift, Distance* (Published 2008)
    —*DSSU Cosmic Redshift-Distance Relation –Converting the Cosmic RS into Distance for our Cellular Universe* (2005-rev.2014)
    —*Cosmic-Redshift Distance Law Without c and Without H*, Galilean Electrodynamics Vol.**25**, No.3, pp.43-55 (May/June 2014)

What do the experts believe?

Experts Martin White (of the Enrico Fermi Institute, Chicago) and Professor Wayne Hu (of the Institute for Advanced Study, Princeton) have published considerable research on what is called the Sachs–Wolfe effect —an effect specifically dealing with photons traversing cosmic gravity wells. They believe, if the gravitational potential does not change (that is, the depth or shape of the gravity well stays the same) while the photon (or lightpulse) is crossing it, then there will be a cancellation "between the infall blueshift and the outclimb redshift."[20] Their assessment is unambiguous and expresses no doubts, the two shifts cancel.

The no-net-redshift view is the prevailing wisdom and is treated as an

---

[20] M. White and W. Hu, *The Sachs–Wolfe Effect*, Astronomy and Astrophysics Vol.**321**, pp.8-9 (1997). Posted at: http://background.uchicago.edu/~whu/pub.html

established fact. A check of any encyclopedic work on astronomy or astrophysics yields the assertion that photons gain energy entering a gravity potential and lose it while climbing out. If the gravity potential remains stable, then the gain and loss expectantly cancel.

So believe the learned professors.

Discussion of validity

How compelling is the proof of the Principle and its applicability to the cosmic redshift? How valid is all this?

In addressing the question of validity, consider the Law itself and the underlying elements. Say one wishes to refute the Principle of velocity differential propagation and expose it as some unscientific fabrication. There are several ways to attempt this.

• One might claim that quanta of light do not have wavelengths. Surely, no one would ever make such an assertion.

• One could claim there is no space medium. To my knowledge, no serious researcher has ever done this except in presentations for uncritical popular audiences or in an elementary level discourse. (For instance, see the quote, below, by Sheldon Glashow.)

• One could claim there is no intimate connection between light and the space medium, no interaction, no embeddedness, no relationship. Aether, or whatever, is not luminiferous. This amounts to an assertion that light is not some kind of disturbance or excitation in, or of, the space medium. But if light is not a disturbance/excitation of the vacuum, then what, one must ask, is? And if nothing disturbs/excites the medium, then why bother with it at all? In other words, if one refutes the connectedness between light and aether, one is effectively denying the very existence of the space medium. And as stated above, no researcher is seriously willing to do this.

• One could claim there is no velocity differential of the space medium — no gradient in the flow of aether. This will not alter the validity because the proof does not depend exclusively on the velocity differential concept. One is still confronted with the proof based solely on the indisputable property of gravity as an inverse-squared effect —the gradient of a force-like effect.

• One might maintain there is no stretching of lightwaves nor of the gaps, but then one is essentially saying that gravity does not influence the motion/propagation of light. This amounts to asserting that gravity accelerates mass particles but not energy particles. In that case, one is automatically rejecting the phenomenon of gravitational lensing!

• Lastly, one could, maybe as an act of desperate frustration, simply reject objective reality (as seems to be the modern trend). Needless to say, one would then be outside the realm of science. To which nothing more need be said.

• But I should add a post script: There are, however, dishonest people who will, without hesitation, reject the argument (hence, the reasoning supporting it) solely on the basis of the conclusion. They hinder the advancement of science while exposing themselves as frauds. This, I suspect (for good reason) is probably the most common factor.

As for challenging the validity of the Law's applicability to the cosmic redshift, the Law's involvement in its veritable cause: The only possible objection here is to declare there are no gravity wells. The universe does not consist of gravity fields. *That*, assuredly, is absurd.

Revisiting the unavoidable question

How could scientists have missed the Principle?

The main factors were the influence of relativity (as a way of interpreting reality) and the total lack of restraint in the adoption and application of a totally unverified hypothesis.

First and foremost, scientists of the 20th century missed this Law because of the single-minded effort to interpret, whenever possible, Nature and its phenomena in terms of relativity theory. It was applied wherever a phenomenon was in some way dependent on the perspective or the motion of the observer or the measuring instruments. Gravity, too, was interpreted in terms of relativity theory. The theory's application extended to ever greater time scales and ever more extremes of size scales. From cosmic beginning of time to the unimaginable future, from the scale of the whole universe down to the scale approaching singularities, explanations were sought within Einstein's conceptual framework. However, the Principle having pivotal relevance here involves something that happens to electromagnetic radiation (and neutrinos) at a level entirely independent of any observer. Here was a phenomenon no one thought of investigating.

One might be tempted to assert that Einstein's rejection of aether played a major role. He had, in his early and rebellious years, rejected the very existence of aether. Although he was wrong about aether, the extraordinary success of the relativity theories and the other great theories (such as the quantization of light and the photoelectric effect) led to the wholesale acceptance of what became the "new physics," including, unfortunately, the nonexistence of aether.

However, the mature Einstein made it quite clear, as expressed in his 1920 Leyden University lecture and in his 1922 book Sidelights on Relativity, Ether and the Theory of Relativity, that aether does exist. In fact, he stated that light would not be able to propagate without it. "According to the general theory of relativity, space without aether is unthinkable; for in such space there ...

would be no propagation of light," He further stated "As to the part which the new aether is to play in the physics of the future we are not yet clear."[21]

And that is pretty much where he left the issue of aether. Einstein never exploited "the new aether" beyond the confines of relativity theory, and, for the most part, neither did anyone else. But why!? Einstein and his followers insisted: *the idea of motion may not be applied to the aether of the new physics.* Without aether motion, of course, there can be no aether-flow differential.

The aether, whether old or new, was neglected for the rest of the century, with the definitive experiments of Dayton Miller being a notable exception. Even as late as 1988, Nobelist (1979 Physics) Sheldon L. Glashow would dismiss aether in these terms: "What we call light consists of electromagnetic oscillations, periodically changing electric and magnetic fields, but they do not need aether or any other medium to propagate themselves. They are perfectly capable of propagating through empty space."[22]

Secondly, there was the adoption of an unverified hypothesis and its outrageous extrapolation.

The hypothesis was that the apparent recession of distant galaxies was caused either by the galaxies actually flying away through the vacuum or by the expansion/growth of vacuum between here and the distant galaxies.

The theoretical backing came from Einstein's general relativity, which allowed the space medium, the vacuum, to expand. Additional backing came in 1932 with the introduction of the Einstein-deSitter model of the universe. It was a mathematical construction of a universe that expands and over time gradually slows down its rate of expansion.

It seemed there was no alternative to the hypothesis of the recession of galaxies. Edwin Hubble was truly perplexed. In 1937, he wrote, "There must be a gravitational field through which the light quanta travel for many millions of years before they reach the observer, and there may be some interaction between the quanta and the surrounding medium." But what medium was Hubble referring to? He did not specify. No matter. He immediately summed up the situation in these words, "The problem invites speculation, and, indeed, has been carefully examined. But no satisfactory, detailed solution has been found. The known reactions have been examined, one after the other —and they have failed to account for the observations."[23]

---

[21] A. Einstein, *Sidelights on Relativity, Ether and the Theory of Relativity*, translated by: G. B. Jeffery and W. Perret (Methuen & Co. London, 1922); republished unabridged and unaltered (Dover, New York, 1983) p23 & 20; posted at http://www.gutenberg.org/ebooks/7333

[22] Sheldon L. Glashow, *Interactions* (Warner Books Inc., New York, N.Y., 1988); p43.

[23] E. P. Hubble, *The Observational Approach to Cosmology* (Oxford University Press, Oxford, 1937)

ONE• Cosmic Redshift and the Velocity Differential Propagation of Light 47

And so it happened that the experts settled on the expansion idea (medium or no medium), which was immediately extrapolated into a hypothetical expansion of the entire visible universe. It was even back-extrapolated to a hot-and-dense primordial state —some sort of tiny cosmic egg.

Hubble had laid out the available options: "this dubious world, the expanding universe of relativistic cosmology," on the one hand, or "some unrecognized principle of nature," on the other. A practical man, Edwin Hubble, refused to commit himself and simply gave up on any further investigation of the causality issue; and was quite content to let the experts figure it all out and maybe decipher the true meaning. What he got was a consensus interpretation.

The accredited experts contrived and promoted the expanding-universe paradigm —a supreme masterpiece of misconception. Whether the particular version hypothesized a steady-state expansion or an explosive growth, made no difference, each was exquisitely crafted and equally unrealistic.

The only evidence to be had consisted of those *apparently receding* galaxies —misinterpreted as actual receding motion. There was simply no real evidence. None whatsoever. Model variations and alternative theories were explored. The situation was desperate. Then it happened. In 1965, a pair of Bell Lab researchers Arno Penzias and Robert Wilson measured the background radiation believed to be coming from deep space; essentially it was treated as the background temperature of the universe. Some actually considered it to be a major discovery, worthy of no less than a Nobel Prize. The only thing "major" about this temperature measurement was its potential for supporting the *big bang* hypothesis; theorists realized they could exploit this temperature as the critical evidence, otherwise lacking. *Here*, they declared was the proof of expansion, evidence that the cosmic climate had changed from a sunny-hot 5000 kelvin down to a sub-freezing 3 kelvin. At last the universal-expansion adherents had something to serve as evidence. When, in 1978, the Nobel was awarded to Penzias and Wilson, everyone knew the big-bang believers were definitely in control.

It became orthodoxy that the universe expanded from a hot and dense matter-dominated state and, during the expansion, transitioned to the diluted state with a cool 3K background temperature evident at the present time. This radical change, supposedly took place during the last 13.8 gigayears. Although completely wrong, there is no doubt that the scenario is a masterpiece and was conceived by brilliant minds. For instance, the time scale of 13.8 billion years just mentioned was established over many years of research and is claimed to be accurate to within about 1 percent. In all seriousness. "[T]he age of the Universe is now known with better than 1% accuracy"! That's what is quoted in the report issued by The Nobel Committee for Physics of the Royal Swedish Academy of Sciences: *Scientific Background on the Nobel Prize in Physics*

2019 (8 October 2019). The good authority for this degree of accuracy is the world's foremost cosmologist James Peebles (yes, another big-bang Nobel laureate).

(In accordance with their flawed interpretation of the cosmic redshift evidence, the cosmic climate change believers believe the universe is heading into the ultimate ice age, the mother of all deep-freezes. But in the fashionable obfuscating logic of academics they refer to their grave prediction as "the heat death of the universe.")

There is, however, nothing special about this temperature value. It is just the temperature that the Universe happens to be —at least in the regions away from hot spots like stars, quasars, and astrophysical jets. The background temperature is what it is and quite worthless as evidence of some hot-and-dense big-bang beginning. If the temperature had been different; if it had measured 15 K or 50 K, the claim that it was evidence to serve as the supporting pillar for the big-bang paradigm would still have been advanced.

The resolution of Olbers' Paradox is often elevated to the status of a supporting pillar. Again, one finds a finely-crafted argument. Unfortunately for the big-bang defenders, *that* pillar was toppled in 2016.[24]

In summary, the Principle was overlooked primarily because the leading thinkers of the time were preoccupied with the new physics of relativity and, moreover, became obsessed with the paradigm of universe-wide expansion. Others merely followed; naturally, that included following the money. Much more could be said on this matter, but the point has been made, the question answered.

Momentous misinterpretation

Theorists of the last century, along with all their many hangers on —journal editors, book publishers, videographers, financial funders, educators, and others— had committed themselves to a false narrative, and they did so to the depth and extent where there was no way out. The current century has inherited this dilemma. When the "collective wisdom" has it wrong to such a degree, there really is no easy way out. Don't expect anyone to turn traitor to the collective. Instead, expect the misinterpretation to sink ever deeper. Expect more awards such as the 2011 Nobel Prize; that's the one given for the "discovery" of ever-faster cosmic expansion! Expect ever more abandonment of objectivity and disconnect from reality. How, then, can this ever be resolved? What fate for the masterpiece of misconception? ... I see only two

---

[24] C. Ranzan, *Olbers' Paradox Resolved for the Nonexpanding Infinite Universe*, American Journal of Astronomy and Astrophysics Vol.**4**, No.1, pp.1-14 (2016).
(Doi: https://doi.org/10.11648/j.ajaa.20160401.11)

outcomes. It will either suffer a prolonged dissolution by decay, or it will undergo a replacement by revolution!

✠ ✠ ✠

First Interlude

Reviewers Confronted with a New Exciting Theory

> *On the one hand, the professional duty of a scientist confronted with a new and exciting theory is to try and prove it wrong. That is the way science works. That is the way science stays honest.* –Freeman J. Dyson, *Infinite in All Directions*

> Either the professionals have failed in their duty or their attempt to disprove the theory of differential propagation has failed. –C.R.

Needless to say, the discovery of a new law of physics is a rare occurrence. When it does happen, the discoverer must publish as soon as possible in order to establish chronological priority. He must submit his research to a professional journal where it undergoes quality evaluation, where it is assessed by "peer" reviewers. Physics journals exist primarily for the purpose of publishing innovative findings, new processes, advances in techniques, and infrequent fundamental discoveries. They pride themselves with bold statements of being dedicated to presenting works of ground-breaking significance, of wide interest, or of at least possessing sufficient originality. But in today's world very little is what it seems.

The journal Physics Letters A defines itself in these terms and in bold print: "a publication outlet for **novel theoretical and experimental frontier physics**."

But when the manuscript detailing the *velocity differential propagation of light* was submitted for publication to Physics Letters A, I received an eye-opening response.

The work, detailing a revolutionary discovery, was rejected because it lacked "the level of novelty, urgency, sufficient originality, wide interest and/or significance to justify its publication." That assessment came, "Sincerely," from Dr. Carina Arasa Cid, the Managing Editor of Physics Letters A.

Understand that the topic was relevant to their own policy statement: It welcomes the submission of new research in general physics and cross-disciplinary physics. It seeks "Articles focusing on break-through research in physics."

If a self-proclaimed physics journal is not interested, maybe an astrophysics journal would be. Considering the revolutionary cosmic implications of the discovery, surely it would appeal to an astrophysics journal.

An anonymous reviewer for the International Journal of Astronomy and Astrophysics reported, "The ideas proposed by the author are inconsistent with General Relativity and Special Relativity."

Well, what do you know! Of course, this new law of physics is not consistent with General Relativity. Otherwise there would be no reason to consider it "new." The new law is able to explain the Taurus A observational evidence and the Viking mission anomalies; while General Relativity fails utterly! And to say there is inconsistency with Special Relativity is just nonsense. The new law has practically nothing to do with the Special theory. There is no violation of the observed speed of light and no violation of physical laws (they remain the same, as they should, for all uniformly moving observers).

The reviewer merely re-stated what is obvious. The Article's conclusion does not agree with the conventional thinking! But of course, it is *not* intended to agree —it is, after all, a *new* interpretation.

And what about the supporting proofs so carefully presented? ... Ignored. Not a word.

The editor's decision, "After careful consideration," based on one reviewer's incomplete and incompetent assessment and no customary feedback from the Author: Reject.

Turning to another journal and another review: An anonymous reviewer for Physics Essays journal objected to the description of the neutrino as being a composite of linearly propagating photons: "This is speculation. The Author has given no proof to substantiate this character advanced for the neutrino. I recommend that the Author revise this section to reflect its speculative nature or cite experimental evidence for it."

Not sure why this would be considered speculative. The neutrino *does*, in fact, travel at the speed of light. Theorists *do*, in fact, treat it as "a superposition of two wave packets." The neutrino *does*, in fact, very nicely hide (i.e., internalize) its electromagnetic fields/effects and manages to almost disappear from the rest of the electromagnetic world. Scientific American Special Edition (2003) had this to say: "Described quantum-mechanically, the neutrino is apparently a *superposition of two wave packets* of different mass."

The same Reviewer recommended including additional scientific details on

First Interlude. Reviewers Confronted with a New Exciting Theory 53

the interpretation of the cosmic microwave background (CMB):

> *I draw to his attention that it has been proven, beyond any doubt, that the alleged CMB does not exist. It is a scientific fact that the CMB monopole signal has never been detected beyond ~900 km of Earth; precisely because Earth is the source of the signal, from its oceans, via the hydrogen bond in water* [1][2]. *In 1988 professor Paris Herouni, in Armenia, sampled the Cosmos using the most sensitive radio telescope ever built (ROT-54/2.6, self-noise 2.6 K), of his own design, construction financed by the Soviet Union. The 54 m diameter antenna returned 2.6 K, i.e. the self-noise of the antenna, hence 0 K for the Cosmos. Professor Herouni's measurement is direct and definitive. It too has been ignored by the astronomers and cosmologists for reasons unscientific. I recommend that the Author revise his discussion of the CMB.*

My response: I should point out that the purpose of the discussion of the CMB is to address the question: *How could scientists have missed the Principle?*

To go into the level of detail suggested by the reviewer, I believe, would be inappropriate.

It was the interpretation of the CMB as the remnant radiation of the alleged birth of the universe that was used to support a preposterous cosmology and consequently led to the dismissal of other interpretations. In other words, the background-temperature interpretation was an important factor in diverting attention from *the Principle* and the true cause of the cosmic redshift. (And, no doubt, the same "official" interpretation also caused the Earth-ocean-sourced signal interpretation to be ignored as well.)

The interpretation that the CMB was Earth sourced was really not a factor in explaining how scientists missed the Principle.

The details relating to the CMB are not too important. The importance of the CMB in the discussion rests simply with the fact that it was adopted as the main pillar of a preposterous model. This is why I stated that the background temperature, whatever it may be, is not important.

(It later occurred to me that maybe there are two "background" signals, one cosmic-sourced and one Earth-sourced. The problem with the theory that the microwave background is an Earth-sourced signal —the signature from the Earth's oceans, via the hydrogen bond in water— is this: It only explains a monopole signal, but does not explain a dipole signal, where the signal is

---

[1] Pierre-Marie Robitaille, WMAP: *A Radiological Analysis*, Progress in Physics, Vol.1, pp.3-18 (2007). (http://www.ptep-online.com/2007/PP-08-01.pdf)

[2] Pierre-Marie Robitaille, *Water, Hydrogen Bonding, and the Microwave Background*, Progress in Physics, Vol.2, pp.17 (2009 April). (http://www.ptep-online.com/2009/PP-17-L2.pdf)

different in opposite directions. Only a cosmic source can do that.)

Calling it a "very interesting paper" the reviewer's assessment concluded as follows:

> I consider this paper suitable for publication in Physics Essays. The Author has delivered his arguments in terms understandable by specialist and non-specialist alike, and reveals the fallacies that currently pass as explanation of the redshift of light due to gravity, cogently arguing the said redshift in the form of classical physics. This paper deserves the attention of the scientific community.

"This is an excellent Paper. A photon red-shifts when both entering and leaving a gravity field because a photon is an extended object that gets stretched when two ends of it are pulled differentially by a gravity gradient."
–reviewing Physicist

A second reviewer for *Physics Essays* had this to say:

> This is an excellent Paper. A photon red-shifts when both entering and leaving a gravity field because a photon is an extended object that gets stretched when two ends of it are pulled differentially by a gravity gradient: greater force pulling the end that is deeper in the gravity field. ... The stretching increases the wavelength, and hence incurs the redshift. Once stretched, the photon retains the stretch after it emerges from the gravity field: it is not a spring action with a restorative force. As I see it, the velocity of the photon going through the gravity field is only incidental to the mainly positional effect. All the author's proofs and discussions reinforce this picture in my mind. The citation of experiments and other theorists further bolster's the author's idea.

However, this reviewer had an issue with the algebraic signs associated with equations (1), (2), and (3) (see Chapter 1, *Proof based on a dynamic space-medium*). The "quibble" was

> ... with the clashing of the algebraic sign (negative velocity) in Equation (1) with the assertion that the v's are positive in Equations (2) and (3). The author should repair this blemish.

My response: I think I can best justify the correctness of the equations as they are, by making the following comparison:

Just as it would be wrong to write equation (1) as: $-v_{(@r)} = -\sqrt{2GM/r}$

(with the extra negative sign in front of the velocity v); so too would it be wrong to write equation (2), *with the extra negative signs*, as:

(Relative velocity between ends of lightpulse)
= (vel. of higher end) – (vel. of lower end)

$$= \bigl(-c + (-v_2)\bigr) - \bigl(-c + (-v_1)\bigr).$$

This equation will then give wrong results. The same goes for equation (3).

In conclusion, the v parameter does not need to carry a sign. The sign is carried into the equation when a specific value is substituted into (2) or (3). And those values *are* negative.

To my amazement, the reviewer was not convinced.

> *I think the author's positive/negative problem remains. To resolve it will require strict interpretation of velocity as a vector—albeit a one-dimensional vector. ... The only way the radius can be positive in both directions [both right and left in the graphs] is for* r *to be a scalar, not a vector.*

To which I responded: There is another way. One could use a cylindrical coordinate system —and confine oneself to a thin slice through the center of the gravity well. The r and v could then still be vectors. The radius r-axis shown in the figures can be thought of as a diameter of the gravity well.

The Reviewer also suggested treating r and v as scalars in equation (1). The alleged sign problem can then be fixed by "removing the negative sign from equation (1) (which would make it a statement about the magnitude of v) and then use the vector v (with its implicit algebraic sign) in the other equations."

I responded by reiterating my original point: The signs in equations (2) and (3) cannot be changed: simply because they will give the wrong results! The symbols v_1, v_2, etc., are placeholders for actual values. For example, v_2 is the placeholder for the velocity at r_2. One does not need to put signs on placeholders; the sign is part of the actual value and enters the equation upon substitution.

Following the change to a cylindrical coordinate system, the Paper was approved and published: C. Ranzan, *Law of Physics 20th-Century Scientists Overlooked (Part 1): The Velocity Differential Propagation of Light*, Physics Essays Vol.33, No. 2, pp.163-174 (2020 June).

Proof using Cartesian coordinate system

The finding presented in Chapter 1 is so important that it should be checked and confirmed using a regular Cartesian coordinate system.

The rule is very simple. Rightward motion is always positive and leftward always negative. The rule applies to the inflow component of aether's motion *and* to the lightpulses. So, for a thin slice through an isolated gravity sink, the aether inflow on the left-hand side (see **Figure 1-8a**) is positive and tracked above the horizontal axis; the aether inflow on the right-hand side is negative and tracked below the horizontal axis.

For the inbound lightpulse shown: At the instant that the pulse is located at

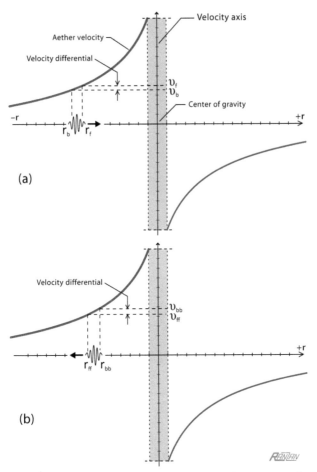

Figure 1-8. Using a Cartesian coordinate system to demonstrate the velocity differential redshift produced by gravity wells, or aether inflow sinks. With this kind of coordinate system, the rule is: rightward motion is always positive and leftward always negative. The rule applies to the inflow component of aether's motion and to the lightpulses. Lightpulses "experience" a flow differential between front and back ends, both upon entering a gravity sink, as in Part (a), and upon exiting, as in Part (b). In other words, because the medium (aether) through which a pulses is traveling has a speed gradient between the two ends of the pulse, the pulse ends are being conducted with two slightly different speeds. As a consequence, the pulse tends to elongate; the pulse undergoes redshifting. (Lightpulses are greatly exaggerated.)

the radial position indicated as r_f and r_b, the leading and trailing pulse-ends will have velocities $+c+v_f$ and $+c+v_b$ respectively.

The motion differential of the two ends of the lightwave or lightpulse, then, is

(Front-end velocity) minus (Back-end velocity);
which equals $(+c+v_f) - (+c+v_b)$;
and simplifies to $(v_{front} - v_{back}) > 0$.

Which, by definition, means the pulse ends are moving apart. The lightpulse is therefore being redshifted.

Now if the same pulse were going in the outbound direction to the left (see **Figure 1-8b**), then

(Front-end velocity) − (Back-end velocity)
$= (-c-v_{ff}) - (-c-v_{bb})$;
$= (v_{bb} - v_{ff}) > 0$.

Again, the pulse ends are moving apart. The lightpulse is being redshifted.

Arguments for the right-hand side of the graph will, likewise, result in a positive differential. Inbound or outbound, the redshifting process is unavoidable.

Velocity differential effect holds the key to an amazing positional effect

The second reviewer for *Physics Essays* made an astute observation. "*As I see it, the velocity of the photon going through the gravity field is only incidental to the mainly positional effect.*" The effect on the photon traveling through the gravity field is essential to the explanation of the cosmic redshift. That is now well understood. But the "positional effect" stands in importance far beyond anything this reviewer could have imagined.

The wavelength altering attribute of the *velocity differential law* is, as we have seen, a propagation effect. But it can also manifest as a positional phenomenon —an ongoing alteration of light due merely to its location (inside the gravity sink). If a photon or a lightpulse can be "held" in place within the gravity well, locked into one radial position, then the velocity differential effect will continue to progressively alter the particle's wavelength. And there is one place where this can, and does, happen. There is one place —within the most extreme type of gravity wells— where something truly amazing occurs.

The positional effect underpins Law number 2 and will be explored in the next chapter.

✠ ✠ ✠

*Law of Perpetual Energy
Generation/Amplification*
OR
Law of Blueshift Accrual
(by virtue of location)

2. Energy Generation via Velocity Differential Blueshift

Energy generation based on the velocity differential propagation of light

> *[E]ssentially all physical processes that we observe in the Universe are finite and nonrenewable. ... The supply of material for new stars is limited.*
> –Paul Davies & John Gribbin, *The Matter Myth*

Not so!

We now understand how wrong the experts have been. Contrary to the statement by mathematical physicist Paul Davies, the supply of material for new stars is *not* limited. It turns out that there are stars, *Terminal neutron stars*, capable of expelling far, far more energy than they take in.

Remarkably, the *law of velocity differential propagation of light* is responsible for two opposite processes. One produces the energy loss associated with the cosmic redshift. The other produces the *energy gain* associated with the most powerful energy generators of the universe. This is the type of law that scientists dream about —a law that unifies extraordinarily diverse phenomena. Essentially, the Propagation Law encompasses the Law of

Redshift Accrual (detailed in Chapter 1) *and* the very opposite —the Law of Blueshift Accrual. See **Figure 2-1** for the conceptual unification flowchart.

One manifestation of the Propagation Law was used in Chapter 1 to solve the century-old problem of the cause of the cosmic redshift. Its other manifestation will now be used to solve the mystery of *astrophysical jets* —the mystery of the underlying driving mechanism. This chapter explores a law of Energy Amplification requiring no conventional-energy input and reveals the limitless energy source that empowers those cosmic beacons.

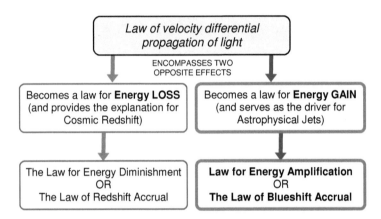

Figure 2-1. The Law of velocity differential propagation of light is responsible for two opposite effects. One effect manifests as the *energy loss* associated with the Cosmic Redshift. The other manifests as the *energy gain* associated with the mechanism that empowers Astrophysical Jets. Essentially, the propagation law branches into the sub-laws shown. The focus in this chapter is on Energy Amplification —the Law of Blueshift Accrual.

The discussion begins with a review of the Law of Velocity Differential Propagation of Light.

The Propagation Law governs the ongoing interaction between electromagnetic radiation and ambient gravity gradients. It provides a rule for in-flight activity between lightwaves and the gravitational environment; and applies to photons, lightpulses, and neutrinos. The Law has two essential requirements: One, light propagates through a nonmaterial medium (called *aether* for convenience); light, after all, cannot travel through nothingness. Two, photons are spread-out entities; photons, definitely, are not point particles.

The Propagation Law affects all electromagnetic radiation, quanta of light and the in-between gaps, and the photons that constitute neutrinos. It applies to freely propagating light and neutrinos; it does not apply to confined photons, such as the self-orbiting photon that makes up the electron. As detailed in

Chapter 1, its most easily understood manifestation is the redshift associated with crossing gravity wells. Light waves and neutrinos traversing the external portion of a gravity well will inherently lose energy. They will lose energy during the inbound propagation and during the outbound propagation. In short, light and neutrinos lose energy —*throughout the entire journey*. The effect accrues, without limit, over multiple gravity wells; and it is the integral of all the increments of wavelength stretching that is observable as the cosmic redshift.

The other manifestation of the Propagation Law occurs in the interior of a gravitating body. Lightwaves/pulses and neutrinos propagating radially in the body's interior will have their wavelength *shortened* during both the inbound direction *and* the outbound direction. In other words, lightwaves and neutrinos *gain* energy. This effect is extremely important for the physics of end-state gravitational collapse (misleadingly called "black-hole physics").

The second manifestation is, in truth, Nature's primary law of energy acquisition. It is appropriately called the **Law of Energy Regeneration/Amplification** —or equally appropriate, the **Law of Blueshift Accrual**.

The process of Energy Amplification, or Blueshift Accrual, is defined this way: Photons and neutrinos trapped within any *end-state neutron stars* (also known as *Terminal-state stars*) continuously gain energy. These particles undergo amplification (gain energy by wavelength contraction) while propagating *in place* in a zone of negative velocity differential. It is the "positional effect" anticipated at the end of the First Interlude.

The proof of the Energy Regeneration/Amplification process makes use of two features of light and one feature of the space medium (aether):
- Light quanta possess wavelengths. Photons are spread out.
- Light propagates as an excitation in (and *of*) the aether.
- The space medium —DSSU aether[1]— is dynamic. The medium's dynamic nature manifests as the familiar gravity effect.

We again make use of a representative gravity well and its characteristic aether flow profile; but this time the focus is on the interior portion, the spherical mass region.

Gravity well and aether flow profile

Consider a gravitating mass body of uniform density. Also assume it is not rotating and is completely at rest within the space medium. As was explained

---

[1] Like Einstein's aether, DSSU aether is *nonmaterial* and *dynamic*; but unlike Einstein's aether it is not a continuum. Rather, it consists of discrete units; and it is kinetic. It turns out that the DSSU aether is the first ever dynamic aether consisting of nonenergy, nonmass, discrete entities. In other words, the aether has the ability to manifest energy —yet its discrete units (when in the unexcited state) do not!

earlier, the body has an associated aether-flow field —a symmetrical pattern of aether flowing inward. The continuous flow sustains the very existence of the mass body. The rate of this external flow, in accordance with the derivation in Appendix A, is:

$$v = -\sqrt{2GM/r}, \quad r \geq R \qquad (1)$$

where G is the gravitational constant and r is the radial distance (from the center of spherical mass M, radius R) to any external point of the gravity well. This expression was used to construct the "external aether velocity" curves in **Figure 2-2**.

Now for the flow rate in the interior portion: logically, it is maximal at the surface (where r equals R) and diminishes to zero at the center (where r equals 0). The rate depends on the quantity of mass inside a particular radial position. In other words, M must be treated as a function of the radial position. Based on the volume of a sphere and on the density ρ of its mass, the mass function is easily derived:

$$M_{(r)} = (4/3)\pi r^3 \rho, \quad r \leq R \qquad (2)$$

When this is substituted into equation (1) we obtain

$$v_{(r)} = -\sqrt{\tfrac{8}{3}\pi G \rho r^2}; \quad r \leq R \qquad (3)$$

Except for the radius, everything on the right hand side remains constant (recall, the mass density ρ is assumed, for the sake of simplicity, to be uniform). The expression, therefore, reduces to

$$v_{(r)} = (\text{constant}) \times r. \quad r \leq R \qquad (4)$$

Thus, for the interior portion, the velocity is a simple linear function (represented by the "internal aether velocity" segments in **Figure 2-2**).

Note that the aether is actually *accelerating* inward; in the exterior region the acceleration is in proportion to the inverse-square law, in agreement with Newtonian gravity; in the subsurface region the acceleration is directly proportional to the radius. For presenting the proof, however, the velocity function is most convenient and agreeably intuitive.

In the previous chapter, it was proved that when light traverses a gravity well there will occur an energy loss during the inbound leg of the journey as well as during the outbound leg. An intrinsic redshift will accrue throughout. What follows is a simple proof of an opposite effect, of *energy gain* that occurs in the subsurface portion of any gravity well. It will be shown that light and neutrinos propagating radially within the interior portion of a gravitating body will undergo wavelength contraction, that is, they will undergo

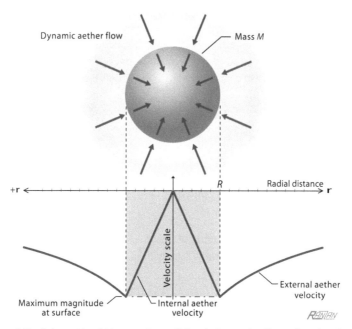

Figure 2-2. Schematic of the gravity well (top) shows the flow of aether into a spherical central mass. It is assumed the body is not rotating, is not moving (with respect to the surrounding space medium), and has a uniform density. The aether flow profile (bottom) is linear for the interior of the structure and inversely proportional to the square root of *r* for the exterior region. The velocity scale in the graph is arbitrary, since specific values are not important for proving the Law of Blueshift Accrual.

blueshifting. Moreover, the contraction will occur regardless of radial direction. Both inbound and outbound particles will become blueshifted. Granted, it sounds counterintuitive; and, hence, may well be the foremost reason that the phenomenon was missed by 20th-century scientists.

Rest assured, the following proof will, again, use nothing more than elementary algebra.

Process of Blueshift Accrual –the proof

Imagine that a tunnel has somehow been drilled clear through the center of our mass body (as shown in **Figure 2-3**). Lightpulses are beamed through the tunnel.

As was done in Chapter 1, we follow the intuitive approach of using a radial coordinate system. The radius axes on the left side and on the right side are both *positive*. Thus, any motion away from the center of gravity is considered positive and motion toward it is negative. (A regular Cartesian coordinate system will, of course, give the same proof results.)

Lightpulse during subsurface inbound journey

Consider the lightpulse propagating into the gravity well. It is moving in the same direction as the aether. By simple inspection (**Figure 2-3**), one can see that the front end of the pulse must be moving inbound slower than the back end. It is a straightforward matter to show that the two ends are moving closer together.

At the instant that the lightpulse is located at the radial position indicated as r_1 and r_2, its two ends will have velocities $-c+v_1$ and $-c+v_2$, respectively.

That is, the velocity of the pulse-end *lower down* in the gravity well is $-c+v_1$; while the velocity of the pulse-end *higher up* the well is $-c+v_2$. Next, subtract the two velocities: From the one *farther out* of the gravity well, subtract the one *deeper* in the well. An expression for the end-to-end relative velocity, then, follows:

(Relative velocity between ends of lightpulse)
= (velocity of farther end) − (velocity of deeper end)
= $(-c + v_2) - (-c + v_1)$
= $(v_2 - v_1) < 0$,　　　　　　　(5)

where v_2 and v_1 are the radial velocities of the aether flow. Both, of course, are

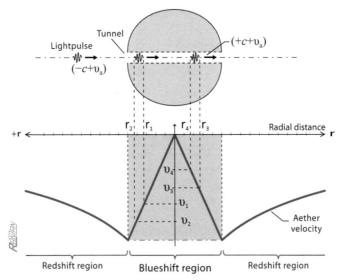

Figure 2-3. Lightpulse located between r_1 and r_2 "experiences" a flow differential between its front and back ends. This differential tends to contract the pulse; the pulse undergoes an elementary blueshifting process. The lightpulse positioned between r_3 and r_4 experiences the same effect —despite the fact that it is propagating AGAINST the flow of the aether medium. Blueshifting occurs during the entire interior journey. (Note, the graph uses a radial coordinate system; the radius axis is positive in both directions.)

negative; but, clearly velocity v_2 is more negative than is v_1. Therefore, the expression must be negative. Hence, there is a velocity of approach between the two ends of the pulse.

The pulse inside the tunnel and heading toward the center of gravity is having its wavelength shortened and, hence, is gaining energy[2]. No surprise here. Standard physics makes the same prediction. But watch what happens during the outbound leg.

Lightpulse during subsurface outbound journey

During the lightpulse's propagation through the ascending half of the tunnel, it is moving in the opposite direction of the aether flow.

And we find again, it is the front end of the pulse that is moving slower than the back end (in the outbound direction). But the reason is different. The front end is slower because it encounters a stronger "headwind." This may be easily confirmed by using the configuration and symbols in **Figure 2-3**:

(velocity of back end) − (velocity of front end)
$$= (+c + v_4) - (+c + v_3)$$
$$= (v_4 - v_3) > 0 \,. \qquad (6)$$

By inspection, it is seen that v_3 is more negative than v_4; which makes the expression positive; thus, proving the back end propagates faster than the front (with respect to the coordinate system). So the back end is gaining on the front end.

Relative motion method. It can also be shown that the two pulse ends are moving closer together in a relative sense. This simply means we construct the proof so as to show that the two ends have a negative *relative* velocity.

At the instant that the lightpulse is located at the radial position indicated as r_3 and r_4, the two ends of the lightpulse have velocities $+c+v_3$ and $+c+v_4$, respectively.

Subtract the two velocities: From the one *farther out* of the gravity well, subtract the one *deeper* in the well.

(Relative velocity between ends of lightpulse)
$$= \text{(velocity of farther end)} - \text{(velocity of deeper end)}$$
$$= (+c + v_3) - (+c + v_4)$$
$$= (v_3 - v_4) < 0 \,, \qquad (7)$$

where v_3 and v_4 are the aether velocities from **Figure 2-3**. Since v_3 is more negative than v_4, the expression must be negative; which means, the two ends are tending toward each other. There exists a negative end-to-end *relative*

---

[2] The shorter the wavelength, the greater the energy. This is why, for example, x-rays carry more energy than Sun rays.

motion.

This confirms the existence of an intrinsic energy gain. Moreover, the gain occurs during the entire cross-transit through the subsurface portion of the gravity well.

Proof based on gravity as a force effect

The Scientific community has long understood that the influence of gravity — its ability to accelerate things— applies to electromagnetic radiation. Gravity can cause a change in the direction of propagation and the spacing between pulses and the actual wavelength of light itself. For the evidence of the change in the propagation direction, there is *gravitational lensing*. For the evidence of the change in the gaps between light emissions, there is the time delay imprinted in the light signatures (radiation profiles) of distant supernovas. For the evidence of the change in wavelength, there is the proven phenomenon of *gravitational redshift*.

The "force" argument, then, depends only on self-evident factors: Light quanta (photons) are extended entities, in that they possess wavelengths; a photon can undergo a change of its dimension, its extension, unlike a mass particle; and further, that gravity "pulls" on photons (and neutrinos).

Turning to **Figure 2-4**, the peak effect of gravitational acceleration is at the surface of the body. Pass into the imaginary tunnel and the effect decreases. The greater the depth, the smaller will be the gravity effect. And for a constant density structure, the force-effect decreases linearly. At the center, there is no gravity; gravity equals zero.

For any spherical body, the standard expression for gravitational acceleration is: $a = -GM/r^2$.

For the external-to-body portion of this acceleration function, the mass M remains fixed.

But for the internal portion, the mass that actually determines the acceleration varies with the distance from the center of gravity. The applicable function is found by expressing mass in terms of the volume times the density ρ. This means replacing M with $(4/3)\pi r^3 \rho$ to obtain

$a_{(r)} = -\frac{4}{3}\pi G \rho r$, which, for an homogeneous body, becomes

$a_{(r)} = $ (constant) $\times r$. $r \leq R$ \qquad (8)

Simply put, the acceleration magnitude of the internal portion of the gravity well (assuming uniform density) is a function directly proportional to the radius r. While for the external portion, the gravitational acceleration is a function proportional to $1/r^2$. (See the schematic graph **Figure 2-4**.)

When the lightpulse descends into the tunnel, the trailing end is "experiencing" a stronger force (a greater acceleration magnitude) than is the leading end. This difference, or differential, in the acceleration exists throughout the descent journey. It follows that, in the frame of the pulse itself

TWO • Energy Generation via Velocity Differential Blueshift

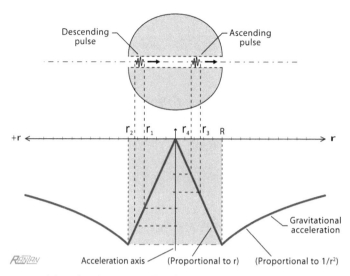

Figure 2-4. *Lightpulse transiting subsurface portion of a gravity well.* During the descent into the tunnel, the gravitational acceleration acting on the trailing end of the pulse is slightly more intense than the acceleration on the leading end. This differential in the acceleration manifests as a wavelength foreshortening. During the ascent, the situation is reversed; the gravitational acceleration acting on the fore end is slightly *more intense* than what is experienced by the back end. In other words, the front end is being "dragged back" more than is the trailing end. Consequently, there is again a wavelength contraction. Light acquires a blueshift throughout its subsurface propagation. (Pulses are greatly exaggerated.)

(and in the frame of the coordinate system), there will occur some degree of shrinkage between the two ends. A contraction of the wavelength will accrue.

Then, when the pulse ascends the tunnel, it is the leading end that experiences a stronger force (a greater acceleration magnitude). The gravitational pull on the front end is ever so slightly more intense than is the pull acting on the back end. There exists a gravitational acceleration differential as evident in **Figure 2-4**. The leading pulse-end "feels" a stronger backwards pull throughout the outbound propagation. Again, it follows that there will be shrinkage between the two ends —manifesting as wavelength contraction.

The argument is equally valid for a train of lightpulses. It also applies to the spacing between mass objects (aligned along the tunnel axis and undergoing inertial freefall). Objects falling in tandem will, as a matter of fact, experience a decrease in their vertical separation. Such a scenario exhibits a basic effect due to a gravitational potential differential and serves as an analogy of a gravity differential *blueshift* —a blueshift going into the tunnel as well as coming out. The conventional view holds that the shift (gap closure or wavelength contraction) going in is cancelled by the shift (separation or wavelength expansion) coming out.

Clearly, we have here another powerful principle that the experts had overlooked.

Environment most favorable for Blueshifting-mode propagation

Consider a couple of questions: Do conditions actually exist where blueshifting manifests in a major way? Are there mass structures to be found in the real world where the energy-gaining Blueshifting effect is significant? And how significant?

To answer the questions, it helps to first consider the location where the more familiar process of redshifting occurs. There are places where it manifests in the extreme. At or near the exterior surface of a neutron star, the phenomenon can radically alter the wavelength. For the penultimate situation, in which aether streams into the neutron star with a speed approaching lightspeed, any outbound photons (or lightpulses) would struggle to escape and would, in the process, undergo severe elongation. The subsurface region of such a structure would certainly provide the aether flow (a flow with a large velocity differential) capable of producing significant Blueshifting; except that the neutron mass is surely far too dense to permit the passage of light particles no matter how energetic (no matter how far beyond x-rays and how deep into the gamma range of the electromagnetic spectrum).

But now consider the truly ultimate situation. Once the neutron star has accreted sufficient additional mass, it enters a critical state; it becomes an *end-state neutron star*, also called a *Terminal-state star*. In the process of transforming into a Terminal star, the additional mass causes the aether inflow velocity to increase, in accordance with equation (1). The speed is always maximal at the surface. When this speed attains the speed of light, the surface of the neutron star undergoes a transformation. The neutron mass at the surface, compelled by the laws of physics, transforms into pure energy. A thin layer of photonic energy forms and encloses the star. In order to comply with the rule of special relativity (according to which mass and aether can never have a lightspeed relationship relative to each other), the so-transformed surface consists ONLY of particles that travel at the speed of light. Only two such particles are known to exist —photons and neutrinos. Thus, the end-state neutron star is enveloped by a lightspeed boundary, a buffering thin layer of radially propagating photons and neutrinos.

The formation of Terminal stars is a remarkable story in itself. Chapter 3 of this book delves into some of the details and the underlying physics. For the

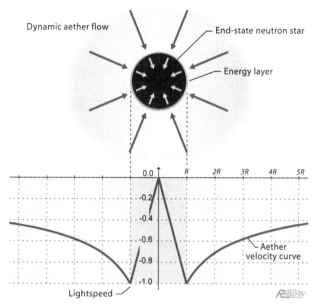

Figure 2-5. End-state neutron star (or Terminal star). It is the stable structure that results when gravity compresses a sufficient quantity of mass to its maximal density state. It has a surface aether-inflow speed equal to lightspeed; and it is enveloped by a thin layer of pure energy. Within that energy layer, photons and neutrinos propagate "in place" —*trapped within a Blueshifting environment*. Notice that the aether velocity attains lightspeed only at the neutron star's surface. The velocity then rapidly decreases to zero —all within a radial distance of about 10 kilometers. (Aether-flow scale is marked off proportional to lightspeed.)

journal articles[3,4], visit www.cellularuniverse.org .

For the present, however, a few comments are in order. When a sufficiently massive body collapses to the Terminal state shown in **Figure 2-5**, a very small quantity of the original mass undergoes a total conversion to energy. Only the surface mass is so affected. The 100 percent conversion of surface mass to energy should not come as a surprise, for it is simply the logical outcome of the fact that all mass particles are nothing more than trapped/confined energy particles (gamma photons and neutrinos) —albeit in highly intricate patterns which have yet to be deciphered. The electron is the exception. Its photon confinement configuration is reasonably well under-

---

[3] C. Ranzan, *The Nature of Gravitational Collapse*, American Journal of Astronomy and Astrophysics. Vol.**4**, No.2, 2016, pp.15-33. (Doi: http://dx.doi.org/10.11648/j.ajaa.20160402.11)

[4] C. Ranzan, *Natural Mechanism for the Generation and Emission of Extreme Energy Particles*, Physics Essays Vol.**31**, No.3, pp.358-376 (2018). (Doi: doi.org/10.4006/0836-1398-31.3.358)

stood[5]. And so, the compelled mass-to-energy conversion is but the constituent energy particles being prevented from following subatomic looping patterns and, instead, having to propagate linearly (aligned perfectly along, and into, the direction of aether flow).

The really important point is that these surface particles, photons and neutrinos, are propagating in-place at top speed —while going *nowhere*!

Bringing all the discussed elements together, here is what we have: With the Terminal star, we have a subsurface zone where gamma particles travel "in place" endlessly. And we have a subsurface region where the inflowing aether, by its dynamic deceleration, produces Blueshifting (**Figure 2-5**). These are the necessary and sufficient conditions for the amplification of energy by the velocity differential Blueshifting mechanism. These are the conditions for producing staggering amounts of energy.

The gravitational collapse of a gaseous star to form an end-state neutron star is not the only possible way to produce the "ultimate situation." It is possible for a massive dwarf star in a binary star-system to accrete sufficient mass over time and attain the critical state. Another scenario may involve a merger of a pair of orbiting massive dwarfs. The critical state may also be attained as the result of the simple collision of sufficiently massive dwarfs. In other words, the collapse of a gaseous star is not an essential element in the argument. It does, however, serve as an easy-to-understand simplification —a convenient thought experiment of a controlled collapse.

Regardless of how the end-state structure comes about, the environment most favorable for Blueshift-mode propagation is where aether flow attains lightspeed and is decelerating. There is where one finds the environment for the *fundamental energy amplification process*.

The Terminal neutron star and the Amplification process

The size of the Terminal neutron star is easily determined. Start by making the reasonable assumption that matter cannot be compressed beyond the density of nucleons (neutrons and protons) —a density generally agreed to be about 1.6×10^{18} kilograms per cubic meter. Terminal neutron stars are the macroscale embodiment of nature's maximal density state.

Assembled here are the three essential values:
- The mass density, per above assumption.
- The speed of aether inflow at the surface. Which, by definition of a

---

[5] J. G. Williamson, *On the nature of the electron and other particles*, The Cybernetics Society 40th Anniversary Annual Conference (2008) in London.
(https://www.researchgate.net/publication/267370968)

Terminal star, is lightspeed c.
- The Newtonian gravitational constant G.

Plug the numerical values[6] into the earlier derived equation (3) and solve for the radius (which is now given the symbol $R_{Terminal.star}$). The calculation gives the result,

$R_{Terminal.star}$ = 10,000 meters.

Thus, the Terminal neutron star has a diameter of about 10 kilometers. Significantly, this size, once attained, does not change. The end-state neutron star is a stable structure and *cannot collapse further* [7].

Energy Regeneration/Amplification process

Its modus operandi is the growth or regeneration of energy. As long as radiation particles are trapped within the energy layer (**Figure 2-6**), their respective wavelengths will undergo contraction, all in accordance with the definitive arguments presented above. This is the *energy Generation/Amplification process* —a process that can take an ordinary cosmic background photon and boost it into the gamma range. It's a very slow change. It may take a billion years, or longer. No matter. The process runs continuously. This energy boosting process is a manifestation of the law of the velocity differential propagation of radiation.

Both the process and the Law were completely overlooked by scientists of the 20th century.

On the question of layer thickness. The energy layer probably has a depth of a few centimeters, but possibly may extend to several meters. Here is the reasoning. As the aether passes through the energy layer, a portion of it (the aether) is absorbed/consumed by whatever is trapped therein; this absorption/consumption reduces the quantitative flow of aether causing it to slow down to subluminal speed. Since the energy layer holds Nature's densest state of matter in the form of radiation (this density is probably even higher than that of neutron mass), the consumption rate must be staggeringly high. So then the question is: how do lightspeed particles remain stationary within aether flowing somewhat less than lightspeed? Not a problem; photons can and do propagate at less than their normal speed in vacuum. (In standard physics, for instance, the photon's speed is related to the index of refraction of the medium through which it is propagating.) Although the radiation is NOT propagating *through* the energy layer, just the fact of the extreme density

---

[6] Substitute $v_{@surface}$ equals lightspeed (-3.00×10^8 m/s); $G = 6.67\times10^{-11}$ N·m²/kg²; $\rho_{neutron}$ = 1.60×10^{18} kg/m³.

[7] C. Ranzan, *The Nature of Gravitational Collapse*, American Journal of Astronomy and Astrophysics Vol.**4**, No.2, pp.15-33 (2016). (Doi: http://dx.doi.org/10.11648/j.ajaa.20160402.11)

slows their speed —permitting the radiation particles to remain stationary despite a subluminal aether headwind. Unfortunately, the relationship between propagating speed and density is not known, and so neither is the depth of the pure-energy layer.

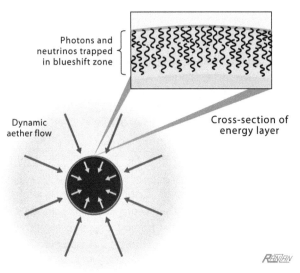

Figure 2-6. The end-state neutron star has a unique surface layer. It is a thin zone consisting of photonic energy (mostly gamma photons) and neutrinos. These particles are propagating outward; but because the aether is flowing inward with the same speed, the photons and neutrinos simply remain stationary within the energy layer. The crucial aspect is that the propagation is happening within a *blueshifting region* (in accordance with the Law of velocity differential propagation). Consequently, the particles undergo energy Amplification —they slowly gain energy.

It is important to realize that there are no mass particles within the surface layer. No particle having mass can travel at the speed of light; not through aether, not through space, not through whatever. The matter particles that existed in the pre-collapsed surface had either become part of the neutron-density region or had undergone a transformation. It was a transformation that required no energy input; it only required an extreme change in the gravitational environment. When Nature imposes the ultimate gravitational environment, mass particles unravel —quite literally. They do so in accordance with the photonic theory of particles, which holds that ALL particles are either free photons or some configuration of one or more photons. No exception. The concept is beyond brilliant; it means there is only one true fundamental force particle.

Here is the picture so far: A stable neutron star with the usual stuff falling in (things like stray atoms, microwave background, thermal photons, gamma particles, neutrinos, space rocks, and so on); but with nothing coming out; nothing whatsoever. The Terminal structure has an energy layer, in which photons and neutrinos continuously gain energy; also, additional energy is constantly being added from the external environment. However, the picture presented up to this point, has *not* included rotation.

What happens to this ever-growing bottled-up energy? For the answer we must take rotation, and its accompanying magnetic effect, into account.

Rotating Terminal neutron star

Magnetic channels

Generally speaking, all rotating structures are surrounded by magnetic fields. Rotating neutron stars possess extremely powerful external magnetic flux. When sufficiently massive stars collapse to the neutron state and further collapse to the Terminal state, they do not lose their magnetic fields. The external magnetic fields do not participate in the end-state collapse. The explanation for this is straightforward.

Two things happen during the collapse that morphs a mass body into the end-state. One, rotation rate speeds up, as a consequence of angular momentum conservation. Two, the magnetic field becomes ever more twisted and collimated. The faster the collapsing neutron star rotates, the more collimated will be the polar lines of magnetic force (**Figure 2-7a**). The result is a pair of polar cones, or columns, possessing extraordinarily high energy density. It is these energy-filled columns that prevent the inflowing aether from attaining lightspeed and "sealing off" the surface.

As aether streams down these energy-intense columns, a significant proportion (of that aether) is consumed. Remember, all energy and matter sustains its existence by the absorption/consumption of aether. Naturally then, the inflow acceleration and speed will be less than they otherwise would be (**Figure 2-7b**). This state of affairs is in effect during the collapse and continues after collapse. Therefore, the aether inflow at the surface —at the column bases— is always very much less than lightspeed. Meanwhile, the rest of the structure continues to experience the maximum inflow speed that Nature allows.

Recapping, when a rotating high-density structure transforms to the Terminal state, the polar magnetic channels remain in place. An energy layer forms and finalizes the collapse. The neutron star, thus, becomes sealed off

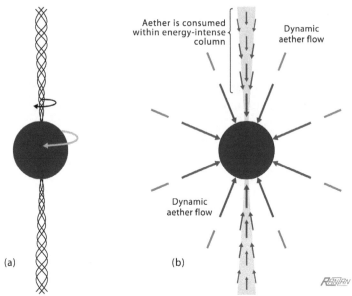

Figure 2-7. Rotation causes collimation of the magnetic force-lines of the Terminal neutron star, as shown in part (a). The rotating structure, thus, possesses two beams (or columns) with very high magnetic energy density. Part (b): As aether streams down these energy-intense columns, a large proportion is consumed. As a result, its acceleration and speed are seriously affected. Although acceleration and inflow speed still *increase* with proximity to the structure, these are greatly attenuated. Hence, the aether velocity at the surface —at the column bases— is very much less than lightspeed.

everywhere over its surface —everywhere *except* at the polar magnetic channels.

Escape mechanism

The *energy escape mechanism*, like everything else in the DSSU Worldview[8], is perfectly natural.

The energy layer just underneath (and touching) the lightspeed boundary contains photons and neutrinos —including nature's most energetic of such particles. The layer holds nature's densest state of radiation. Here is a domain absolutely saturated with electromagnetic energy waves —a domain totally inaccessible to investigation from the outside world, so that the enormous density can only be imagined. Extreme density, naturally, entails extreme lateral pressure. The lateral pressure pushes the particles toward the two places where the stationary energy layer is absent. As shown in **Figure 2-8b**, the

---

[8] **DSSU theory** is, by far, the most successful problem-free cosmology. For instance, it does not require so-called *dark matter* —not for galaxies, not for galaxy clusters, and not for the Cosmos.

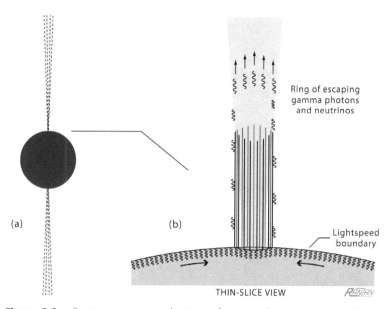

Figure 2-8. Energy escape mechanism of Terminal neutron stars. Gamma particles and neutrinos make their escape through the polar portals and outward along the magnetic channels. Part (a) shows the bipolar emission beams. Part (b) is a thin-slice schematic of the emission portal and its cylindrical beam of escaping photons and neutrinos. The extreme density of the energy layer produces a lateral pressure that continuously pushes surface-embedded particles towards the portals where they then escape at lightspeed (speed with respect to the aether medium).

radiation particles are pushed toward the edges of the polar portals. Once they reach the opening, they escape into the external environment (**Figure 2-8**). They shoot out at lightspeed; but note, this speed is not with respect to the surface but, rather, with respect to the inflowing aether. Collectively, they form a curtain around the opening —a cylindrical shaft of ultrahigh energy blasting into deep space.

Thus, photons and neutrinos make their escape through the polar portals and along the magnetic channels.

The lateral displacement of photons and neutrinos and their escape at the portals never diminishes the surface-layer's energy density. There are two reasons. One is the ongoing intake of new particles; the other is the ongoing energy Amplification. Lost photons are endlessly being replaced by infalling mass undergoing mass-to-energy conversion, as well as by the capture of incoming radiation. There exists the cosmic background radiation (including microwaves and a wide range of other wavelengths); it provides a staggeringly abundant stream of relatively low energy photons. Neutrinos, too, are replaced; as the cosmic background radiation of neutrinos provides for an equally abundant source for replenishment. Preposterously plentiful, neutrinos are said to outnumbering atoms a billion to one. And then there is the process that actually pumps energy into all those surface-trapped particles. This is the process whereby energy is generated via the velocity differential Blueshift — the blueshifting of the trapped photons and neutrinos. As a natural mechanism, it plays the key role in fueling the emission system.

Thus, thanks to constant replacement and ongoing energy Amplification, the streaming of escaping energy from any Terminal neutron star is a *continuous phenomenon*.

The question of energy conservation

Does the Blueshifting process violate thermodynamic laws?

It may be argued that the described process violates the tenets of energy conservation in the sense that the Terminal structure behaves as what is known as a "perpetual motion machine of the first kind" —one that produces more energy than it consumes or absorbs. Here is a system where low energy photons are constantly streaming in, while, at the same time, high energy photons are streaming out at those polar portals. Viewed in isolation, the system stands as a serious violation of energy conservation. Moreover, it also violates the Second Law of thermodynamics. The mechanism, as it has been described, is an entropy lowering process! But this is only the case if this Amplification aspect is treated in isolation.

However, these unique structures in which energy Amplification occurs are but components of a much larger system. And within that larger system, there is no violation of the conservation law and no violation of the entropy rule[9].

It should be pointed out that within cosmology theories the conservation of energy is handled differently. Most physicists and philosophers assert that such restriction does not apply to cosmic regions; while others treat it as something unknowable or simple evade the issue altogether. Cosmologist Edward Harrison, for instance, claimed outright, "[it] is obvious: Energy in the universe is not conserved."[10]

DSSU theory, however, does have a unique way of assuring compliance to the rules. How energy conservation is achieved and how natural processes manage to maintain entropy stability will be discussed in some detail in Chapter 6.

✠ ✠ ✠

---

[9] C. Ranzan, *Nature's Supreme Mechanism for Energy Extraction from Nonmaterial Aether*, Infinite Energy Vol.25, Issue#144, pp8-16 (2019 March/April). Reprint posted at: www.CellularUniverse.org/

[10] E. R. Harrison, *Cosmology, the Science of the Universe* (Cambridge University Press, Cambridge, UK, 1981); p276.

The Blueshift-accrual-and-release mechanism solves some of the most perplexing problems that plagued scientists of the previous century and considerably limited their understanding of the Universe. One is the mystery of the driving mechanism associated with *astrophysical jets*. Another is the complete bafflement with regard to the source of ultrahigh energy particles. A third involves a deep philosophical question about ultimate origins.

The energy Amplification process as the driver of astrophysical Jets

The energy Amplification process is the heart of the mechanism driving *astrophysical jets*. Here is a mechanism whereby energy actually escapes from the interior of a gravitationally collapsed body —a body that categorically cannot collapse further. Moreover, it escapes in prodigious limitless amounts. The superiority of this mechanism over all others is that it requires neither an accretion disk nor rotation[11]. To repeat, the Terminal star's mechanism powering astrophysical jets requires neither a mass-source disk nor self-rotation. The combination is nothing less than revolutionary: enablement of escape from total gravitational collapse, no dependency on rotation, no reliance on infalling mass.

The no-reliance-on-infalling-mass should be self-evident from the nature of the mechanism. But the no-dependency-on-rotation may not be. Consider this "what if" scenario: What if the rotation were somehow stopped, or canceled, would not the magnetic field (and the collimated force lines) simply collapse, thus causing the polar portals to close and seal off any further emission? No. If two Terminal stars collide or combine in such a way so that their separate rotations cancel each other and the combined structure is left with no rotation or negligible rotation, the energy escape mechanism and the driving force behind the astrophysical jets would in principle be unaffected. Take away the collimated magnetic field. Throw it aside. What remains are two columns of energy —beams of electromagnetic energy and neutrinos. They consume aether; they attenuate the aether inflow; they keep the portals open. Thus, the polar energy outflow continues. It is unstoppable.

Let us briefly examine the old 20th-century view. In particular, there is the failure of the astrophysics community to recognize the reality of emission jets coming from inside a totally collapsed structure. This failure can be blamed on three misconceptions.

---

[11] C. Ranzan, *Natural Mechanism for the Generation and Emission of Extreme Energy Particles*, Physics Essays Vol.**31**, No.3, pp.358-376 (2018).
(Doi: http://dx.doi.org/10.4006/0836-1398-31.3.358)

- They believed, and their theory demanded, that the space medium on the inside (the interior side of the event horizon) must be flowing inward FASTER than the speed of light. The assertion of space-medium flow faster than lightspeed is, in itself, not a problem. It does not violate Einstein's relativity. However, it automatically leads to the conclusion that nothing from the inside can escape. Academic physicists really had no inkling of how anything below the event horizon can possibly escape to the outside world. (But they tried. They did speculate about escape through white holes and passage into parallel universes.)
- They failed to grasp the full nature of the lightspeed boundary/horizon. The view that their event-horizon boundary divides two regions, where the medium flow on one side is less than lightspeed and on the other side is greater than lightspeed, is quite valid; it may, just maybe, apply to overly-massive regions (noncontiguous supermassive aggregations). However, *it does not apply to end-state neutron stars*. In the latter context, they failed to appreciate its nature as a photonic surface, a physical energy surface, *a perpetual generator of gamma photons*. The restrictive view of the boundary caused them to miss the source of the energy feeding the jets.
- The biggest misconception was the belief that the mass hidden deep inside their event horizon is point-like. The notion is so outrageous and outside the realm of natural physics that it really doesn't need elaboration.

As a broad critique, under the Old Physics view, there is no plausible mechanism for linking the interior mass —mass which everyone knows is causing the gravity, but is wrongly believed to reside in a point-like "structure"— with an external magnetic field! Since magnetic lines-of-force cannot travel faster than light, the magnetic field of the point mass cannot possibly manifest, cannot extend to the event horizon, and cannot reach through it. Even an escape hole in their horizon can't resolve this problem.

Energy Amplification process as the source of ultra energy particles

The remarkable thing about the Blueshift accrual taking place within the energy layer is that it has no limit. As long as they remain embedded in the surface layer, the photons and neutrinos will gain energy. There is no upper limit on how much energy can be conferred to the trapped particles. This provides the natural explanation for ultrahigh energy particles —particles that have been repeatedly detected, particles whose energy is far beyond what can be produced by any known mechanism and any theoretical process and any imagined action.

The energy Amplification process perfectly explains the PeV neutrinos often detected by the IceCube Neutrino Observatory located on the Antarctic continent. These are neutrinos in the peta-electron-volt (or quadrillion electron volt) range, corresponding to about a million times the mass-energy of a proton! As American physicist Spencer Klein has pointed out, "These neutrinos have energies more than a thousand times higher than any neutrinos that we have produced in particle accelerators." Canadian astrophysicist Ray Jayawardhana perspicaciously stated, "we may have to look to distant celestial sources to uncover the violent origins of these neutrinos."[12]

Another example. The basic mass energy of a proton is about 10,000,000,000 electron volts (or 10 GeV). Yet particles, assumed to be protons, have been detected with energy *ten billion times greater*! That's 10^{10} times greater.[13] How does Nature generate such ultra-extreme energy protons? The answer is that it probably doesn't. Without question, Nature does generate cosmic-ray particles with an astonishing energy of 10^{20} electron volts —just not in the form of protons or neutrons, not with such level of energy. So, what's going on here?

The total energy of any incoming particle is determined by analyzing the splash of particles produced when it collides with a subatomic mass particle. This means adding up the masses and kinetic energies of all the products produced in the collision. There is, for the most part, no problem with determining the source-particle's energy. The procedure and resulting values can be trusted. The problem is with the difficulty of distinguishing between a pure energy particle and a subatomic mass particle as the source entity.

Some years ago, Discover magazine (in an article on cosmic rays and collisions) discussed the dilemma. In the detection process that researchers use (recording and analyzing ultrahigh-energy air showers) there is no way of conclusively identifying whether the source particle is a proton (or even an alpha particle) or a gamma particle. No way to distinguishing between a cosmic-ray particle of mass or a particle of pure energy![14]

Unless the track of the incident proton is actually observed/detected (as when its positive electric charge highlights its path through a cloud chamber), there can be no conclusive identification. The source particle could just as likely be a gamma photon or even a neutrino. Moreover, the higher the cosmic particle's energy, the greater will be the probability of it being a gamma or a neutrino.

Incidentally, the neutron is rejected as a candidate cosmic-ray particle

---

[12] Ray Jayawardhana, *Neutrino Hunters* (HarperCollins, Toronto, 2013); p22-23.

[13] J. Linsley, *Evidence for a Primary Cosmic-Ray Particle with Energy 10^{20} eV*, Physical Review Letters, **10** (4), pp146-148 (15 Feb 1963). (Doi: http://dx.doi.org/10.1103/PhysRevLett.10.146)

[14] Discover, September 2003, page 47.

because a free neutron has a rather short life; it readily disintegrates within a matter of minutes.

What about the transfer of momentum, from the particles of the Terminal star's emission beams, to the hydrogen nuclei (protons) present in the immediate or interstellar environment? Photons, as is well known, possess momentum and can transfer their momentum. Could we assume that the greater the photon's energy, the greater transferred of momentum energy? The problem is the primary emission particles have too much energy, too much momentum. An ultrahigh-energy photon or neutrino striking a hydrogen nucleus will, undoubtedly, produce a fireworks of secondary particles including new protons. Some of the freshly minted protons may possess considerable kinetic energy, but nowhere near the ultrahigh range.

Now let us make the connection with the spectacular cosmic jets observed by astronomers. The emission beams, shown in **Figure 2-8**, contain the source energy particles. The *transfer of their momentum energy* is what produces the visible jets extending out to great distances from the collapsed structure. And it is the *fireworks of secondary particles* that constitute the considerable mass content of those jets.

We come to the following conclusion regarding ultra-energy radiation: It makes reasonable sense to treat the evidence of ultra-range gamma photons and extreme neutrinos as the experimental proof of energy amplification outside the constraints of the ordinary rules of energy conservation. What other choice do we have? There really is no other mechanism —proven or theoretical or conceptual— capable of the energy levels discussed here. No collision, no interaction, no nuclear process has ever been proposed for such mind-boggling levels of energy.

The energy Amplification process provides a bonus feature

There is a truly profound implication associated with of the Law of Blueshift Accrual. The deep meaning of the *velocity differential Blueshifting* mechanism is that an energy-particle creation process is not a necessary feature of the Universe. The energy Regeneration/Amplification process —the Blueshifting process— is the only ingredient needed. It alone in the quantum world, along with aether emergence in the subquantum realm, ensures that the Universe will never run down. It is the unique process wherein the material and the nonmaterial realms of existence intersect, generate energy, reenergize radiation particles, and help to maintain our Universe as a perpetual steady state cosmos.

In other words, if one's theory of the universe includes a fundamental process of re-energization, then a separate hypothetical *process of creation* becomes unnecessary. The spontaneous creation of mass or energy particles is then entirely non-essential! Ponder the implications for the efforts to resolve some of the deepest, most challenging, philosophical mysteries that have long barred a full understanding of the Cosmos.

Blueshift Accrual may turn out to be the most significant of all the DSSU discoveries.

Summary

Scientists of the 20th century failed to recognize an underlying law of physics governing the propagation of light. Linearly propagating electromagnetic radiation (and neutrinos) is subject to the *principle of velocity differential propagation.*

The Principle has two manifestations: the Redshifting of radiation in the external portion of a gravity well; and the Blueshifting of radiation in the internal portion. There are three underpinning factors:

(i) The fact that light quanta are extended entities. Light has an associated wavelength.
(ii) The fact that aether is the conducting medium of light. Put another way, light is embedded in the aether medium.
(iii) The fact that aether is not static but is involved in a dynamic flow, in accordance with the DSSU aether theory of gravity.

Combine these factors and you end up with a velocity difference within a lightwave. This velocity difference along a wavelength is the consequence of the constancy of the speed of light with respect to the conducting medium *whose own velocity is not exactly the same at the front and back ends of the energy particle.* In general, any gradient in the motion of aether, will impart a spectral shift.

The most unexpected aspect of the Principle is that lightwaves are inherently redshifted when entering a gravity well. (The view had always been that redshifting only occurred when climbing out of the gravity well.)

The most practical aspect of the Principle, identified as the Law of Blueshift Accrual, is its ability to solve the mystery of astrophysical jets. The Law drives the jets that have baffled physicists for many decades, the jets they associate with their flawed concepts of stellar black holes, the jets for which they have no plausible explanation.

The most profound implication. The Blueshifting process leads to what the previous century would have deemed impossible: the limitless expansion/amplification of energy.

The primary source of the energy comes *directly* from inside a totally collapsed mass structure. It does not come from the energy of the magnetic field; and it does not come from the energy of the rotation. The Blueshifting, the energy Amplification process, is a perpetual activity taking place in a surface layer —continuously fed by mostly low energy photons and neutrinos.

The high energy end-product of the process is blasted out through the poles. Yet the supply of gamma photons and neutrinos never diminishes. The energy flow is without limit.

This is revolutionary.

✠ ✠ ✠

Second Interlude

Praise, Flip-Flop, Rebuke

In the course of attempting to publish material directly related to the previous chapter, I encountered a wide range of opinions and judgements.

Scholarly and erudite

One reviewer had this to say:

> This is a scholarly and erudite article that deserves to be published in recognition of its tutorial nature and conviction of the author in his ideas and concepts. But, having been equally impressed by another submission, of which I was a reviewer, whose author also convinced me, for the same reasons as now, of the validity of his proposals, I am disturbed by the contradiction of what is proposed now by the present author with what was proposed earlier by the previous author.
>
> Specifically, the present author states in the abstract: *"The Terminal star converts mass to trapped radiant energy ..."*; whereas the previous author* convincingly showed that the principle of conservation of mass and the principle of conservation of energy *deny* the possibility of converting mass into energy.
>
> I therefore pass on to the present author the invitation to address, with the same professionality and tutorial attitude, the above dilemma. I insist that a summarily dismissal of such contradiction will not satisfy me, this leading to not recommending his paper for publication.

* Ling Jun Wang, *A critique on Einstein's mass-energy relationship and Heisenberg's uncertainty principle*, Physics Essays **30**, 75 (2017).

So, the reviewer was faced with the dilemma of convincing arguments leading to contradictory conclusions. In a situation like this, one must turn to fundamental concepts. One needs to appreciate the significant difference in the domain of the underlying theories being used to reach opposing conclusions. The short answer that resolves the contradiction is that Ling Jun Wang in his denial of converting mass to energy is using an *incomplete theory of mass and energy*, while DSSU theory is a broader theory incorporating a fundamental process of energy and a unifying principle linking mass and energy particles.

Einstein was a mathematical physicist. Ling Jun Wang, the denier of the possibility of converting mass into energy, no doubt, is also a mathematical physicist. As such, they construct *possible* worlds and sub-worlds based on logical explanations. The emphasis is on the "possible" aspect of the construction. A mathematical theory may be valid over here, but not over there. It may explain this phenomenon, but not some other phenomenon. Relativity theory, as Einstein admitted, is an incomplete theory. Furthermore, Einstein's *geometric* gravity theory mathematically describes the phenomenon —but does not explain it. Exactly the same thing may be said of Newton's *force* gravity theory. DSSU theory, on the other hand, *does* explain the phenomenon of gravity. Moreover, the Theory explains more phenomena, more aspects of physics, than any other previous cosmology/astrophysics theory.

In any case, the above argument along with some additional supporting evidence must have convinced the reviewer that the *primacy of processes* trumps the *primacy of mathematics*. Approval to publish was forthwith granted.

Acceptance turns into rejection!

The International Journal of Astronomy and Astrophysics gave a favorable assessment. They reported that the energy-generating model was presented and explained using good scientific arguments rooted in astrophysics and cosmology. "The Paper is accepted."

Surprise, surprise! Twelve days later, on the 24th of April 2018, I received this email notice:

> *On behalf of the International Journal of Astronomy and Astrophysics (IJAA) Editorial Board, we are sorry to inform you that your manuscript is not accepted by our journal. The referees' comments are as follows:*

> The author has tried to reinterpret the words and statements of scientists, taking them out of context to create an argument. The model that is presented has no value to science, since it just makes interpretations more intricate which is not the nature of science. There are simply too many assertions for which there is no justification [**and yet the referees give not a single example!!** – CR]. The paper should have been and should now be rejected.

> *Thank you for considering International Journal of Astronomy and Astrophysics for the publication of your research. We hope the*

outcome of this specific submission will not discourage you from the submission of future manuscripts.

Please feel free to ask us if you have any question.
Best regards,
Connie Zhang, IJAA Editorial Assistant

How nice, I've been given an opportunity to respond, a chance to find out about those *"many unjustified assertions!"* and also point out a simple fundamental truth: When assertions have been validated —that is, when they have been shown to be supported by evidence and rational arguments—they become facts of reality.

But instead of responding to my polite queries, Connie Zhang revealed why one referee flatly refused to do the review; and then had my access to the Journal's website (and author's account) blocked! (Presumably, it was to prevent me from submitting new research.)

According to editorial assistant C. Zhang, one of their reviewers refused to evaluate the article because he did not agree with the conclusion! So here is what DSSU theory is up against: For him the factual aspects, the conformity to the laws of physics, and the logic of the argument were irrelevant —the conclusion was not to his liking, and that was that. The conclusion did not fit his venerated version of how the world works. It did not agree with the determined-by-consensus "truth."

As for those *many unjustified assertions*? They remain a mystery.

Strange reasons for rejecting the Law of Energy Regeneration/Amplification

Then there was the reviewer whose fault-finding had me scratching my head. How was I to respond to someone who claims gravitational self-compression does not occur, gaseous and neutron stars do not exist, and there is no such thing as Cosmic Microwave Background Radiation?

The main objection was that my presented argument depends on two things that the reviewer, I'll call him Reviewer Strange, claimed do not exist: gaseous stars and neutron stars. "Stars are not gaseous" ... "Neutron stars do not in fact exist."

Quoting the reviewer: "The Author's arguments ultimately rest on the premise of 'gravitational collapse' of a gaseous star to form first a neutron star and thereafter an end-state Superneutron star [Terminal star]."

My response:

The argument does NOT depend on how the collapse occurs. The key requirement is that there is a limit to the aggregation of contiguous mass; that at some stage a critical state is reached; and it occurs when the surface inflow of aether reaches lightspeed.

The collapse of a massive gaseous star was used merely as a convenient easy-to-understand example. Other modes of collapse are not excluded. In order to clarify this point for the Journal readers, the following paragraph was added:

> The gravitational collapse of a gaseous star to form an end-state neutron star is not the only possible way to produce the "ultimate situation." It is possible for a massive dwarf star in a binary system to accrete sufficient mass over time and attain the critical state. Another scenario may involve a merger of a pair of orbiting massive dwarfs. The critical state may also be attained as the result of the simple collision of massive dwarfs. In other words, the collapse of a gaseous star is not an essential element in the argument. It does, however, serve as an understandable simplification —a convenient thought experiment of a controlled collapse.

The argument does *not* depend on the existence or nonexistence of neutron stars. There are just two requirements:
- The dynamic aether theory of gravity ("aether" may, if one wishes, be replaced with "universal space fluid");
- The existence of some maximum density state of mass;

And of course there must be compliance to the 2nd postulate of relativity as it applies to the relative motion between mass/matter and aether.

"Neutron star" or *neutron density* is simply a convenient shorthand for saying "this is the maximum density any stable particle or aggregation of particles can have." If someone wants to call it something else, that's quite alright —just don't call it a black hole. (I entirely agree with the Reviewer, black holes, i.e., singularity-type black holes, do not exist.)

Nuclear density was selected (and used in the various calculations) because it is the maximum that is known to exist in nature. But the argument remains valid regardless of what the actual stable density may be. The surface layer/boundary must still consist of pure radiant energy —radiating while staying stationary with respect to gravitating mass M.

All in all, a most reasonable approach.

However, it did not satisfy the reviewer. The objection remained; and so it went for three rounds.

Reviewer Strange (R. S.) in the 2nd report stated:

> "Self-compression, by gravitational or other means, does not exist."

"All physical evidence attests that the stars are condensed matter. This is proven by the thermal spectra of stars alone, since only condensed matter can emit a thermal spectrum ... The Author now includes condensed matter in his Paper and permits condensed matter to self-compress, again by means of gravitational collapse."

"Condensed matter is essentially incompressible, so neutron stars do not exist, ..." "Condensed matter too cannot compress itself by any means and cannot therefore 'collapse'..."

My response:

The Paper's argument does NOT depend on the occurrence of gravitational collapse or of self-compression (and not even of the existence of neutron stars). There is no dependency on HOW the end-state with its critical surface comes about.

Consider a sphere of your "incompressible condensed matter" having a lattice structure with the same density as water. If the sphere has a radius of about 400,000,000 kilometers, it would be a critical-state structure (by definition). Without invoking any sort of collapse or compression to greater density, it would acquire a critical-state surface simply by virtue of its size and mass quantity. And if it could maintain that same density indefinitely, that is, if water-density were the maximum density Nature allowed, it would be an **end-state structure** (by definition).

In order to avoid violating the 2nd postulate of relativity, the structure must possess (that is, it must have acquired) a surface energy layer. The Law of Blueshift Accrual is then automatically in effect. The proof has been given.

Again, it really does not matter how the end-state is attained —be it a textbook conforming collapse or the accretion of condensed matter (as favored by the Reviewer). I have presented this in a way that most people should have no trouble understanding and no difficulty in appreciating the possible ways of how the end state comes about. They can then focus attention on what happens at the surface of the end-state structure.

How matter aggregates is entirely irrelevant to the main subject — the process of energy amplification/generation by the Blueshifting mechanism of the Paper's title.

Before long, I received a third-round report.

Again, R. S. objected to the possible scenarios for gravitational collapse and gravitational aggregation. And reiterated was the rejection of the use of neutron stars in the discussion.

I was scratching my head. Why would someone deny the fact that gravitational collapse (and gravitational aggregation) does exist; it does so within DSSU theory and within the evidence-based real world. Gravitationally collapsed bodies abound. For goodness sake, we even live on one.

Anyway, this was my response:

> I have presented the collapse-and-aggregation aspect in the way it is conventionally understood, the way it is commonly described in textbooks. I have done this so as to provide a recognizable framework for the benefit of the widest readership; thus, setting the stage for the Paper's main argument. I believe this to be the optimum strategy when one is trying to introduce a new and unfamiliar concept. The main focus of the Paper is on the energy layer of a critical quantity of mass, regardless of how the critical quantity came together. The Blueshifting that occurs within *that* layer is the central issue of the Article and an important overlooked law of physics.
>
> Since mass collapse/aggregation does in fact occur —we *do* live on an aggregation of low-temperature mass (Earth) and we *are* supplied with energy from an aggregation of high-temperature mass (Sun)— it is surely reasonable to bring it into a theoretical discussion.
>
> Regarding the other objection: My use of neutron stars and neutron density has already been explained. This form of matter is cited because it is the densest known, or theorized, to exist.

Obviously this was just going in circles.

Urged withdrawal

Reviewer Strange concluded the first report by bluntly urging the Author to withdraw the Paper.

My reply: "The Process described and explained in the Paper does not stand in isolation, but is an integral part of a much larger theory. The Blueshifting mechanism is one piece of a jigsaw puzzle in which all the available pieces fit together and the overall picture, after some years of research, has finally emerged. Withdrawal of this work is not an option. DSSU theory/cosmology is a complete picture. Without the Blueshifting process, *that* jigsaw assembly would be left with a gaping hole!"

The 2nd report from R. S. had this: "The Author's Paper remains constructed upon the non-existent self-compression by gravitational collapse, now, of gases and condensed matter. I cannot see how the Author can salvage anything from his paper once gravitational collapse is expunged, as it must be."

So, R. S. still insisted that gravitational collapse needs to be removed. Once it is removed, *nothing of the Paper can be salvaged*. In other words, there is then no point in publishing; the Author might as well just withdraw the Paper.

The 3rd report again insisted that because the Paper invoked things that the reviewer said do not exist (namely, gravitational compression, gravitational collapse, neutron stars, and the cosmic microwave background), R. S. opposed its publication.

No CMB radiation!?

Reviewer Strange objected to my statement, *There exists the cosmic background radiation (including microwaves and a wide range of wavelengths); it provides a staggeringly abundant stream of relatively low energy photons*; to which R. S. countered with the assertion, "The so-called Cosmic Microwave Background Radiation does not exist." Strange pointed out, "Dr. Pierre-Marie Robitaille has proven that what astronomers and cosmologists call the CMB is in fact microwave emission from Earth, from its oceans, via the hydrogen bond in water."[1]

I responded by drawing attention to the fact that in my Article, the statement, "cosmic background radiation" was not capitalized. Included in this *generic cosmic background radiation* is, as I have stated "a wide range of wavelengths." This includes dispersed starlight, which undeniably exists; and which "falls" onto the Terminal star's *energy layer* just as it "falls" onto the Earth.

I did not receive a 4th report. Perhaps, the editor sensed the futility of additional rounds and ended the discourse, for nothing more was heard of this reviewer. I surmise that none of my above explanations were accepted. Reviewer Strange conceded nothing.

✠ ✠ ✠

---

[1] P.-M. Robitaille, *Water, Hydrogen Bonding, and the Microwave Background*, Progress in Physics Vol.2, April (2009). (http://vixra.org/pdf/1310.0129v1.pdf)

Law of Noninteraction Mass-to-Energy Conversion

3. Noninteraction Mass-to-Energy Conversion

Introduction

There are three basic ways by which mass can be converted directly to energy.

The most unambiguous mechanism is the one presented by the annihilation reaction between an ordinary mass particle and its antiparticle. For example, when an electron encounters an antielectron (a *positron*) a mutual annihilation takes place resulting in the emission of high-energy light —a pair of gamma photons. The conversion is expressed as: $e^- + e^+ \rightarrow \gamma + \gamma$.

Such particle-antiparticle pair annihilation represents a *total* mass-to-energy conversion.

A second way is through *nuclear fusion*. Under this process only a small portion of the interacting mass particles is converted to energy. Energy is released when small atomic nuclei fuse together to form larger nuclei. The fusion process (with energy release) applies when the reactants and the product are classified below iron within the periodic table of the elements. Fusion is the nuclear reaction that takes place within stars. It converts a very small fraction of the mass, of the combining smaller nuclei, into pure energy —electromagnetic radiant energy.

Lastly, there is mass-to-energy conversion through *nuclear fission*. Again, only a small portion of the mass particle is converted to energy. Energy is released when large atomic nuclei split apart to form smaller nuclei, as happens in the decay of radioactive elements, or as occurs within laboratory particle accelerators, or as takes place within specially-constructed nuclear reactors.

The mass loss corresponds to the mass difference, or deficit, between the pre-reaction component(s) and the post-reaction component(s). And the energy equivalence of this "mass deficit" is called the *binding energy* of the pre-reaction nucleus or nuclei.

The mass-to-energy conversion concept goes back to the very early part of the 20th century. The essential feature was succinctly expressed by British physicist James Jeans in one of his popular books written for a general audience.

> *It follows that any substance which is emitting radiation must at the same time be losing weight. In particular, the disintegration of any radio-active substance must involve a decrease of weight, since it is accompanied by the emission of radiation in the form of gamma-rays.*[1]

Those are the conventional three. There is, however, another mechanism for mass-to-energy conversion —one that is total and requires no interaction whatsoever. Here was something quite unconventional —a law of physics completely unrecognized by the 20th-century experts.

Conventional motion versus "stationary" motion

As a preliminary to the explanation of interaction-free conversion of mass to energy, it will be helpful to examine what is required to drive a mass particle towards lightspeed and also appreciate what happens to the particle during its acceleration to such extreme speed.

What makes this somewhat challenging is that there are actually two distinct kinds of motion —the two being fundamentally different from each other. In other words, there are two ways to bring about the approach to lightspeed. We will refer to them as **Scenario 1** and **Scenario 2**.

Scenario 1 involves interaction

The conditions applicable for this approach lie within the realm of conventional motion and 20th-century physics. When a particle is forced —say by repeated bombardment or by powerful electromagnetic fields— to travel closer and closer to the speed of light, it will gain energy. Under the right condition, whether natural or artificial, particles can and often do gain a truly extraordinary amount of energy.

The energy that a particle gains as it is accelerated has a certain *mass equivalence*. That is, energy can be measured in terms of conventional units of joules (or in units of electron volts) or in terms of mass (grams or kilograms).

---

[1] Sir James Jeans, *The Universe Around Us,* 3rd ed. (Cambridge University Press, London, 1933); p139.

THREE • Noninteraction Mass-to-Energy Conversion

When a particle is accelerate close to the speed of light, the gain in energy, the mass equivalence so produced can far exceed the original "resting" mass of the particle.

Consider this dramatic example from the CERN laboratory near Geneva, Switzerland. Within the Large Electron-Positron Collider, which operated during the 1990s, electrons and positrons (antielectrons) were accelerated to velocities within about one part in a hundred billionth (10^{-11}) of the speed of light. Speeding around in opposite directions, the particles were then smashed into each other, producing a lot of subatomic particulate debris. A typical collision might produce ten pions (*PI mesons*), a proton, and an antiproton. Let's compare the total masses, before and after:

Before collision: one electron + one positron: 2×10^{-28} gram
After collision: ten pions + proton + antiproton: 6×10^{-24} gram

Remarkably, what comes out weighs about *thirty thousand times* as much as what went in.[2] The electron and positron must have gained an amount of energy that is equivalent to thirty thousand times their rest mass!

Other particles are likewise affected. As particle physicist Frank Wilczek (Nobel Physics Prize, 2004) noted, "If you bang protons together really hard, what you find coming out is ... more protons, sometimes accompanied by their hadronic relatives. A typical outcome would be, you collide two protons at high energy, and out come three protons, an antineutron, and several PI mesons. The total mass of the particles that come out is more than what went in."[3]

One may confidently conclude that adding ever more collisional momentum and transferring ever more energy to a particle causes it to gain more and more kinetic energy. And all the while it is brought closer and closer to the speed of light. But no matter how far the process is pushed, the particle can never attain lightspeed.

In the case of charged particles, the source of the energy driving the acceleration is the electromagnetic fields generated by extremely powerful electromagnets or by natural magnetic fields generated, for instance, by rotating planets and stars.

What about other particles, like gas molecules and dust motes, charged or uncharged? It turns out that the acceleration of particles can also be achieved by simple photon bombardment. This, in fact, is the mechanism by which stars are able to clear away the enshrouding gas and dust of the cloud from which they originally coalesced; and in the process stars make themselves visible to

---

[2] Frank Wilczek, *The Lightness of Being* (Basic Books, New York, NY., 2008); p16.
[3] Ibid., p31.

the rest of the surrounding stellar neighborhood. The mechanism works whether the particle is charged, ionized, or neutral. And, in accordance with conservation rules, the higher the energy of the bombarding photons, the greater will be the transfer of momentum upon impact. The proviso is that the incident photon's energy must not be so great that it "destroys" the target particle and instead of merely transferring its momentum causes the formation of new particles.

All of the preceding was well-understood.

Scenario 1 centers on this key feature: It always involves a transfer of energy from one to the other of the interacting particles; or a transfer of energy from the electromagnetic field to the charged particle.

Scenario 2 involves no interaction

Let us look at the scenario in which there is effectively no interaction between particles. When a particle is compelled to travel ever closer to the speed of light without any particle interaction, *it does not gain energy.*

This counterintuitive situation arises passively through nothing more than a change in the gravitational environment in which the particles find themselves. What we are talking about is the change that occurs during any gravitational contraction. As a massive star contracts, its surface gravitational acceleration increases AND the inflow speed of aether at the surface also increases. Surface particles, of course, are subjected to this speed of inflowing aether. This speed increases as the star's gravitational collapse progresses. For the ultimate situation, one that follows an extreme change in the gravitational setting, surface particles (having lost their mass attribute) actually attain lightspeed. This type of collapse requires sufficiently massive stars. Such total collapse may also come about when low-mass dwarf stars acquire significant amounts of additional mass.

Let's take a closer look at this.

Gravity is always most intense at the surface of any gravitating body. The surface is special. It is special for two reasons. There the gravity is most intense; and there the inflow of the space medium has its maximum speed. For example, consider an isolated Earth-like body (at rest within the surrounding aether medium): The effect of gravity is strongest at the surface and the inflow of the space medium (aether) at the surface is 11.2 kilometers per second. For a similarly isolated star the same size as our Sun, the surface inflow is 617 kilometers per second. All else being equal, the more massive the body is, the greater will be the surface inflow speed of aether. This is all in accordance with the aether theory of gravity. Similarly, but this time by holding the total mass constant, the smaller the body is, the greater will be the surface inflow speed of aether (and also the greater will be the surface gravity).

Stated another way, increasing a planet's or star's density increases the surface inflow speed. And here is the point: As the structure collapses towards Nature's maximum density state, the surface inflow speed tends towards lightspeed.

Our focus of interest is with the surface mass —its embedded particles. Surface particles/objects travel through the aether as explained and illustrated in Appendix A. Those particles, by virtue of location, are in a very real sense moving through the space medium. Although, apparently "stationary" at the surface, they are speeding upward through the aether flow. As an astronomical body undergoes collapse or merely acquires more mass, the speed of such *in-place motion* increases. This collapse process is explained in more detail below under the heading "Mass-to-energy conversion mechanism."

Key feature: Scenario 2 does not involve significant transfer of energy from one particle to another. (The transfer of energy that does occur is due to comparatively small *gravitational heating*.) There is of course the passive pressure of the underlying mass supporting the surface particles; but very little transfer of energy as such. During an actual stellar collapse (end-stage collapse), there are no supporting layers since everything is "falling" inward. By the time the structure has collapsed to its final maximum density state, the "surface" mass will already have transformed to pure energy.

Recapping: There are, within the realm of established physics, two ways to accelerate mass. One requires the transfer of energy, the other does not.

Under Scenario 1, the acceleration of the subject particle(s) is caused by some type of physical interaction.

Under Scenario 2, the acceleration of the particle(s) is associated with gravity itself. It is the acceleration that a person experiences while merely standing still at the Earth's surface. It is the acceleration that stationary surface particles "experience."

The energy triangle

For an understanding of the motion of objects and particles at a fundamental level, several forms of energy must be taken into account, namely the *kinetic* energy and the energy associated with *momentum*, as well as the *total* energy and the *mass* energy. The mathematical expressions for these energies have been shown to be valid (if properly interpreted) for all speeds, from zero to lightspeed.[4][5] Moreover, the four energies are mathematically related; in

---

[4] L. B. Okun, *The Concept of Mass*, Physics Today, Vol.**42**, No.6, 31 (1989); p32.
(Doi: http://dx.doi.org/10.1063/1.881171)

[5] C. Ranzan, *Mass-to-Energy Conversion, the Astrophysical Mechanism*, Journal of High Energy Physics, Gravitation and Cosmology Vol.**5**, No.2, pp.520-551 (2019).
(Doi: https//doi.org/10.4236/jhepgc.2019.52030)

fact, they are related in such a way that they can be configured into a most useful triangle —the *mechanical energy* triangle (**Figure 3-1**). It is also referred to as the relativistic energy triangle.

The base of the triangle represents the energy of mass itself. Usually called the *rest energy*, it is defined by the well-known expression $E = mc^2$.

The long side of the triangle represents the total mechanical energy, which includes the particle's rest-mass energy and its energy of motion —kinetic energy. The total energy is just their sum:

(Total energy) = (Rest energy) + (Kinetic energy).

The height of the triangle stands for the *momentum energy* of the object or particle.

The relationship that combines all the components of mechanical energy is

$$\left(\begin{matrix}\text{Total}\\ \text{energy}\end{matrix}\right)^2 = \left(\begin{matrix}\text{Rest}\\ \text{energy}\end{matrix}\right)^2 + \left(\begin{matrix}\text{Momentum}\\ \text{energy}\end{matrix}\right)^2.$$

The fact that this energy expression has the basic Pythagorean form means that the energy graphic is actually a right-angled triangle as shown in **Figure 3-1**. Most University-level physics textbooks present the topic in a similar graphic way.

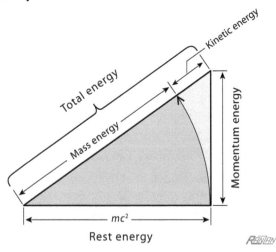

Figure 3-1. The energy triangle represents the Pythagorean relationship among the energy components associated with bodies or particles. It captures the relativistic relations among the total energy, the rest energy (or mass energy mc^2), the kinetic energy, and the momentum energy. It can be applied to elementary particles —be they mass or massless, be they stationary or moving at the speed of light. (Since the focus is on the energy associated with motion, the "Total" here refers to total mechanical energy.)

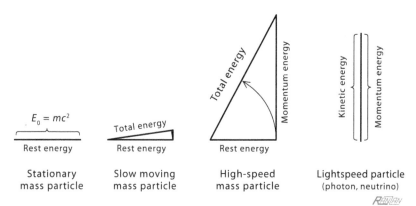

Figure 3-2. Energy triangle works for all situations pertaining to mass and speed that are found in nature. The flattened "triangle" on the left pertains to a mass particle with zero speed. The short triangle applies to a "slow" moving mass particle. The tall triangle applies to a high speed mass particle and shows its full complement of energy components. The right-hand graphic is for energy particles, which, by definition, have no mass and propagate in a vacuum at the speed of light; since energy particles have no rest energy, the two other sides of the triangle come together making it clear that the two energies of motion must be equal.

The energy formula, and its graphic representation, works for both mass and massless particles. Moreover, the equation and the triangle are applicable to all speeds —from zero through to lightspeed. This is not something new; it is all part of standard physics. **Figure 3-2** demonstrates how the triangle can be manipulated to represent these diverse situations in terms of speed.

This broad applicability is of particular importance as it means that the energy triangle can be used to explain both of the scenarios discussed earlier.

Energy triangle applied to Scenario 1

The first thing to note is that regardless of how much energy is applied to a mass particle, there is no way to get to the pure energy state (as represented in the right-hand graphic of **Figure 3-2**). More broadly, there is absolutely no non-collisional way to transform a subatomic particle to pure energy with 20th-century physics. Restated, there is no way except through particle-antiparticle collision to bring about a transformation to pure energy within the old physics.

Next, it is important to note that for several decades of the 20th century there was some confusion over the fundamental nature of mass. In the development of the Special Theory of Relativity, a thought experiment of a ballistic nature arose. It was a situation in which if two relatively moving observers agreed quantitatively on the mass of a moving object, then they would not be able to agree on its momentum. Similarly, if they instead decide to agree on the momentum (as viewed from their respective frames), they would not be able to agree on the object's mass.

At this point there were two choices available to theorists. They could assume that momentum principles —in particular, the conservation of momentum— do not apply at large velocities. Or alternatively, they could look for a way to redefine the momentum of a body in order to make momentum principles applicable to Special Relativity. And the simplest way to do this was to allow mass to change with its speed. And this is the alternative that was chosen by Einstein. He showed that all observers will find classical momentum principles to hold if the mass m of a body varies with its speed v according to[6]

$$m = m_o \Big/ \sqrt{1 - \left(v^2/c^2\right)},$$

where m_o, the *rest mass*, is the mass of the body measured when it is at rest with respect to the observer.

And so it was, Einstein and his followers, including the influential Sir Arthur S. Eddington, for a good number of years claimed that mass increases with speed —a belief that became a signature feature of special relativity theory. But, as with several of his other early viewpoints, Einstein reconsidered and changed his stance. In a private letter written in 1948, he made it clear that he had abandoned the idea, stating,

> "It is not good to introduce the concept of the mass $M = m \big/ \sqrt{1 - v^2/c^2}$ of a moving body for which no clear definition can be given. It is better to introduce no other mass concept than the 'rest mass' m. Instead of introducing M, it is better to mention the expression for the momentum and energy of a body in motion." –A. Einstein, in letter to Lincoln Barnett [7]

Gradually, but not without years of uncertainty and opposition, the fixed-mass view prevailed.

The "modern" view, the now-well-established stratagem among physicists is to treat mass as being special and keep it constant —regardless of its state of motion. The generally accepted view is that there is no distinction between *rest* mass and *relativistic* mass. When an object is accelerated, its gain in energy goes entirely into increasing its kinetic and momentum energies. Strange as it may seem, there is no theoretical upper limit. Let me emphasize

---

[6] Ronald Gautreau and William Savin, *Schaum's Theory and Problems of Modern Physics* (McGraw-Hill Inc., U.S.A., 1978); p39.

[7] As in L. B. Okun, *The Concept of Mass*, Physics Today Vol.**42**, No.6, 31 (1989); p32. (Doi: http://dx.doi.org/10.1063/1.881171)

this, a particle's motion energy has no theoretical limit.

However, uncertainty persists.

In the CERN accelerator experiment cited earlier, the kinetic energy of the electron and positron pair was great enough to be converted into mass particles equivalent to 30,000 times the actual mass "carried" by the original pair! The question has to be asked, *what form did this energy of motion take?* Sure, it's easy to say this vast energy is manifest in the momentum and quantified in the kinetic energy's mathematical expression. But, undoubtedly, there must be something deeper.

Evidently something is being transferred during the interaction that causes the acceleration. Something is given-up by the initiator in the interaction and taken-up by the recipient. The charged particles gain and the magnetic field loses. Yes, the result is a gain in *total* and *kinetic* energy —a neat and clean accounting quantity. But what is it in a nuts-and-bolts sense? What is this something that has been shown to be many thousands of times more substantial than the naked mass particle? ... Does the particle's constituent photon (or photons) shorten its wavelength, thereby shrinking the particle as it gains energy —thus becoming more massive (since more energy is equivalent to more mass)? Or do ancillary energy particles (photons, neutrinos) somehow attach themselves to the recipient particle? Possibly, the kinetic energy manifests as wavelets riding atop the main carrier wave(s) of the mass particle. It is an interesting speculation.

Based on cognizance of the fundamental process of energy [8][9][10], one thing is for certain. As the kinetic energy of a particle increases, there is a corresponding increase in aether absorption/consumption. And for all intents and purposes, this is equivalent to an increase in mass! (However, this does not happen with *noninteraction conversion* via Scenario 2.)

The question remains: At the most fundamental level, what is it that "carries" the extreme energy when Scenario-1 particles approach lightspeed? ... It remains one of the deepest mysteries in physics.

Let us return to the energy triangle. The established approach in applying the relativistic energy equations (represented by the energy triangle) is to treat the mass energy as a fixed quantity. This means that the base of the triangle does not change in length and makes it easy to demonstrate what happens as a particle or object gains energy as it is accelerated. As can be seen in

---

[8] C. Ranzan, *The Fundamental Process of Energy, Part I*, Infinite Energy Issue **#113** (Jan/Feb 2014). (http://www.infinite-energy.com/iemagazine/issue113/index.html)

[9] C. Ranzan, *The Fundamental Process of Energy, Part II*, Infinite Energy Issue **#114** (Mar/Apr 2014). (http://www.infinite-energy.com/iemagazine/issue114/index.html)

[10] C. Ranzan, *DSSU Validated by Redshift Theory and Structural Evidence*, Physics Essays Vol.**28**, No.4, pp.455-473 (2015). (Doi: http://dx.doi.org/10.4006/0836-1398-28.4.455)

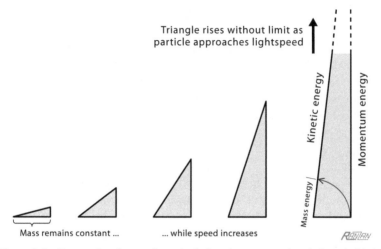

Figure 3-3. Conventional way of manipulating the energy triangle is to hold the base constant (in keeping with the assumption that mass is invariant) and then adjust the triangle's height. There is no theoretical limit to this process of pumping kinetic energy into a particle, and as the energy grows without restriction, the speed approaches c. However, the particle never attains lightspeed and, therefore, can never become massless. The sequence captures the essential aspects of Scenario 1.

Figure 3-3, the height of the triangle increases in proportion to the energy gained. All the while, the base, representing the mass, must remain unchanged.

Underscoring the main point of Scenario 1: When particles are driven toward lightspeed, the triangle representing the energy becomes stretched above a fixed base. There is no theoretical limit to this process.

In contrast, the triangle for Scenario 2 has a well-defined height limit, as will be explained next.

Energy triangle applied to Scenario 2

No matter how much energy and momentum is pumped into a particle, one can never bring it up to the full speed of light in a vacuum. It is an incontrovertible fact.

But what happens when a mass particle acquires motion and is brought to ever higher speeds without any interaction —no collisions, no energy fields, no exotic force? Nature has a way of doing this, as will be explained in a moment. But first, we need to recognize a simple fact. In the absence of some interaction, there is no logical way for the particle to gain energy; consequently, *the total mechanical energy of the particle can then never be greater than the original mass energy* (its rest energy).

For example, an electron has mass energy equal to 0.511 mega electron

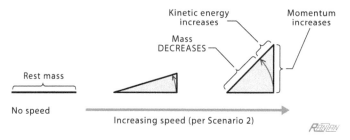

Figure 3-4. Manipulation of the energy triangle for the situation when there is no interaction and, therefore, no energy gain. As the speed of a particle/object increases, as a consequence of a change in the gravitational environment, its energies of motion increase; while simultaneously its energy of mass decreases. The triangle gains in height, *while the base contracts*.

volts (corresponding to its mass of 9.11×10^{-31} kilogram); and a proton has 938 mega electron volts (corresponding to its mass of 1.67×10^{-27} kilogram).

When at rest, this is the only energy they possess.

When these particles acquire speed under Scenario 2, they naturally also acquire kinetic energy and momentum. And naturally, the greater the speed, the greater will be the momentum, etc.

Now for the crux of the matter. Since no interaction —no energy field, no impacting— is involved here, where then, does the energy for the momentum come from? ... There is only one place. It comes from the particle's own mass. There simply is no other source. What this means in terms of the energy triangle is shown in **Figure 3-4**.

In order to accommodate situations of extreme gravitational environments as occur under Scenario 2, the basic energy triangle needs to be reinterpreted. Under the special-situation interpretation, the *total mechanical energy* remains constant, mass itself varies, and motion ("stationary" motion) is referenced to the universal medium (aether). See **Figure 3-5**.

Noninteraction mass-to-energy conversion

The stop-action sequence shown in **Figure 3-6** represents a greatly simplified gravitational collapse of a massive star. The contraction sequence ends with the formation of a Terminal star. Recall, this means that the star becomes enveloped in a lightspeed boundary. It also means that some "surface" mass must convert to pure energy.

Let us now see how this surface mass conversion can be interpreted with Einstein's relativistic mass equation,

$$m = m_o \Big/ \sqrt{1 - \left(v^2/c^2\right)} \ .$$

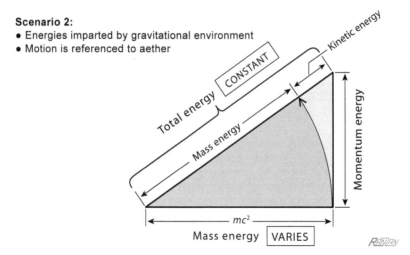

Figure 3-5. Energy triangle applicable to the Scenario-2 situation —the gravitational situations in which particles/objects suffer practically no interaction. Since there is no interaction and, therefore, no energy gain, the "Total energy" stays constant. The length of the hypotenuse remains unchanged, while the values of all the energy components will change with the velocity as referenced to the universal space medium. Here is the important difference with the earlier energy triangle: While the energy components for the Scenario-1 triangle use relative-to-observer velocity, the equations for the Scenario-2 triangle use relative-to-aether velocity.

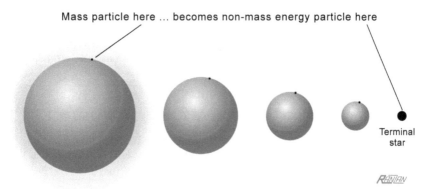

Figure 3-6. Idealized gravitational collapse of a massive star, ending up in the Terminal state. During the collapse, a typical particle on its surface undergoes total conversion to energy. The resulting energy particle, a photon or neutrino, becomes trapped in the Terminal star's surface. No interaction is involved; the particle only "experiences" a change in its gravitational environment.

THREE• Noninteraction Mass-to-Energy Conversion

Recall, in Einstein's original equation for mass, mass m changes with the *relative* motion between observer and the object of interest. The greater the velocity magnitude, the greater will be the predicted mass. Mass m_o is a constant representing the conventional rest mass; it is the same as the mass used in Newtonian mechanics. Before making modifications to this equation, let's be clear on the meaning of the original terms here. The moving mass is the *apparent mass* when there is relative motion; the rest mass is the *apparent mass* when there is zero relative motion; and unsubscripted υ is the *relative* speed. The radical term is called the Lorentz gamma factor[11] and symbolized by the Greek letter γ. Expressed as a word equation we have

$$\begin{pmatrix} \text{MASS moving} \\ \text{with respect to observer} \end{pmatrix} = \begin{pmatrix} \text{MASS at rest} \\ \text{with respect to observer} \end{pmatrix} (\text{Gamma factor}).$$

In other words,

(VARIABLE mass) = (CONSTANT mass)×(Gamma factor).

Now for Scenario 2, it is the rest mass, *the mass resting on the surface* in **Figure 3-6**, that actually varies. And what was previously the "variable" mass (variable because of relative motion between mass and observer), now changes to serve as the "constant" mass (constant because we are now taking aether into account). To encode this situation, we need to arrange the terms this way:

(CONSTANT mass) = (VARIABLE mass)×(Gamma factor);

$$\begin{pmatrix} \text{MASS at rest} \\ \text{with respect to aether} \end{pmatrix} = \begin{pmatrix} \text{MASS at rest} \\ \text{with respect to surface} \end{pmatrix} (\text{Gamma factor}).$$

Here is the same equation just restated:

(**Mass** constant-value) = (**Mass** at surface) γ_a.[12]

Rearranged in accordance with the simple rules of algebra, it is equivalent to

$$(\textbf{Mass}_{\text{surface}}) = (\textbf{Mass}_{\text{constant-value}}) \sqrt{1-(v_a/c)^2}.$$

---

[11] Lorentz gamma factor, γ, is $\dfrac{1}{\sqrt{1-(v^2/c^2)}}$, where υ is the relative speed between observer and observed.

[12] This time the gamma factor, γ_a, is referenced to the aether: $\dfrac{1}{\sqrt{1-(v_a/c)^2}}$. The velocity term is appropriately subscripted to indicate this important distinction.

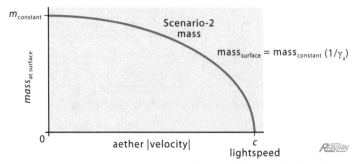

Figure 3-7. How Scenario-2 mass varies with speed. The graph shows how mass varies when the cause of the change in speed is not an applied force but, rather, a change in the gravitational environment in which the particle or body finds itself. Mass$_{Surface}$ can diminish from a maximum value of mass$_{constant}$, its value as measured in an aether rest-frame, down to a minimum of zero. (Symbol γ_a is the aether-referenced gamma factor.)

Note carefully, the velocity has now been referenced to the aether medium and is subscripted with the letter "a" to indicate this. Also, the "constant-value" mass is the mass as measured in an aether rest-frame. (The latter corresponds to the Newtonian mass.)

The graph of this relationship is presented in **Figure 3-7** and shows how mass *decreases* with increasing speed of the aether inflow. In conformance to conservation laws, the mass is converted to non-mass energy.

While examining the graph in **Figure 3-7**, it is worth reflecting on how radically different the two Scenarios are. Scenario 1 is about what someone observes or what some instrument measures. But Scenario 2 is about what actually happens to an object irrespective of any observer.

Returning to our example particles, let us consider their total conversion. An electron with its 9.11×10^{-31} kilogram of mass will convert to 0.511 million electron volts of pure energy. It will, according to standard-physics equations, emerge from the conversion as a photon of wavelength 2.42 picometers (or 2.42×10^{-12} meters).

For a detailed treatment on the structural nature of the electron and how it undergoes conversion, see the journal Article[13].

Turning to the proton. We know that a proton with its 1.67×10^{-27} kilogram of mass will convert to 938 million electron volts of pure energy. But what is not known is the proton's photon configuration. Is the proton an intricate self-looping configuration of a single photon or is it a complex linkage of possibly

---

[13] C. Ranzan, *Mass-to-Energy Conversion, the Astrophysical Mechanism*, Journal of High Energy Physics, Gravitation and Cosmology Vol.5, No.2, pp.520-551 (2019); Section 5. (Doi: https//doi.org/10.4236/jhepgc.2019.52030)

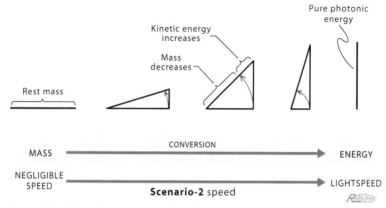

Figure 3-8. Schematic demonstration of mass-to-energy conversion. Since under the Scenario-2 conditions described in the text no energy transfer occurs, there is no reason to expect any change to the total energy (mass energy plus kinetic energy). Thus, the energy triangle's hypotenuse remains constant (while the base shrinks) during the conversion. Conservation law is obeyed; with nominally no loss or gain in energy. The conversion ends with the photon(s) as the carrier(s) of non-mass kinetic energy (more properly called momentum energy).

3 photons? In any case, once the conversion is complete, only linearly-propagating photons remain. The conversion process changes all self-looping into linear propagation.

The Scenario-2 conversion, from "stationary" mass to photonic energy, may be summarized in terms of the energy triangle. See **Figure 3-8**. Ignored is the comparatively small thermal energy gained from the loss of gravitational potential energy.

For the full mathematical treatment of how the mass-to-energy conversion works, see the original journal Article[14].

Mass-to-energy conversion mechanism

> *The ideal is to reach proofs by comprehension rather than by computation.* –Georg Bernhard Riemann (1826-1866)

Simplified stellar collapse and the end-state structure

Consider a star with a mass equivalent to about 3.4 times that of our Sun. It is at the end of its normal lifespan and it has no rotation. Imagine the fate of this massive star as it undergoes a simplified gravitational collapse. No sudden

---

[14] C. Ranzan, *Law of Physics 20th-Century Scientists Overlooked (Part 3): Noninteraction Mass-to-Energy Conversion*, International Journal of High Energy Physics Vol.7, No.1, pp.19-31 (2020). (Doi: http://dx.doi.org/10.11648/j.ijhep.20200701.14)

implosion, no rebound ejection, no nova event, and no supernova explosion — just an uncomplicated slow-motion contraction.

Depending on the theory being used, there are two radically different outcomes predicted. One aligns with the long-established Einstein theory of gravity, which conspicuously lacks a causal mechanism; the other accords with the validated DSSU aether theory of gravity, and which *does* have a causal mechanism. The one conforms to the 20th-century thinking on extreme gravitational collapse, which, if taken to its logical conclusion, ends in a physical impossibility; the other produces a perfectly natural end-state.

Let us briefly look at the traditional general-relativity view. Upholding the conventional view is the notion that there is no force in nature that can prevent the star's complete implosion. Given the aggregated quantity of 3.4 Suns, there is simply too much mass. As the star gradually uses up its nuclear fuel, it collapses. As the star collapses, the mass density rises. It rises until it reaches the ultimate density —nuclear or neutron density. The conventional belief is that stationary maximum-density stars cannot exceed 2.16 solar masses. *"With an accuracy of a few percent, the maximum mass of non-rotating neutron stars cannot exceed 2.16 solar masses."*[15] What this means is that if the collapsed object's neutron-density mass exceeds 2.16 Suns, then Nature's ultimate pushback, the *degeneracy pressure of neutronium matter* as it is called, is not sufficient to sustain the structure. If the mass exceeds such limit, which it clearly does in the hypothetical example of 3.4 Suns, then there is no known force to prevent *total* collapse. The collapse proceeds as shown in **Figure 3-9a**; and is predicated on there occurring a catastrophic physical breakdown ending in a mathematical entity called a singularity. Such is the conventional view.

As expert Paul Davies, a professor of mathematical physics, states in one of his popular books, "We now know, from relativity theory, that no force in the Universe can prevent the star from continuing to collapse, once it has reached the light-trapping stage. So the star simply shrinks away, essentially to nothing, leaving behind empty space —a hole where the star once was. But the hole retains the gravitational imprint of the erstwhile star, in the form of intense space and time warps."[16]

In other words, the star becomes a point mass that is somehow able to retain all of the erstwhile star's gravitational influence. Furthermore, it is surrounded by a lightspeed boundary —what in relativity theory is called an *event horizon*. Anyway, at some point in history, this hypothetical type of star

---

[15] Goethe-Universitat Frankfurt am Main (2018) *How massive can neutron stars be?* ScienceDaily (January 16, 2018). (https://www.sciencedaily.com/releases/2018/01/180116093650.htm)

[16] Paul Davies & John Gribbin, *The Matter Myth* (Simon & Schuster, Touchstone, New York, 1992); p265.

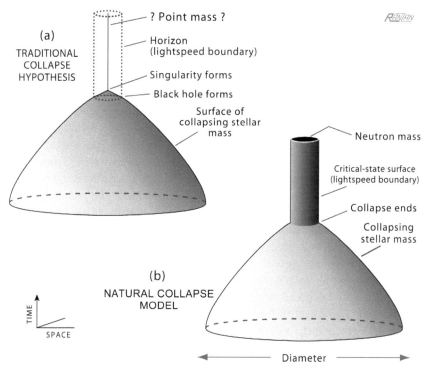

Figure 3-9. Contrasting views of stellar gravitational collapse. (a) Schematic of traditional unstoppable gravitational contraction; no traditional force can prevent it, so it is believed; the result is a singularity-type black hole. Its critical boundary is an event horizon located in free space. (b) Natural end-stage collapse results in a critical-state neutron star. Its critical boundary is a pure-energy layer located at the physical surface.

became known as a *black hole*.

Admittedly, the nature of the point mass was a mystery; and worse, a paradoxical mystery; but physicists knew it was there, somewhere deep below the event horizon; and they knew what happens to anything that succumbs to its gravitational influence.

Professor Davies again, "The balance of opinion among the experts is that all matter entering a black hole eventually encounters a singularity of some sort."[17]

Decades earlier a professor of astronomy and astrophysics, Herbert Gursky, had stated, "*Unless some new laws of physics intervene, the matter*

---

[17] Paul Davies & John Gribbin, *The Matter Myth* (Simon & Schuster, Touchstone, New York, 1992); p272.

will shrink down to a singular point."[18] Yes indeed, Professor Gursky, a new law of physics does intervene.

We now know the singularity was completely unnecessary. There was something the 20th-century experts failed to recognize. It turns out that no additional force is needed to prevent "catastrophic physical breakdown"!

Turning our attention now to Natural gravitational collapse: The key to understanding the Natural collapse mechanism is in recognizing the convergent flow of aether into mass —any and all mass. The defining characteristic of mass is its ontological need to continuously absorb/annihilate aether [19]. (This characteristic also applies to energy particles.) The very existence of the mass depends on a steady inflow. Now, the nature of fluid dynamics requires that, as the medium converges towards a mass body, the speed of the flow increases. From the two facts just cited, it follows that the maximum flow speed occurs at the surface of the mass structure —at the surface of the collapsing stellar structure, in this case. After penetrating the surface, the aether is absorbed/annihilated as it continues its convergent course, but with steeply diminishing speed, towards the center of mass.

The important point is this: the inflow speed is maximal at the surface *and* as the area of the surface decreases (due to the contraction of the stellar mass to ever greater density) the aether flow speed must increase. In order to sustain the same quantity of mass, the same volume of aether needs to flow through a smaller surface area. It is somewhat analogous to the *venturi* effect found in the spraying end of an ordinary garden hose. Therefore, as the structure collapses, and this is still being visualized in our minds as a gradual non-violent contraction, the flow increases until eventually it becomes critical. The aether inflow attains the speed of light; the structure has acquired a critical-state surface (**Figure 3-9b**).

It is at this stage that the collapse comes to an end. The mass has reached its peak density (neutron density 1.6×10^{18} kilograms per cubic meter); it now has a one-way lightspeed boundary; yet it does not violate the speed rule of special relativity (as will be confirmed shortly). The final product is an *end-state neutron star* with a radius of 10 kilometers. It simply cannot contract further —*even if more mass is added*. The reason it cannot contract further is explained in Chapter 4.

The significant difference is with the critical boundaries: With a

---

[18] H. Gursky, *Neutron Stars, Black Holes, and Supernovae*, as in Frontiers of Astrophysics, Eugene H. Avrett, editor, (Harvard University Press, London, England, 1977); p158.

[19] C. Ranzan, *The Nature of Gravity —How one factor unifies gravity's convergent, divergent, vortex, and wave effects*, International Journal of Astrophysics and Space Science Vol.6, No.5, pp.73-92 (2018). (Doi: http://dx.doi.org/10.11648/j.ijass.20180605.11)

singularity-type black hole, the critical boundary is an event horizon located in free space. With a Natural collapse to a Terminal neutron star, the critical boundary is a pure-energy layer located right on, or at, the physical surface. The Terminal neutron star's critical-state surface is a *physical* one-way boundary.

Surface mass transitions to energy state

The mechanism of conversion of mass to energy is confined to the surface region —a rather thin surface layer. This, as we have noted, is where gravity is always most intense; and where aether flow speed is greatest. The surface region provides the gravitational environment where the transformation takes place. Essentially, what happens is that the surface material of the pre-collapsed star transitions into pure energy.

The stellar mass is the 3.4-solar-mass spherical body introduced earlier. The mass constituting the surface —be it gaseous, plasmic, solid, or a crushing superfluid degenerate state— is subjected to an aether "wind" in accordance with the following equation (as derived in the Appendix A),

$$v_{\text{aether@surface}} = -\sqrt{\frac{2GM_{3.4\odot}}{R_{\text{surface}}}}.$$

As the structure contracts to a smaller and smaller radius, this equation traces the increase in the flow speed at the surface. **Figure 3-10** shows a graph of the change in speed. The simplest way to interpret the graph is to picture a stop-action sequence of the collapse. The sphere and its surface is shrinking; for each radius reached by the surface as indicated along the horizontal axis, at those points, the speed of the aether flow is calculated. The collapse necessarily ends when the flow reaches lightspeed —when the radius bottoms out at 10 kilometers.

For example, at the stop-action point where R_{surface} equals 1000 kilometers (beyond the right-hand edge of the graph), the boundary-layer mass will experience an aether flow of one-tenth lightspeed (or 30 000 kilometer per second). When the radius reaches 250 kilometers, the aether flow will be one-fifth the speed of light. Continuing, when the aether speed is calculated at radius 40 kilometers, the surface flow will be one-half lightspeed (150,000 kilometers per second). And if the collapse is checked when the radius shrinks to 15.7 kilometers, the aether flow will be eight-tenths lightspeed (240,000 kilometers per second). Finally, at 10 kilometers from the center of mass, it all ends; a critical velocity boundary then forms, structural stability is attained, further collapse is precluded. Four of the "stopping points" are indicated in **Figure 3-10**.

One of the physicists who reviewed the *noninteraction mass-to-energy law* had trouble understanding the role of the surface stuff, or so he claimed. So,

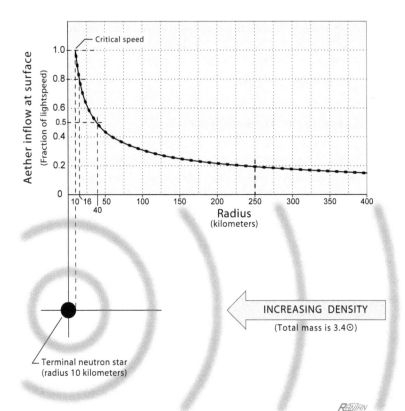

Figure 3-10. Gradual collapse of a 3.4 solar mass (without undergoing any mass loss during the size shrinkage) results in a Terminal star (an *end-state neutron star*). The graph plots the aether inflow speed at the surface of this 3.4☉-mass structure. The equation for the surface-inflow curve is given in the text. The collapse necessarily ends at the point where the flow reaches lightspeed —when the radial contraction reaches 10 kilometers. The density of the terminal structure is the maximum that Nature permits —nuclear density, 1.6×10^{18} kg/m^3.

for the benefit of any overly-specialized experts, let me clarify the meaning of "surface mass/matter." It is that region of the collapsing structure (in **Figure 3-9b** and **Figure 3-10**) which —when collapse stops— experiences lightspeed inflow of aether (popularly called the vacuum or the quantum foam). This surface matter is what becomes the pure-energy skin of the Terminal star. Obviously, "surface mass" only has meaning during collapse; the final structure does not, and cannot, have a mass surface. As for the gravitational process being considered: Think of the described slow-motion collapse as a convenient thought experiment and as an understandable simplification of what actually happens.

Let us examine the course of the collapse in terms of the density (assumed to be uniform). After the star's nuclear energy processes have run their course, with such fuel depleted, the star will contract to become a white dwarf. At this stage, it has a radius of 54,450 kilometers and its matter exists in the *electron degeneracy state* with a density of 10^7 kg/m^3. The contraction continues. Upon reaching the radius of approximately 2528 kilometers, the density has increased to 10^{11} kg/m^3 and the electron degeneracy state is replaced by the *neutron degeneracy state*. This neutronium form of matter increases in density to the ultimate that nature permits[20]. The radius has shrunk to only 10 kilometers; and the surface is in the critical state. The neutron star now has a lightspeed boundary.

During the collapse sequence, any mass particles embedded in the surface experience the aether flow as a headwind through which they must "propagate" in order to maintain their surface position. As the collapse progresses and the aether wind increases, the mass attribute of the particles undergoes its gradual conversion to pure energy. The photons that constitute those particles are compelled to unravel, so to speak. For the vectorial treatment of the transition as it applies to the electron, see the journal Article[21].

Consider the purely intuitive perspective. Mass objects or particles cannot exist in (or on) the critical-state surface; this is simply because mass cannot travel at lightspeed. During the collapse process, the structure does have a mass surface; upon collapse completion, *that* mass will have lost its "mass" quality. But it is not lost instantaneously. An electron, for instance, does not suddenly go from 9.11×10^{-31} kilogram down to zero the moment lightspeed is attained. No, the mass loss is a transitional process —the process accompanying the surface transition to criticality. (Keep in mind, the stellar collapse sequence often involves long periods of stability between very short periods of contraction.) The intuitive answer is found in the *principle of mass variance* as depicted in **Figures 3-7** and **3-8**.

The end result is a surface layer of photons (and neutrinos) propagating radially outward but never actually escaping. The layer of radiating particles trapped in this manner is probably only a few centimeters thick (and a few meters at most).

Collisional mass-to-energy conversion

Although the main theme of this chapter is the noninteraction conversion to pure energy, some attention should be given to the unavoidable situations

---

[20] Nature's ultimate density is estimated to be 1.6×10^{18} kg/m^3.

[21] C. Ranzan, *Mass-to-Energy Conversion, the Astrophysical Mechanism*, Journal of High Energy Physics, Gravitation and Cosmology Vol.**5**, No.2, pp.520-551 (2019); Figure 11. (Doi: https//doi.org/10.4236/jhepgc.2019.52030)

involving collisions.

The collapse described above ends in a Terminal neutron star. The question now is what happens when a compact object, say on the scale of small or large space rocks comparable to meteors or asteroids, falls onto its energy layer? Obviously, this involves a significant collisional interaction. Since the Terminal star, by definition and physical laws, cannot change its size or its matter content; only three things can happen. (i) The incoming mass and its kinetic energy undergo conversion to photonic energy along with the polar ejection of a more-or-less equivalent amount. (ii) A corresponding quantity of mass undergoes extinction/annihilation, as explained in Chapter 4. (iii) Some combination of the first two takes place.

But what if the infalling particle striking the energy layer is a subatomic particle, say an electron? ...

Consider the case for such a particle comoving with the aether. Flowing *with* aether means it has no kinetic energy with respect to aether. But what about the kinetic energy with respect to the neutron star? If the electron is comoving with the inflowing aether (meaning it is at rest with respect to aether), it will strike the Terminal star's critical boundary with the full speed of light (**Figure 3-11a**). The interaction aspect of this would be equivalent to the collision between two electrons each traveling at half the speed of light. Our single electron, during the impact (during the transformation to photonic energy), should gain as much energy as an electron-positron pair together would gain from a collision in which each particle is traveling at one-half lightspeed within a particle accelerator.

The calculation of this collision energy is a straightforward textbook exercise; it turns out to be 0.158 million electron volts (MeV). This energy becomes available during the instant of conversion. It seems reasonable to expect that the collision energy and the mass energy would both participate in the conversion. Meaning, the resultant photon would be more energetic (shorter wavelength), since, as discussed earlier, the particle's *total* energy is the determining factor. See **Figure 3-11b**.

When the collision energy (0.158 MeV) is added to the electron's basic mass energy (0.511 MeV), the energy of the resultant photon is expected to be 0.669 million electron volts. This represents a 31-percent gain in the energy conversion process.

Thus, when comoving particles are involved, it is surmised that a 31-percent increase in the photonic energy will occur in the conversion event (the event of the particle striking the critical surface).

What about a particle that is moving THROUGH aether and *does* have significant motion energy? If a mass particle is speeding *through* the aether and heading towards the end-state star, then this through-aether speed must be added to the speed of the aether itself. Say the particle is moving with 90

THREE • Noninteraction Mass-to-Energy Conversion

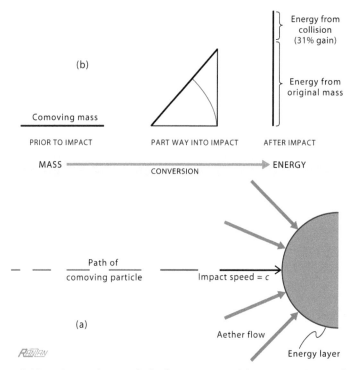

Figure 3-11. Conversion analysis for mass particle —comoving with the inflowing aether— that falls onto a Terminal neutron star. The particle has no kinetic energy with respect to the aether itself. Part (a): Since aether is carrying the particle along with it, the speed of the collision must be equal to lightspeed. Part (b) gives the energy-triangle schematic of the mass-to-energy conversion. As described in the text, the energy gained from the collision is equal to 31 percent of the original mass.

percent lightspeed, then the speed at the instant of impact must be 190 percent lightspeed (with respect to the critical surface, and with respect to background Euclidian space). The mass will actually strike the critical boundary with far more than the full speed of light. The interaction aspect of this would be equivalent to a head-on collision between two identical particles each traveling at 95 percent the speed of light. The end-state neutron body is, naturally, far, far too dense for anything to penetrate. Most certainly nothing penetrates deeper than the surface layer —not even neutrinos.

The outcome of the almost instantaneous conversion to pure energy is predicted to be a photon with considerably more energy than in the previous comoving example. Again, it is the particle's *total energy* that undergoes conversion. For the details and a numerical example, see the journal-published Article[22].

Some relevant aspects of mass-to-energy conversion

The question of further collapse. Can it happen that the end-state neutron star, over time, will collapsing further —say, when it runs out of internal fuel or nuclear energy? ... No. All the nuclear energy has long been exhausted. No fuel remains. The neutron degeneracy pressure mentioned earlier is what prevents any further collapse. This degeneracy pressure is unique. It cannot be changed in any way. It is unaffected by temperature. Even at absolute zero the pressure persists. It can remain in equilibrium with gravity forever. While the pressure generated from some form of internal energy (chemical or nuclear) lasts only so long, degeneracy pressure continues forever.[23]

Gravitational heating. During gravitational contraction, the potential energy (of gravity itself) is converted to thermal energy. Given that gravitational contraction generates heat, how does this thermal interaction affect the conversion process? When surface particles gain gravity-induced thermal energy during final collapse, the process is not altered; it is still by virtue of being subjected to a lightspeed environment that the particles must transform to pure energy. Thermal energy is added to the total energy (the hypotenuse in the energy triangle); and it is this higher total energy that converts to photonic energy.

The heat generated by extreme gravitational collapse can be quite significant. The fact remains whatever the energy a particle gains (and however it may have been gained), all of its energy (mass, thermal, kinetic) must transform into elementary photonic energy.

Amplification of surface-trapped energy. Nature, as explained in Chapter 2, has a remarkably simple process for amplifying the energy within the surface layer and an equally remarkable process for expelling it.

Escape of surface-trapped energy. The natural mechanism by which the surface energy continuously escapes to the external world was explained in Chapter 2. Additional details may be found in the journal Article[24].

The question of excess matter. Given that a contiguous mass of less than 3.4 solar masses is insufficient to form a lightspeed boundary; and given that this same mass is just sufficient for the task; the question then is *What happens*

---

[22] C. Ranzan, *Law of Physics 20th-Century Scientists Overlooked (Part 3): Noninteraction Mass-to-Energy Conversion*, International Journal of High Energy Physics Vol.7, No.1, pp.19-31 (2020). (Doi: http://dx.doi.org/10.11648/j.ijhep.20200701.14)

[23] G. Greenstein, *Frozen Star* (Freundlich Books, New York, 1983); p197.

[24] C. Ranzan, *Natural Mechanism for the Generation and Emission of Extreme Energy Particles*, Physics Essays Vol.31, No.3, pp.358-376 (2018). (http://doi.org/10.4006/0836-1398-31.3.358)

if the collapsing body is greater than 3.4 solar masses? And, say, none of the material is expelled. This contingency is the subject of Chapter 4.

No violation of Relativity theory. It should be pointed out that the mass-to-energy conversion mechanism, and the associated gravitational collapse, is not a contravention of relativity. As is well-known, *general relativity* is not a complete theory. Einstein himself admitted this. And probably the best evidence of incompleteness is that the theory does not say what happens to the matter on the inner side of the critical boundary, the so-called spacetime *event horizon*. And if one pushes the equations too far, general relativity predicts outrageous nonsense —a singular point of infinite density. It predicts a mathematical object with no connection to reality.

Furthermore, general relativity says nothing about the interior except that something inside there *somehow* produces the gravitational effect —*somehow* causes spacetime curvature. But it does not say how. It does not say how the "interior something" can reach through that ultimate one-way barrier and influence the outside world. Therefore, our end-state neutron structure, as it has been presented, cannot be in violation of a theory that breaks down for the interior region —that fails for the most important domain.

Also, there is no violation of special relativity. Not with the gravitational collapse; and not with the Terminal state structure. Nowhere does mass travel at lightspeed with respect to aether; and nowhere do photons or neutrinos travel faster than the *c*-constant, likewise with respect to aether.

A powerful argument can be made that it was the slavish conformance to Einstein's relativity theories that prevented scientists of the last century from discovering several fundamental laws and structural features of the Universe.

Implications

Nature has a way of compelling matter to travel at lightspeed (with respect to the universal medium). It is a situation, as scripted in Scenario 2, of not merely striving but of actually attaining the ultimate speed. And in the process of doing so, matter loses its attribute of mass and converts to pure energy — totally and naturally. Dependent only on a radical change in the gravitational environment (as illustrated in **Figure 3-10**), this process is essentially the *noninteraction conversion of mass to energy*.

With the failure to recognize this underlying law of physics governing mass-to-energy conversion, Scientists of the 20th century unwittingly had to forego its profound implications.

Three things about the conversion mechanism stand out and point to far-reaching consequences for physical science: Its patent naturalness, its relevance to black-hole physics, and its cosmological implications. These are the three salient aspects in more detail:

ONE. The noninteraction conversion of mass to energy stands as a perfectly natural mechanism. It is a 100-percent conversion process that requires no new force, no new particle, and no *radically* new physics. Nor does it require changing any existing force. Its validity is rooted in the photonic theory of particles and the natural aether theory of gravity.

TWO. It is of game-changing importance for research into black-hole physics. Crucially important to the study of gravitational collapse, this overlooked process circumvents the breakdown of theoretical physics that plagues the conventional 20th-century view of terminal collapse.

The new interpretation avoids the paradoxes associated with singularity-type black holes. Consider the following:

Black holes, by definition, preclude the existence of any form of energy between the central gravity-causing singularity and its surrounding event horizon. Any energy present in the gap between those two must be absorbed by the point mass. But at the same time, and also by definition, there is a gravitational field surrounding the singularity and extending out to the event horizon and beyond! So why isn't this energy-possessing gravity field sucked into the singularity? There is no answer —and therein lies the paradox.

Then there is the angular momentum paradox. Black holes, it is claimed, inherit the angular momentum possessed by the pre-collapsed structure. But here's the problem. Angular momentum, most definitely, requires a radius for the material that is present; however, the radius of a singularity, regardless of how much matter it supposedly contains, is always zero. No radius, no angular momentum. Hence, a paradox.

One more self-contradiction worth mentioning. It can be stated bluntly as the outright paradoxical notion of having a vast quantity of matter "inside" a spatial speck of nothing!

Needless to say, there were 20th-century experts on this subject who abhorred the contradictory consequences and strongly suspected something was missing. Sir Arthur Eddington and Lev Landau thought this sort of outcome was ridiculous and repeatedly argued that there must be some law of nature, some law as yet unknown, that would prevent such collapse.[25]

THREE. The implication for cosmology are many, but the highlight is probably what the conversion process means for our understanding of *astrophysical jets*. The process —the process of mass-to-energy conversion of infalling particles and debris— serves as the key element of the mechanism that drives astrophysical jets. These jets are beams of matter being pushed outward at high speed and extend to enormous distances from a central object such as a neutron star or a hypothetical "black hole." They are usually paired and are aligned along the star's spin axis. Although they have been observed

---

[25] John Gribbin and Martin Rees, *Cosmic Coincidences, Dark Matter, Mankind, and Anthropic Cosmology* (Bantam Books, new York, 1989); p157.

for decades, astronomers, we are repeatedly told, are still not sure what produces them, what they are made of, or what powers them. Under the conventional wisdom, the true source of the energy behind these emission beams is a major unresolved mystery.

According to Wikipedia's entry for *Black Holes*, these jets are the ejection of matter, often at relativistic speed, along the polar axis and carry away considerable energy. *"The mechanism for the creation of these jets is currently not well understood."*

The new insight goes a long way in resolving this long-standing mystery. For additional details, see Chapter 2.

In conclusion, the described conversion mechanism is reasonable in its modus operandi, relevant to a proper understanding of gravitational collapse, and revolutionary in its implications for cosmology.

✠ ✠ ✠

Third Interlude

The Unhinged Theoretical Physicist

> *To say that the established scientific world is prejudiced against new ideas is an understatement. It is paranoid about them.* –Bruce Harvey, dissident physicist, 2002[1]

Great fun was had with this overlooked law of physics —*noninteraction mass-to-energy conversion*. The fun was in the attempts to get the article published. It was sent to a journal focusing on advances in physics, gravity, and cosmology. Perfect! My article was relevant to all three of the journal's subject categories, as implied in the submitted brief summary.

> Critical scientific information: The submitted Paper details a new mechanism of total mass-to-energy conversion. It describes a process whereby 100 percent of the mass is converted to radiant energy *without* involving particle-antiparticle annihilation. Using only basic physics and relativistic motion —in a way that exploits the distinction between *apparent* and *innate* levels of reality— the article follows a logical sequence in explaining how mass converts to energy and how that conversion leads to an explanation of the mechanism driving astrophysical jets. And as a bonus, the mechanism provides the resolution to the long-standing paradox of stellar black holes. ▫

The Paper was sent out for peer review. Unfortunately, it turned into a surreal review process. Although it was quite entertaining, the unfortunate aspect was that it wasted considerable precious time in getting the article published. There was something odd about the person assessing the Paper. At first I thought I was dealing with a failed scientist or maybe even a senile one,

---

[1] B. Harvey as quoted in, Juan Miguel Campanario & Brian Marten, *Challenging Dominant Physics Paradigms*, Journal of Scientific Exploration Vol.**18**, No.3, pp. 421-438 (2004); p429.

possibly driven to distraction by his failure to find the flaws with this "crackpot" theory he had undertaken to review. (The *crackpot* term was his descriptive for the Paper, according to an email inadvertently forwarded to me.) He turned out to be distinctly brash. I gradually realized, during the three rounds of review, that he just had not learned to think critically and to question long-standing assumptions.

First review

In a nutshell, the reviewer wanted a rewrite of the Paper with the inclusion of material on gravitons (the hypothetical particles that carry gravitational force); and something about "How light travels slower in a crystal, as opposed to its speed in a vacuum"; and stuff on general relativity. The implied message was that my theory should be altered to be more like his preferred theory.

Author explains why no discussion of the properties and propagation of gravitons:
I should point out that I am NOT using a *force* theory; rather, I am using a *process* theory of gravity. Therefore, force carrier particles (gravitons), are simply not needed and do not appear in my Paper. Their absence is entirely consistent with the fact that gravitons have never been detected.

I should also point out that the Paper is based on DSSU theory, which is a comprehensive and fully-validated model. See the article, *DSSU Validated by Redshift Theory and Structural Evidence*, Physics Essays Vol.**28**, No.4, pp.455-473 (2015), posted on the CellularUniverse.org website. □

Author's response to Reviewer's claim, "There is no transition to pure energy, as to making 'particles' go at lightspeed":
One needs to understand that only the "surface" mass converts to pure energy. If it fails to convert, it would stand in violation of special relativity's requirement that no mass particle can travel at lightspeed in a vacuum. This feature is clearly presented in the present Paper.

Check out this article published in the very same journal with which you are associated: *Mass-to-Energy Conversion, the Astrophysical Mechanism*, JHEPGC Vol.**5**, No.2, 2019. It gives the basic-physics details on how particles are compelled to transition to pure energy. □

Second review

The second report still insisted on radical change of the presented theory. Also, it stated "this reviewer has found substantial errors in the manuscript." But instead of specifying those alleged substantial errors, the Reviewer advised me to "Learn some humility" and encouraged me to abandon my theory.

The ball was in my court; I now had to deal with a lengthy list of objections and criticisms.

Author's response to the report's opening point, "I urge the author to review the connection between gravitons and gravity which the author was so quick to dismiss. Discuss it and put it in.":

I am presenting a model that does not use, and does not need, gravitons (whether massive or massless). Since I wish to present my argument as clearly as possible, I see no point in including something quite unrelated. Best to stay focused; make things easy for the reader. Besides, there already is a plethora of papers on gravitons in the literature. □

Author's response to Reviewer's claim, "This review has found substantial errors in the manuscript ... the author has violated Einstein relations several different ways.":

The equations were checked and agree with standard textbooks. They also agree with Professor Lev Okun's cited work (*The Concept of Mass*, Physics Today, June 1989). The conversion mechanism I have presented is purposely designed so that it does not violate special relativity. □

Reviewer: "I did not wish to be so blunt, but I do not accept this [cited *DSSU Validation*] article.
I wish you to junk that [DSSU theory] construction."

Author's response: So, the Reviewer considers the Physics Essays article, *DSSU Validated by Redshift Theory and Structural Evidence*, to be irrelevant. If someone considers theoretical and observational evidence as not being relevant within a scientific discussion, then there is really not much that can be said or done. Really.

Regarding the "wish" to have that DSSU construction junked, the success of the DSSU theory is so overwhelming that it is simply not possible to destroy and discard it. For a quick summary of its success, please see the two Tables comparing the DSSU with the conventional Worldview:
- www.CellularUniverse/Educators/CosmologyTestingPt1.htm
- www.CellularUniverse/Educators/CosmologyTestingPt2.htm □

Reviewer objects to a remark made in the 1st round: "The remark that there have been 'no errors' in an offered manuscript is unprofessional. Every PhD physics researcher knows that errors are part of the process. There is no theory which exists which is a perfect 1-1 connection to phenomenology. At best each theoretical idea is an approximation, albeit good if successful to the actual physical fact."

Author's response: Stating that I was pleased "that no errors were found in the Manuscript" is not the same as saying "there are no errors."

Speaking from my own experience, I have been working on DSSU theory for 19 years with over 35 published papers and have never encountered an error in its framework and predictions. (Of course there have been rejections; but never on the basis of an error in the physics. And certainly a theory can always be improved and even extended.) All I can say is that the model works! It works spectacularly!

The historical evidence suggests there are some theories with a perfect connection to phenomenology. Recall how Galileo conclusively proved the Heliocentric model with the powerful evidence of the seasonal phases of Venus. On the placement of the Sun at the center of the system of planets, Copernicus' assertion and Galileo's discovery were 100% correct. And yet, there are individuals who can never be convinced. For example, Harvard scholars at the time refused to examine the evidence —they adamantly refused to look through the telescope. They were exercising the Bellarmino bias (named after the contemporary Italian Cardinal Bellarmino).

DSSU theory has a *checkmate proof* that is as incontrovertible as was Galileo's. □

Reviewer critiques surface mass: Originally, the Reviewer claimed that there is no transition of surface mass to pure energy. Now he questions the meaning of "surface mass" itself.

"You have used the term "surface" mass in a way which is unprofessional. What is SURFACE mass? The term is NONSENSE as you presented it."

Author's response: "Surface mass/matter" is that region of the collapsing structure [in Figure 3-9b and Figure 3-10, Chapter 3], which —when collapse stops— experiences lightspeed inflow of aether (or the vacuum). This surface mass is what becomes the pure-energy skin of the end-state neutron star. Obviously, "surface mass" only has meaning during collapse; the final structure does NOT have a mass surface.

The reason the collapse must stop is because Nature allows only a certain maximum density to exist. Einstein himself believed that collapse halts at some point and is somehow prevented from going further. As he expressed it, a body can never collapse through its Schwarzschild radius. Sir Arthur Eddington also believed that some unknown mechanism halts the collapse. These historical items are commonly known.

Furthermore, rather than over-explain this stuff (of stellar shrinkage), I prefer to respect the intelligence of the Journal readership. I think they would recognize the described slow-motion collapse as a convenient thought experiment and as an understandable simplification of what actually happens.

In any case, I have added a clarifying paragraph ... [just before Figure 3-10, in Chapter 3].

The Reviewer also wanted to know "Where is the EVIDENCE ?"

Third Interlude. The Unhinged Theoretical Physicist

Yes, it is always good to present evidence.

All neutron stars having contiguous mass of approximately 3.4 Suns have surfaces of pure energy (some of which escapes via polar emission beams). There is abundant evidence of these end-state objects readily observable. They have long been detected as pulsar-type neutron stars. (Mention of this now appears in the Paper.) □

Reviewer expressed concern about the density calculations. [When I first saw this, I thought, *Maybe, finally, a real error!*]
Author's response: I checked the formula and the density calculations and they are correct.

Radius of a sphere (as a function of density) is: $R_{(\rho)} = \left(\dfrac{3}{4\pi} \dfrac{M}{\rho} \right)^{1/3}$, where ρ is uniform or average.

When solved for total mass 3.4 Suns and $\rho = 1.6 \times 10^{18}$ kg/m^3 (neutron density reasonably assumed to be Nature's maximum), the equation gives a radius of 10.0 kilometers. This agrees with what is stated in the text.

I should caution against sinking into unimportant detail here. My presented argument does not fundamentally depend on what the actual density value may be. Just that there is a maximum, whatever it may be. (It cannot be otherwise, if one wishes to stay within the bounds of physical reality.) □

Reviewer suggested reviewing the works of Professor Wolfgang Rindler and kindly attached a pdf copy of his book.
Author's response: In an initial check of the Rindler textbook *Relativity Special, General, and Cosmological*, I noticed the claim about "the Michelson–Morley null result." I have a copy of the 1887 published paper; and it states that they measured an aether wind of 5-7 kilometers per second — this is certainly far beyond a "null" velocity! Even more remarkable is Rindler's complete omission of the Dayton Miller experiments. Those experiments were definitive and actually determined the direction of aether flow.

However, I very much like the book's opening quote. An excellent dictum:

> *The ideal is to reach proofs by comprehension rather than by computation.* –Bernhard Riemann

Reviewer commented about presenting in international physics conferences. Not sure, but if I understand the Reviewer correctly, the suggestion is that my work *"would be vaporized say at Rencontres De Moriond Cosmology, and Marcel Grossman conferences."*

Author's response: I should point out, DSSU theory was presented at the international physics conference —*2002 ESA/ESO/CERN Astrophysics*

Symposium— in Munich (where it made its inaugural appearance in the scientific literature). It was not, as you say, vaporized. □

Reviewer, in closing, made some interesting comments on quantum electrodynamics and the connection to a deeper reality. ... Theories are only approximations ... Quantum electrodynamics works but as Richard Feynman admitted is based on a "swindle" (renormalization). ... The author's theory should not be considered "an airtight theory."

Author's response: In quantum theory, renormalization is only necessary because the theory is founded on a false assumption —the assumption that particles are zero-dimension points. Change the assumption and renormalization can be discarded. In other words, the choice of assumptions is critically important for connecting to the deeper reality.

For 20 years I have been attacking my own theory, testing for leaks, checking the "air-tightness," repeatedly testing DSSU's basic assumptions. With every test, the results have been better than had been expected. Prediction after prediction proved correct —all validated by the available evidence. In a nutshell, DSSU theory works AND is founded on correct assumptions (correct until shown to be otherwise).

Richard Feynman also made the following comment, words aimed at any would-be theorists or anyone who thinks the process of original reality-based ideation is easy:

> *In physics there are so many accumulated observations that it is almost impossible to think of a new idea which is different from all the ideas that have been thought of before and yet agree with all the observations that have already been made.*
> –R. P. Feynman, The Meaning of it All, p22

Feynman wisely left the door unlocked. Applied to the Cosmos, to construct a theory to explain the Universe that agrees with "so many accumulated observations ... is almost impossible." As it turned out, it was difficult but not impossible. □

Author's closing comments: I have just finished reading the book *Faster Than the Speed of Light* by João Magueijo. Physicist/cosmologist Magueijo, in order to make his variable-speed-of-light theory work, had to take the extraordinary steps of
- Violating the principle of Lorentz symmetry!
- Allowing laws of physics to change over time!
- Making the entire universe expand!!!
- And of course, violating the constancy of the speed of light!

And yet, amazingly, his Paper (coauthored with Andreas Albrecht) detailing this unusual model was accepted for publication. João even admitted, as stated on the cover of the book, that the whole thing is a scientific speculation.

Now compare, or contrast, with my work. The Editor and the Reviewer should be able to appreciate why a high degree of confidence in DSSU theory is justified:
- It does not require whole-universe expansion
- It does not violate special relativity (speed of light in vacuum remains constant)
- It does not require time-wise changes in physical laws
- It agrees with the astronomically observed evidence
- And is validated by a checkmate proof.

In light of this comparison and the arguments made herein, any critical reader should feel assured that there is nothing speculative about DSSU theory. Everything is supported by *reasonable assumptions, sound theory, and solid evidence.* □

Now, before getting into the third round, I think it would be interesting to find out something about this guardian of the status quo. From various clues, I managed to determine the identity of the "anonymous" reviewer.

He is a post-doctoral PhD theoretical physicist named Andrea Maselli. According to his website (https://andreamaselli85.wixsite.com/home), Dr Maselli is a Researcher in Theoretical Physics, at La Sapienza University, Dipartimento di Fisica Marconi, in "the most beautiful city in the World," Rome.

I found the following item in his impressive 12-page curriculum vitae (dated 2020-8):

> **General Relativity & Beyond:** My main research interest is Strong Gravity, which plays a crucial role in many astrophysical phenomena involving the most compact objects of our Universe. My work focuses on black-hole and neutron star physics, as isolated and binary systems, as they represent promising sources of gravitational waves for detectors like LIGO and Virgo. I am particularly interested in finding solutions which describe such objects in theories of gravity alternative to General Relativity, and to study their dynamic and stability.

Great! His focus and expertise is on neutron stars and collapsed objects/bodies. Moreover, he is "particularly interested in finding solutions which describe such objects in theories of gravity alternative to General Relativity." (If that were true, here's the ideal candidate to review my work.) If that were true, why would Dr Maselli be so set against a successful alternative theory known as the DSSU? Unfortunately, as it turned out, his mastery of collapsed structures seems to be confined within the limiting framework of

General Relativity. As evident in these reviews, he utterly refused to look outside the 20th-century paradigm.

Third review

Here again Dr Maselli, still unable to point to an actual error, wants major changes to the theory. To begin with, he suggests one of the diagrams [same as Figure 3-11 in Chapter 3] should be removed: "This construction is incorrect. Please remove this figure. I do not accept it." But he does not explain how it is incorrect.

> **Author's response:** The calculations relating to Figure [3-11] have been checked. The energy triangle in the figure is consistent with:
> (1) the principle of energy conservation;
> (2) the energy triangle and relationship explained in Figures [3-5] and [3-8] [see figures in Chapter 3];
> (3) the aether theory of gravity;
> (4) several cited references, including one to the same Journal the Reviewer is connected with. □

Next, Dr Maselli diverted attention to something not in the Paper. "[Y]our statement in your reply about Feynman and QED is incorrect" regarding renormalization. "You are required to defend your statement about Feynman, and his 'incorrect' methodology." (I checked, but could not find mention of *incorrect methodology* in the 2nd response.)

> **Author's response:** But there is nothing in the Paper about Richard Feynman. Moreover, there is nothing said about quantum mechanics and renormalization. In other words, there is nothing here to correct. □

My adjudicator, now for the third time, faulted the surface effects. "Your statement as to surface effects is NONSENSE. I read what you stated and it is NONSENSE." "I insist that you correct your errors in fact." (Please, tell me the errors! or just one.)

> **Author's response:** Again, I was unable to find any errors. I should point out that the key arguments in the Paper are based on the ***process*** concept of physics —and everything is consistent with those processes as specified in DSSU theory. □

Lastly, Dr Maselli claimed that General Relativity is being violated. "You submit[ed] results which contravene GR." (But no hint as to some specific contravention!)

> **Author's response:** The reviewer is insisting on a strict conformance to relativity theory. But general relativity is not a complete theory. Einstein

himself admitted this. If we want to discover something new, we really have to go beyond his restrictive paradigm. In my Paper, I'm presenting a perfectly reasonable way to go beyond general relativity without of course violating its successes and its key principles.

For example, as Edward R. Harrison wrote in his Cosmology textbook, "The surface of a black hole where spacetime in effect falls inward at the speed of light is known as the *event horizon*." ... I do not contravene this; I merely say that *aether* is flowing inward at lightspeed.

But probably the best evidence that general relativity is incomplete is this: GR does not say what happens to the matter on the inner side of the critical boundary, the so-called spacetime *event horizon*. And if one pushes the equations too far, GR predicts outrageous nonsense —a singular point of infinite density. It predicts a mathematical object with no connection to reality. The reasonable approach is to reach beyond GR for a reality-based answer.

Furthermore, GR says nothing about the interior except that something in the interior SOMEHOW produces the gravitational effect —*somehow* causes spacetime curvature. But it does not say how! It does not say how the interior material can reach through that ultimate one-way barrier and influence the outside world. Therefore, with my *end-state neutron structure*, I cannot be in violation of what is not specified. The accusation is like charging someone with an offence without specifying what civil or criminal law was violated! Or worse, just make up a charge on the basis of a made-up law (as now happens in much of the Western world).

I contend that it was the reverent conformance to relativity theories that prevented scientists of the last century from discovering several fundamental laws and structural features of the Universe.

(I have added a brief segment in the "Relevant Aspects" Section of the Paper to point out that there is no violation of the key aspects of relativity theory.)

A point worth considering. If one slavishly adheres to existing relativity theory, one is most unlikely to discover anything new. □

Author's closing comment. In light of the arguments made herein and in the previous responses, one should be able to recognize the reasonableness and naturalness of DSSU theory and the absence of contravention of relativity theory. As stated before, everything is supported by *reasonable assumptions, sound theory, and solid evidence*. □

The dismissal

I do want to mention that at the end of each of my response reports I respectfully thanked the "anonymous" reviewer for the time expended going through the manuscript and preparing a report. (Do not want my readers to get the wrong impression.) I know how much time and effort can be involved; often it's a thankless undertaking. And thanks were extended for the very useful pdf book by Wolfgang Rindler.

Anyway, on the one hand the theory couldn't be changed; on the other, the mind of a fully-indoctrinated physicist couldn't be changed. His mind was sealed, as was the fate of my submitted Paper.

Final referee report, word for word:
Reply to author,
You are dismissed. I have turned down your submission due to your attitude.
I am done with you. You proclaimed you are right and presume to give ME orders. That means you have no idea as to what peer review means.
Good luck. You will need it.
End of reply

Reflections

I think what sent Andrea Maselli into fits of frustration was his failure to find a legitimate flaw in the Article. All his many years of training conditioned him to the ways the Universe is supposed to work and it wasn't anything like the natural DSSU. It is supposed to work in accordance with his venerated General Relativity; as if *that* were the last word on gravity —the immaculate interpretation.

His impressive, lengthy curriculum vitae had not prepared him for what he had undertaken to evaluate. I imagine his initial reaction: *This can't be right! No general relativity, no gravitons, and what's this! Aether! ... Yes, it will be very easy to dispose of this crank theory.*

All he has to do is point out the flaws, expose those violations of the laws of physics. But, on closer inspection, they could not be found. Check and recheck.

Surely there is some error in the arguments, some deficiency in the logic of the presentation. No luck. The reasoning was tight. But something just has to be wrong here!

In desperation Dr Maselli disparages an important reference. Repeatedly. Even in the Third Report, he refuses to look at the evidence! And I quote:

> "The Reviewer considers the Physics Essays article, *DSSU Validated by Redshift Theory and Structural Evidence*, to be irrelevant. I do not accept this article."

He even admitted why. Dr Maselli does not accept it because he considers Physics Essays to be a substandard journal —thereby, wittingly or unwittingly, committing a common logical fallacy.

Third Interlude. The Unhinged Theoretical Physicist

Could it be that the material of the Paper was just too poorly presented, the theory too impenetrable, the ideas too esoteric to be understood? No. Everything was written in absolutely unambiguous terms; with crystal-clear prose, simple math, plain diagrams, helpful definitions, easily visualized processes. In other words, the subject matter was presented so that there can be no chance of being misread. Do not think for a moment that Dr Maselli may have misunderstood what was presented. He understood only too well. But he dared not openly acknowledge its validity. For any untenured academic physicist, that would be an unforgivable act of apostasy.

✠ ✠ ✠

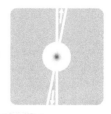

Law of Mass Extinction

4. Mass Extinction/Vanishment

The Total Annihilation of Mass by Aether Deprivation

Understand black holes and you understand the ultimate laws of the universe. –Charles Seife, *Decoding the Universe* (2006)

Introduction

In Chapter 3 it was explained what happens when a star with the mass of 3.4 Suns slowly collapses. The focus was on the mass-to-energy transformation that takes place at the "surface" as the star contracts to a density state beyond which further contraction is fundamentally impossible. The final collapsed structure was described as an *end-state* neutron star —a stable 3.4 solar mass body with extreme nuclear density and a pure-energy surface layer.

The collapse, described more or less as a thought experiment, resulted in a very small amount of mass being converted to photonic energy, which, at the completion of the collapse, remained embedded within the surface energy layer. Importantly, there was no extinction of any of the mass. No matter was lost or expelled. The assumption was that the pre-collapsed star had a mass equivalent to 3.4 Suns and the end-state collapsed structure retained the whole amount. The collapse was presented this way for the sake of simplicity. And it underscores an important point. Regardless of the amount of mass a pre-collapsed star possesses, the 3.4 solar mass is the nominal minimum (or ideal) amount needed at the completion of the collapse, if the end product is to be an end-state structure with a pure energy surface layer.

Now consider a more realistic situation. Say a 10-solar-mass star collapses; passing, most likely, through a supernova stage; ending up as a neutronium remnant with mass greater than 3.4 solar masses. Another realistic scenario is the spiral-type of merger of two orbiting neutron stars with a combined mass significantly greater than 3.4 Suns. And, of course, there is always the remote chance of an outright collision between neutron stars ending in a mass accretion well above criticality.

We want to know what happens under such circumstances, when the collapsed mass is predicted to be greater than 3.4 Suns; or when two stars having already collapsed to the end state try to merge into a single structure. The simple question is *What happens when too much mass aggregates into too small a spatial volume?*

Before addressing these questions, let us review our understanding of a key aspect of the universal space medium. It involves a unique process of critical importance to the most fundamental laws of physics. As was explained in previous chapters, matter —all mass, all radiation, all particles without exception— absorb and consume the universal space medium (aether). The very existence of matter depends on the continuous consumption of aether. Simply put, matter is sustained at the expense of aether. This violates no conventional conservation law because aether (DSSU aether, the universal medium of our Dynamic Steady State Universe) is a *nonmaterial fluid* — which possesses no mass and, in its basic state, possesses no energy.

What this means on the macro scale is that the fluid medium flows into mass bodies and, in the process, produces the familiar gravity effect.

Extreme gravitational collapse

Two foundational properties and a reasonable assumption

There are two foundational properties that play a key role in gravitational collapse. First, there is the axiomatic emergence and existence of an *essence medium* that permeates the Universe. *Aether*, as this medium is called, is defined as a discretized nonmaterial fluid whose discrete entities possess no mass and no energy. Note, however, that although individual *aether units* are devoid of energy, *aether as a bulk fluid* is different. It is then that aether, by way of its inhomogeneous flow, *does* manifest energy. Aether —via its bulk dynamics— *does* produce clearly recognizable forms of energy. This is entirely consistent with DSSU's fundamental process of energy.[1]

---

[1] C. Ranzan, *The Fundamental Process of Energy –A Qualitative Unification of Energy, Mass, and Gravity*, Infinite Energy Magazine, (2014). Posted at www.CellularUniverse.org/Directory.htm

Second, as discussed above, there is the postulated and evidence-supported dependency tying all matter to a continuous supply of aether. All matter particles exist as a continuous process by which aether undergoes excitation, absorption, and consumption. All mass and energy particles exist at the ongoing expense of aether —the volume vanishment of the aether fluid. Without such ongoing absorption of aether, matter simply cannot exist.

Now for a brief discussion of the mechanism of gravitational collapse as it leads to a long unrecognized but crucial assumption about mass. A simplified scenario for the gravitational contraction/collapse of a *sufficiently massive* star involves three stages. Each stage of contraction results in a significant increase in density.

The first stage may be described as the gradual development of a dense iron core. This happens within the densest region of the star as a final reaction in the natural sequence of available nuclear fusions. Technically, atomic iron is the end product of the various steps in the release of the binding energy entrapped within certain lighter elements (namely, helium, carbon, oxygen, neon, and silicon).

The first stage occurs when the star's nuclear fuel has been exhausted (after the various fusion reactions have run their course). The star can no longer resist the gravity-induced inward pressure of its mass and, consequently, contracts until much of its mass is in the *degenerate state* — specifically, in the electron degenerate state in which electrons are stripped from their nuclei and become packed tightly together. This stage of the collapse ends once the star has contracted sufficiently for the gravitational pressure to be in balance with the electron degeneracy pressure. Astrophysicists tell us this requires the density to rise to 10^7 to 10^{11} kilograms per cubic meter[2]. It is now an extremely dense compact star known as a *white dwarf*.

The third and last stage. Over time the white dwarf acquires additional mass; this could come about by accretion, or collision, or merger, or any combination thereof. Additional mass naturally increases the gravitational pressure. When the latter increases beyond the range of the opposing electron degeneracy pressure, the star must collapse (contract) to a still greater density. This stage of the collapse ends when a balance is established between the new gravitational pressure and a new degeneracy state pressure —one produced by the *nuclear degeneracy* state. Probably, this is the final state of degeneracy of mass (although a quark degeneracy state remains a speculative possibility). The final density at this stage may range up to 1.6×10^{18} kilograms per cubic meter —a density that is far beyond normal imagination and comprehension.

---

[2] *The Facts on File Dictionary of Astronomy*, 4th Edition, edited by Valerie Illingworth and John O. E. Clark (Checkmark Books, New York, NY, 2000); p462.

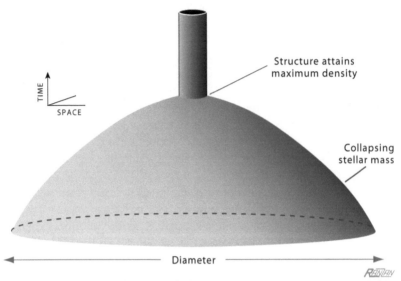

Figure 4-1. Gravitational collapse halts when mass concentration attains the ultimate density. Based on (1) the existence of aether, as defined in the text; (2) the innate absorption/consumption of aether by all matter; (3) the reasonable assumption of the existence of a density limit; and (4) the concentration or aggregation of sufficient degenerate mass having complete absence of thermal agitation; based on those factors, Nature follows a course that ultimately leads to a **final stage of gravitational collapse** (shown in the schematic). The end result is a compact-and-stable structure with maximal density.

But this mutual intensification between density and gravity does not continue indefinitely. Regardless of the amount of mass that may be added to the structure, Nature has its limits.

And here is where a reasonable assumption is invoked: The assumption is that Nature has a maximum density state. There exists a limit beyond which mass cannot be compressed. Since the neutron is the densest stable particle (stable when in the degenerate environment) known to exist, we accept it as the ultimate state of compaction. The ultimate density manifests when matter is in its degenerate state, when mass particles have lost all kinetic energy, when there is a total absence of thermal energy, when neutrons are in direct contact with other neutrons.

Thus, in the context of gravitational collapse, the maximum density is taken to be 1.6×10^{18} kilograms per cubic meter.

Based on (1) the foundational property of the existence of aether, as herein defined; (2) the foundational property of the dependency of aether absorption/consumption by all matter for veritable existence; and (3) the reasonable assumption of unsurpassable mass concentration, of there being a precise density limit; and (4) the availability of a sufficient quantity of mass;

based on those factors, gravitational aggregation must occur and ultimately lead to a final stage of gravitational collapse. This final stage, which results in an end-state compact-and-stable structure, is schematically shown in **Figure 4-1**.

The collapse halts with the attainment of a state of maximum density. But how do we know the structure in **Figure 4-1** has attained the critical degree of concentration? ... How do we know exactly —not approximately but exactly— when the structure attains its maximal state of density?

We check its rate of aether absorption/consumption. In accordance with the aether theory of gravity, every gravitating structure has a characteristic aether inflow profile —a graph of the influx versus radial distance. (Incidentally, the external portion of such a graph has an interesting relationship to conventional gravity theory. The magnitude of the inflow at any radial distance corresponds to the escape velocity there.) The aether's radial-flow profile can be generated with the now familiar expression (derived in Appendix A):

$$v_{\text{inflow}} = \sqrt{2GM/r},$$

where G is the gravitational constant and r is the radial distance (from the center of the mass M) to any position of interest external to the gravitating mass.

If the profile is as shown in **Figure 4-2a**, then we cannot be sure. The density may or may not be the maximum attainable. The graph, however, does tell us that there is room for more mass. The structure still has room to grow (either in density or in overall size). With the addition of mass, the plotted curve could be induced to move upward. There is nothing preventing the structure from acquiring more material and increasing its gravitational potency, its ability to compress matter, and potentially increasing its density. But there is a limit to this process. And when that limit is reached, we can be absolutely certain of having reached the ultimate density.

On the other hand, maybe the total mass is considerably larger and its inflow profile is as shown in **Figure 4-2b**. In this case the structure is in the critical state. The aether influx *at its surface* is the maximum allowable; the velocity curve touches the lightspeed limit-line. In **Figure 4-2b** we definitely have a collapsed structure; but we run into the same problem as before; we still cannot be sure that a state of maximum density has been attained. However, part (b) does illustrate one of the two necessary conditions —lightspeed inflow. The other requisite is that structural collapse must terminate —the radius must establish a stasis. Gravitational contraction must come to a halt. The mechanism that accomplishes this will be explained in a moment.

The essential point. There are two necessary and sufficient conditions for

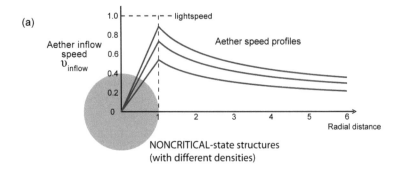

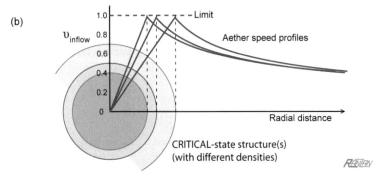

Figure 4-2. Aether inflow profiles relating to gravitationally collapsed structures. Part (a) gives representative profiles of hypothetical same-size structures but with different densities. Although they have undergone serious collapse, or contraction, they are clearly not in the *critical state*. Part (b) illustrates one of the two essential conditions for attaining the ultimate density of mass. Here is a structure that has contracted down to the *critical state*, as indicated by the fact that the profile touches the "lightspeed line." As the collapse progresses in the manner shown here, the density increases —or at least that is the assumption being made. During the ongoing collapse process, there is really no way of knowing when the state of maximum possible density is reached. (The flow speed is given as a fraction of lightspeed.)

attaining the state of ultimate mass density: *lightspeed boundary* and *size stasis*. As will be shown shortly, the lightspeed boundary forms first, the size stability automatically follows.

What about observational considerations. Needless to say, it would be extraordinarily difficult for astronomers to determine whether or not a suspect object meets these conditions. Not only do these objects "appear" totally black, but also they are pitifully small and exceedingly remote.

End-state structure defined

Let us be clear on the precise distinction of meaning between *end state* and *critical state*. "Critical state" simply refers to the presence of a lightspeed

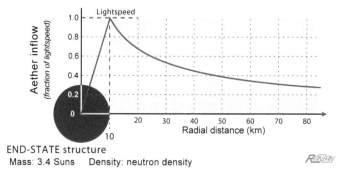

Figure 4-3. Basic properties of the *end-state neutron star* or, synonymously, the *Terminal star*. Based on the assumption of ultimate density being 1.60×10^{18} kg/m^3, the end-state structure necessarily has a total mass-equivalence of 3.4 Suns, a radius of 10 kilometers, a pure energy surface layer, and an aether inflow profile as shown. If the density-value assumption is off the true value, then it would simply change the diameter of the end-state structure.

boundary, where aether inflow speed is about 300,000 kilometers per second. "End state" refers to a structure that cannot collapse further.

An *end-state* gravitating structure is defined as a contiguous mass compressed to the ultimate density (assumed on reasonable grounds to be the density of *nucleons*) and surrounded by a surface-hugging lightspeed boundary (which may be absent at *polar portals*). Manifesting the ultimate density of mass AND possessing the ultimate electromagnetic barrier are the two necessary and sufficient conditions that define an end-state gravitating structure.

The end state is also called the *Terminal state*. The Terminal state exists when we have the greatest quantity of contiguous matter within the least volume (the state of being enclosed by the least surface area); meaning also that the density will be the maximum that Nature will permit. A **Terminal star** is a gravitationally collapsed structure (of neutron mass) that exists in both the *critical state* and the *end state*.

The two terms, *Terminal neutron star* and *end-state neutron star*, are used synonymously.

From previous research[3], the specifics of the *end-state neutron star* are already known and are based on the reasonable assumption that neutron density is the ultimate permissible by Nature. Its basic properties are shown in **Figure 4-3**. But what if the density assumption turns out to be off the mark?

---

[3] C. Ranzan, *Mass-to-Energy Conversion, the Astrophysical Mechanism*, Journal of High Energy Physics, Gravitation and Cosmology Vol.**5**, No.2, pp.520-551 (2019).
(Doi: https://doi.org/10.4236/jhepgc.2019.52030)

Say, compelling empirical evidence reveals a different value. In that case, only the size of the structure would change. Should the density turn out to be higher, then the end-state sphere would be smaller —smaller than the 10 kilometer radius shown in the figure. Moreover, and this is not at all obvious, it would necessarily also have less total mass.

Total collapse of 6-solar-mass body without ejection of mass

Consider a simplified collapse of a 6-solar-mass star. No nova or supernova complication. No mass ejection. No rotation. Once this star gravitationally compresses itself into the neutronium density range, its fate is sealed. Nothing can prevent its almost instant transformation to the Terminal state.

The star's mass equivalence of 6 Suns is more than enough to bring about the neutron degenerate state. This occurs at the end of its normal life. Once it becomes a neutron-density star there is no way to stop the collapse; no atomic process, no nuclear reaction, no thermal activity can alter the inevitable outcome; it follows the sequence shown in **Figure 4-4**. It starts out in the *non-critical state*. Part (a) of the figure shows the structure at the instant when the radius is 49 kilometers and aether inflow (at the surface) is six-tenths lightspeed. A basic calculation gives the density: 2.42×10^{16} kilograms per cubic meter.

An instant later, the structure reaches the *critical state*, **Figure 4-4b**. It now has radius of 17.7 kilometers and a surface inflow equal to 300,000 kilometers per second. The neutronium density is now 0.513×10^{18} kilograms per cubic meter. But this is not the maximum that Nature allows. Any contiguous mass structure that has reached the critical state must continue collapsing until halted by the ultimate density barrier. In terms of the graphical representation, this means the slope of the linear portion (at the interior to the structure) must increase.

Why this is so, is easy to demonstrate.

By inspection of the **Figure 4-4b** graph, $\text{Slope} = \dfrac{v_{\text{surface}}}{R_{\text{surface}}}$.

We know from Appendix A that inflow v_{surface} can be expressed in terms of total mass M and radius R (and a known constant). The mass M, in turn, can be expressed in terms of volume and density. The R conveniently cancels out giving us an expression for the slope in terms of just the density (and the gravitational constant).

Furthermore, we know from **Figure 4-4** (parts b and c) that surface inflow equals lightspeed c.

In other words, we have two expressions for the slope, which can be

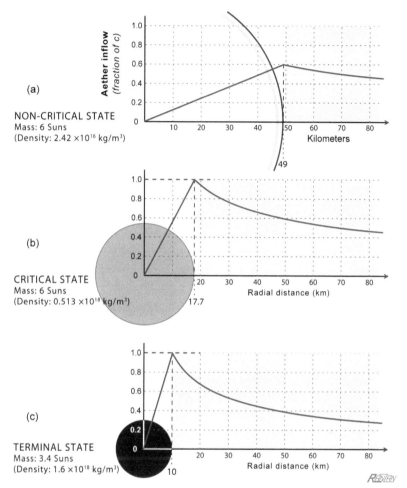

Figure 4-4. Gravitational contraction of a 6-solar-mass body. In stop-action snapshot (a) the structure is in the NON-CRITICAL state. At the instant represented here, the aether inflow (at the surface) is six-tenths lightspeed and the radius is 49 kilometers. Snapshot (b) shows the instant the structure, now shrunk to radius 17.7 kilometers, becomes CRITICAL. The collapse ends in snapshot (c) with the attainment of the ultimate density state —but with the loss of considerable mass! So, what happened to the missing mass?

brought together as: $\text{Slope}_{\text{end-state}} = \dfrac{c}{R_{\text{surface}}} = \sqrt{\tfrac{8}{3}\pi G \rho_{\max}}$. Notice the inverse relationship; if the density increases, the radius must decrease (as happens in the sequence shown in the figure). For the detailed derivation, see Appendix B.

The radius of the final collapsed structure is found by plugging in the known values for c and G (equal to 6.67×10^{-11} N·m^2/kg^2) and the assumed

value for the maximal density ρ_{max}; then solving for $R_{surface}$.
The calculation then gives the Terminal-state radius as 10.0 kilometers.

In summary, the collapse of the 6-solar-mass body —or for that matter any contiguous body— is subject to two inviolate constraints: Aether inflow at the surface can never exceed lightspeed (with respect to that surface); and the body density cannot exceed a natural maximum.

The collapse comes to an abrupt end when both the inflow limit and the ultimate density are present. Size of the end structure depends, of course, on the actual value of that final material density. The higher it is, the smaller the neutronium sphere will be.

But total collapse comes with a strange hidden aspect. Notice what has happened —during the collapse, something truly amazing has occurred.

A 6-solar-mass star has undergone total collapse (as described in the three parts of **Figure 4-4**) without any external expulsion of mass. And yet the post-collapse object has a mass of only 3.4 Suns! Over forty-three percent of the original mass has been lost! *How is this mass loss to be explained?*

Aether Deprivation Annihilation

The loss of mass occurs in conjunction with gravitational collapse ending in the Terminal state. But this is not all; it also occurs when additional matter falls onto (or is absorbed by) the Terminal structure. So, what is going on?

At the instant when the 6-solar-mass star is about 35 kilometers across and acquires its critical boundary, the situation is as shown in **Figure 4-4b** and **Figure 4-5a**. There is lightspeed inflow over the entire surface area. On the inside, there is a spherical quantity of mass totally dependent on this flow.

Obviously, when the density increases (as it must, because collapse ends only when that magic maxi-density is reached), the sphere will shrink and its surface area will decrease. This in turn means that a lower supply of aether will be available for the interior mass. Think about this carefully, why must the supply suffer a reduction? Surely there's some way to increase the aether supply. What about increasing the speed of the flow? No, not possible. The inflow speed cannot be increased; it is already at the special-relativity limit. The inescapable conclusion is this: There simply will not be enough to sustain the entire 6-Sun mass.

Be reminded that all matter is utterly dependent upon a sustained supply of aether —the universal essence. Without a continuous supply, mass and energy particles simply vanish.

Any reduction in surface area (as automatically happens with a rise in density) is equivalent to adjusting a sluice gate of an irrigation system so as to restrict the flow of life sustaining water. Water, being consumed as it flows, will no longer reach the ends of the irrigation channels. Some plants will die.

And so it is with the aether. The reduction in the volume of flow means

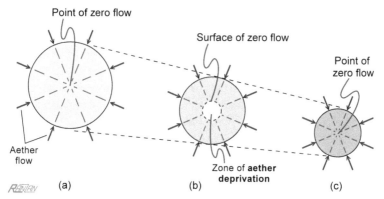

Figure 4-5. If the mass of a neutron star is greater than 3.4 Suns, then the excess mass will quickly be lost. The loss occurs during the gravitational collapse to the Terminal state. The mechanism involves the increase in density, (a) through (c) cross-sections. Simultaneously, the surface area decreases, thus critically reducing the supply of aether —essentially chocking off the flow to the core of the structure, (b). Mass vanishes within the "zone of aether deprivation," as it has been labelled in part (b) and shown greatly exaggerated. Mass literally disappears from the Universe; the reason being that matter simply cannot exist without a sustained supply of the essence fluid. The aether deprivation zone immediately collapses to a point at the heart of what is now the Terminal state structure, (c).

that the aether will be consumed before it reaches the deepest-located mass. The reduction in the volume of flow means that the inflow speed becomes ZERO *before the aether reaches the center of the gravitating body or region*! This core "region" becomes the zone of *aether deprivation*. Matter does not and cannot exist without aether. So this is serious. (See **Figure 4-5b**.)

This "zone of aether deprivation" is where the excess mass vanishes from the Universe —quite literally. Although it has been sketched as a centrally located vacuous sphere, the "zone" is really more of a useful conceptual tool. If a hollow core were to actually form —as a sort of zone of nothingness— it would collapse at near lightspeed. In a real-world collapse scenario the core material *terminates* before any spherical zone of aether deprivation has a chance to develop. In other words, as the critical-state structure contracts, mass vanishes continuously at the core. Whether the collapse is thought of with or without a *deprivation sphere*, the end result is the same; the inflow speed will be zero at the center-of-gravity point (**Figure 4-5c**).

Aether deprivation is not exclusively associated with collapse to the Terminal state. It is also a factor in preventing a Terminal star from changing its size or content. For instance, when a chunk of mass falls onto, or into, the structure, an equal quantity is almost immediately lost at the core —lost via the deprivation of aether. Note, we are here assuming there are no polar escape portals, no axial energy emission beams. It is easy to see how this mass-intake-and-loss becomes a continuous process when there is a steady supply of material, such as when a Terminal neutron star cannibalizes an orbiting vastly-larger gaseous star. In that case, the quantity of material being absorbed will equal the amount undergoing Termination at the heart of the structure. The one is in harmonious balance with the other. (When, however, polar emission beams are present, some proportion of the mass undergoes conversion to pure energy and escapes through the poles and along the beams.)

The most dramatic instance of the aether deprivation process occurs when two Terminal stars come together —either in a collision or orbital merger. While normally the core matter vanishment is able to keep up with any reduction in the supply of aether and prevent the formation of a hollow core (as just described with the continuous accretion and termination process), the coming together of two Terminal stars is truly without parallel. When two end-state bodies combine, a significant region of aether deprivation —a region of nothingness— instantly arises, setting in motion a monumental implosion. But the implosion itself, because it occurs far beneath the lightspeed boundary, has no observable effect on the external world. What *is* observable, however, is a significant burst of energy through the polar emission beams[4] (when present) and a loss of mass equivalent to more than three Suns. Obviously there is a violation of an important conservation law here. This aspect will be discussed and resolved in a moment.

First, some definitions on the ideas covered up to this point.

Aether deprivation: is an absence of aether. It is simply a chocking-off of aether flow. It is the essential condition whereby matter is extinguished —totally. Since matter cannot exist without aether, it vanishes. The condition occurs only in the interior of critical-state contiguous mass.

Aether deprivation annihilation: a process of total destruction of matter that takes place deep inside extreme mass concentrations. It occurs when mass aggregation reaches a state at which an insufficient quantity of aether reaches the core; and since matter cannot exist in the absence of aether, the aether deficiency results in the *terminal annihilation* of the affected matter. (When a neutron star, for instance, gains too much additional mass, its core will become a region of *terminal annihilation*.)

Terminal annihilation: the process of non-interaction vanishment of matter —the total negation of the affected mass/energy. The necessary and sufficient condition that brings about *Terminal annihilation* is *aether deprivation*.

Terminal neutron star (or Terminal-state star): a gravitationally collapsed structure that exists simultaneously in the *critical state* and the *end state*. A neutron star that has acquired a lightspeed surface-boundary. It stands as the Universe's most unusual type of star. Once such a star forms, it can neither grow larger nor smaller. Its

---

[4] C. Ranzan, *Natural Mechanism for the Generation and Emission of Extreme Energy Particles,* Physics Essays Vol.**31**, No.3, pp.358-376 (2018).
(Doi: http://dx.doi.org/10.4006/0836-1398-31.3.358)

volume and mass content remain forever fixed; its density is Nature's ultimate.

Terminal state: The state that exists when we have the greatest quantity of contiguous matter within the least volume. It is the state of bulk matter, with the greatest density Nature will permit, being enclosed by the least surface area (in compliance with special relativity).

Mass Extinction in its supreme manifestation

Mass Extinction can, and does, occur on a scale that is so extraordinary that concern over its connection to the First Law of thermodynamics instinctively thrusts itself to the fore.

The question of mass/energy conservation. What is exceedingly remarkable about a Terminal star is that its size does not vary. This means that when there is a collision merger or orbital merger of two Terminal stars the result is a single Terminal star identical in size and mass to a single original one (**Figure 4-6**). It is a stunning result —a merger accompanied by a magic-like vanishment of wholesale mass. One Terminal star plus another Terminal star equals, not double the mass, but just one of the original!

In terms of the Terminal annihilation of mass-energy, the gravitational merger of two Terminal stars embodies the ultimate energy-changing event that can occur in all nature —an event in which the matter equivalent to 3.4 Suns is suddenly negated. Moreover, this kind of one-plus-one-equals-one merger can occur many times in the course of a Terminal star's lifetime.

Looking at this in isolation, there is obviously a major violation of

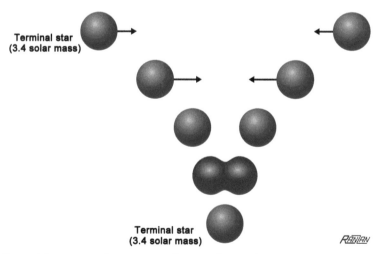

Figure 4-6. Merger of two Terminal stars. The result is a single Terminal star identical in size and mass to just one of the original. Only the rotation rate (and the orientation) is subject to change. Remarkably, the mass equivalence of 3.4 Suns completely vanishes and represents a local violation of the principle of matter conservation. (Polar jets are usually present, but are not shown.)

conservation law. The mass extinction by aether deprivation stands in defiance of the First Law of thermodynamics.

There is, however, the larger system to consider; there is a vast system of Terminal structures. Terminal stars are not only mass-energy destroyers, they are also energy generators. As we saw in Chapter 2, they have the unique ability to amplify the energy of photons and neutrinos. What makes this capability unique is that it is a noninteraction process. The amplified energy is expelled through the polar portals and includes the most extreme energy particles of this type ever detected, such as the ultra-high-energy neutrinos found at the IceCube Observatory located near the South Pole. And so, looking at the larger system, while some Terminal stars are suppressing the existence of mass; others, in fact all of them, are generating fresh supplies of energy. While an individual Terminal star may be, at any one time, a net annihilator of mass energy; another may be a net generator and emitter of energy —energy that is then available for conventional conversion to new mass particles. Applying a strict interpretation, each is a violator of the conservation law. True enough. However, within the grand-scale system (namely, the domain of a great cosmic gravity cell) there exists a dynamic equilibrium between the two. On the cosmic scale, there exists a beautiful *harmony of opposites* between energy loss (via the aether deprivation Annihilating process) and energy gain (via the velocity differential Blueshifting process).

It should be pointed out that cosmology theories of the 20th century handled the conservation of energy differently. Most physicists and philosophers asserted that the usual restriction does not apply to cosmic regions; others treated it as something unknowable or simple evaded the issue altogether. Cosmologist Edward R. Harrison, for instance, claimed outright "[it] is obvious: Energy in the universe is not conserved."[5]

Theory has advanced considerably since then. Under the DSSU paradigm, there is a unique way of assuring compliance to the rules. It is recognized that the end-state structures are but components of a much larger system. And within that larger system, there is no violation of the conservation law; and also, no violation of the entropy rule.[6][7] A more detailed discussion of how energy conservation is achieved and how natural processes manage to maintain entropy stability is presented in Chapter 6.

---

[5] E. R. Harrison, *Cosmology, the Science of the Universe* (Cambridge University Press, Cambridge, UK, 1981); p276.

[6] C. Ranzan. 2021. *Law of Physics 20th-Century Scientists Overlooked (Part 6): Law of Cosmic-Scale Conservation of Energy*, Physics Essays Vol.**34**, No.3, pp.331-339 (Sept 2021).
(Doi: http://dx.doi.org/10.4006/0836-1398-34.3.332)

[7] C. Ranzan, *Nature's Supreme Mechanism for Energy Extraction from Nonmaterial Aether*, Infinite Energy Magazine Vol.25, Issue#144, pp.8-16 (2019 March/April).
Posted at: www.CellularUniverse.org/Directory.htm

Density challenge, black holes, rotation, Higgs

Determining the ultimate density of mass is a challenge. The total mass of a suspected collapsed structure could be determined from gravitational dynamics, provided an observable orbiting companion is present. Basic Newtonian gravity equations work nicely. The real problem lies in measuring the suspected Terminal star's diameter. It's virtually impossible —these end-state structures are just too small, too distant, too dark. Yet, without the diameter, the density cannot be empirically established. And so, a value — either for density or for diameter— must be assumed. Our approach was to make a reasonable assumption regarding the probable ultimate density of mass. Once the assumption was made, the profile of **Figure 4-3** logically followed.

Which comes first, the lightspeed inflow or the ultimate density? Or do they happen together? Although the collapse scenario presented above has the critical (lightspeed) boundary forming first then followed by compression to the end-state density, it may well be that collapse occurs in such a way that aether inflow (with respect to the surface) and mass density both attain their limits simultaneously. Whether the two happen sequentially or simultaneously is an open question.

What about *black holes*? Aren't they supposed to manifest the ultimate collapse of matter? ... Understand that *singularity black holes* are not physical objects —they are mathematical objects, the playthings of theoretical physicists. They are components of mathematical cosmology —the construct of the old 20th-century worldview. These conceptual objects of infinitely dense mass inside an infinitely small "volume" have no place in the Real world. The object-as-a-singularity idea does not pass any reality test, being as it is an affront to common sense and an overextension of an incomplete theory of relativity (Einstein's General theory). The abstract theory that predicts black holes demands that the interiors have "space" flowing inward at a rate far greater than lightspeed; thus, the gravity profile is radically different from that of an end-state neutron star. Turns out, the only thing that black holes and Terminal stars have in common is an enveloping surface where the inflow attains the speed of light.

Something that is not often pointed out: Although a perennial hot topic, black holes are not part of the Standard Model of physics. According to cosmologist Leonard Susskind, "Perhaps the most interesting objects of all, black holes, have no place in the theory."[8]

For more on the contrast between black holes and Terminal stars, see the

---

[8] L. Susskind, *The Cosmic Landscape, ...* (Back Bay Books; Little, Brown and Co.; New York; 2006); p122.

comparison Table in the next section below.

What about rotation? In the earlier analysis, it was assumed to be absent or negligible. But since significant angular motion is almost always present, the question must be asked: How does rotation affect the collapse process and its associated mass loss?

When a pre-collapsed body has significant rotation its shape will be oblate —a normal consequence of the centrifugal effect. Its equatorial diameter will be considerably greater than its polar diameter. This means that the surface area will be greater than would be the area of a perfect sphere of equal mass and density. As the body gravitationally contracts, its spin rate, in accordance with the conservation of angular momentum principle, must increase. Its oblateness grows more pronounced, its surface area expands. The surface area is the important factor influencing further collapse.

More surface area permits more aether to flow into the interior than would otherwise be the case. More aether means that more mass can be sustained; the body can grow larger. It logically follows, more total mass would be required to bring about the end-stage situation. In other words, total collapse would require more than the nominal 3.4 solar mass calculated for a nonrotating body to attain the critical boundary and ultimate density.

With the further addition of sufficient mass the surface will become critical (aether attains lightspeed). At that instant, the body immediately changes from oblate to spheroidal. During this quick transition, an amount equivalent to the extra mass, equivalent to the additional mass required to trigger final collapse, is extinguished. The process of mass extinction by aether deprivation becomes active; any further mass infall can potentially trigger the extinction process.

After total collapse has occurred, the rotation story changes. Final collapse has a remarkable and unexpected side effect. The Terminal star is completely unaffected by self-rotation and must always maintain its spherical shape. This means that its rotation has no effect whatsoever on the mass extinction process.

The collapsed structure is subject to the *principle of centrifugal effect negation*. In other words, the Terminal star's shape is completely unaffected by the rate of rotation. Once a contiguous structure attains the critical state, it becomes immune to the centrifugal effect. No amount of rotation —no limit whatsoever— can produce the expulsion of material. The details of this overlooked law of physics is presented in Chapter 5.

Rotation plays another important role, although it is not related to the extinction process. Rotation is responsible for constricting the polar magnetic fields and facilitating the polar emission beams —thereby allowing surface energy to escape to the external world.

Aether versus Higgs. In light of the prominent role that aether plays in the Law of Mass Extinction (as well as several of the other laws overlooked in the previous century), it is natural to ask *But what about the Higgs field?* According to the 20th-century model of physics, the universe is permeated by a so-called Higgs field. The question then is *How does this field differ from the DSSU aether?* And in particular, one would like to know how the Higgs bestows the property of mass onto particles compared to how the same property is acquired through an aether environment.

Here are the key points:
- The conventional view is that mass particles acquire their property of mass from the Higgs field by interacting with an intermediate particle —the Higgs boson. In contrast, the DSSU aether is not a "field" in the usual sense and, therefore, needs no force carrier. It needs no bosons whatsoever. Let me emphasized, this aether is not a conventional field but rather a *subquantum* universal medium.
- The Higgs mechanism involves extremely massive Higgs bosons; but there is no explanation of where this self-mass comes from! The DSSU mechanism does not have this problem. There simply are no bosons; moreover, the aether, being a *subquantum* medium, possesses no mass.
- Under the DSSU framework, particles acquire the property of mass *directly* from aether. It is accomplished via a combination of processes, namely aether excitation, aether absorption, and aether vanishment —all three occurring almost simultaneously.
- What drives the Higgs mechanism? It is a complete unknown as to what generates the Higgs field. Essentially, it is purely an elaborate mathematical construct. In contrast, the generation of aether, as the essence of the universe, is unambiguous. The process of the steady-state emergence of aether is *axiomatic*. (Unquestionably this is revolutionary. But since aether units are subquantum entities, there is simply no violation of thermodynamic laws.) The existence of a discretized universal essence is the foundational premise of DSSU theory.
- Understand that the Higgs may describe, mathematically, to a limited extent, the mass-acquisition process; BUT it does not explain it. On the other hand, DSSU aether theory provides the explanation; and it does so in clearly understood terms.
- Lastly, DSSU aether has the added ability, lacking with the Higgs mechanism, to literally destroy matter —it accomplishes this via the *aether deprivation annihilation process.*

Summary comments

Mass extinction in perspective. Mass extinction by aether deprivation is but one of six key processes operating in the Universe. For the benefit of readers interested in the broader functional system, here are the other five:
- The excitation/consumption of aether by mass and energy particles. This

foundational process functions as the bestower of the property of mass, the attribute of inertia, and the *primary cause of gravity*. Described on a deeper level, it is the conduction of electromagnetic energy via the excitation-absorption-annihilation of aether. It is the very process by which all matter manifests its existence.
- Emergence of aether; this is what is detectable as the *expansion of the space medium*. It functions as tertiary gravity within the cosmic-scale gravity cells.
- Stress-induced self-dissipation of aether; this is what is detectable as the *contraction of the space medium*. It functions as contractile-type *field gravity*. In DSSU terminology, it functions as *secondary gravity* within any contractile gravity domain. (Self-dissipation is the consequence of the aether's limited ability to sustain stress.)
- Redshifting process (energy reduction); detectable/observable as the *cosmic redshift*.
- Blueshifting process (energy amplification). It functions as the limitless power source behind astrophysical jets, structures associated not only with rotating Terminal neutron stars but also supermassive regions.

(Note: Redshifting and Blueshifting are the consequence of one principle, *the velocity differential propagation of neutrinos and electromagnetic radiation*.)

Notice the common element. Each process involves one or another aspect or property of the universal space medium.

Revolutionary development. The single most important factor responsible for the discovery of the principle of *mass extinction by aether deprivation process*, as well as uncovering several other new laws of physics, is the modern version of aether with its previously unrecognized and underappreciated properties. The developments in aether theory over the last couple of decades have been nothing less than revolutionary. For an excellent timeline of the conceptual development of the universal space medium and the discoveries it has made possible, see the webpage *History of the Aether Theory* (posted at www.CellularUniverse.org/Directory.htm).

Contrasting the old and the new. The Law of Mass Extinction —the process of annihilation by aether deprivation— radically changes the physics of total gravitational collapse. The simple process of aether deprivation entirely avoids the well-known paradoxes associated with the hypothetical black holes that are relentlessly hyped by popular media.

The chart (**Table 4-1**) on the next page provides a quick summary of the ideas presented in this chapter and a point-by-point comparison with the long-held highly-problematic conventional view.

The mass-extinction mechanism is of game-changing importance for research into black-hole physics. Crucially important to the study of gravitational collapse, this overlooked process circumvents the breakdown of theoretical physics in the context of the conventional 20th-century view of terminal collapse.

Table 4-1. Comparison of two views of total gravitational collapse.

	Total Gravitational Collapse	
	20th-century Mathematical View	Natural Process View
Basic collapse:	Self-collapse through the Schwarzschild radius to become a so-called **black hole**.	Self-collapse to become a **Terminal star**. Collapse halts when maximum density is attained.
What happens to excess or additional mass?	Added to the mathematical object called a singularity.	Causes a corresponding quantity to suffer *aether deprivation annihilation*.
Lightspeed boundary?	Yes. A boundary in space called an *event horizon*.	Yes. A pure energy surface (absent only at the polar portals).
Energy escape mechanism:	Black holes are purported to evaporate, via thermal radiation, very *very* slowly.	Powerful polar emission beams (photons & neutrinos).
Problems:	• The singularity absurdity: the paradox of infinite density mass in a zero-dimensional space! • The angular momentum paradox! • The gravity paradox: The gravity-causing singularity sucks in everything EXCEPT the energy of its surrounding gravity field!!	No problems, theoretical or practical.
Relationship to Einstein's view:	Disagrees with Einstein's view that mass does not collapse through its Schwarzschild size.*	Conforms to Einstein's view.
Method for complying with conservation-of-matter law:	• Matter is not permanently lost. Mass never ever dies! • Mass within BHs is mathematically converted to energy and radiated away.	• Local violation, yes. Global violation, no. • Mass extinction by *aether deprivation* is in perpetual cosmic-scale balance with matter-formation process(es).

* In 1939 Einstein published a paper in which he showed that matter could not be so condensed that the Schwarzschild radius would fall outside the physical gravitating body.

The *natural* interpretation avoids the embarrassing paradoxes associated with singularity-type black holes. They were discussed in the "Implications" section of Chapter 3 and included:

• The singularity paradox: The incongruity of having astronomical quantities of matter within a zero-dimensional space! The absurdity of having infinite density mass inside a speck of zero volume!

• The angular momentum paradox! The claim that black holes retain the angular momentum possessed by the pre-collapsed structure contradicts the inability of a zero dimensional entity of actually manifesting such a property.

• The gravity paradox: The gravity-causing singularity is supposed to drag in all the surrounding energy (and matter). But for some reason, it doesn't. It fails to consume the energy of its surrounding gravity field!! Expressed another way, it is the conundrum of extending its influence, its "force" of attraction, out through an impenetrable barrier (the event horizon) and beyond.

As pointed out earlier, some 20th-century experts, most notably Sir Arthur Eddington and Lev Landau, thought this sort of outcome was ridiculous and repeatedly argued that there must be some law of nature, some law as yet undiscovered, that would prevent such collapse.[9] This insight from so long ago was so right.

"There must be some law of nature ... that would prevent such collapse" when there is an excessive concentration of mass. And so there is. With the law of *mass extinction by aether deprivation*, excess matter is never a problem.

The path always seems to lead back to aether.

One wonders, what might have been, if Einstein had not neglected to exploit his own aether. He had, in 1921, acknowledged its existence, but then returned to his purely mathematical interpretations. Yet throughout the 20th century, there it lay unutilized and overlooked: discretized aether and its several associated processes; especially one, the *aether-deprivation phenomenon* —the terminal annihilation of matter. This is the process missing in Einstein's gravity theory.

It is important to note that *aether deprivation* is not an *ad hoc* feature tacked onto a larger theory. Rather, it is something that follows logically from the fundamental premise which deems the existence of matter to be entirely dependent upon the absorption-consumption of aether. Mass is sustained by a continuous flow of aether, when the flow is cut off, stuff vanishes. Where this happens depends on the environment. The process is triggered by extreme gravitational environments —at the core of ultimate mass concentrations.

In conclusion, the DSSU mass-extinction mechanism is perfectly reasonable, logically connected to the larger theory, crucially relevant to a proper understanding of gravitational collapse, momentous to the maintenance of energy balance in the universe, and revolutionary in its implications for cosmology.

---

[9] John Gribbin and Martin Rees, *Cosmic Coincidences, Dark Matter, Mankind, and Anthropic Cosmology* (Bantam Books, New York, 1989); p157.

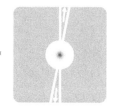

Fourth Interlude

The Key to Decoding the Universe

As the main topic of this interlude, I want to address a couple of questions implicit in Chapter 4's opening quote —*Understand black holes and you understand the ultimate laws of the universe.*

What is it that we need to understand about black holes?

Simply this: Black holes, in the way they are generally conceived, do not exist.

Before proceeding, let's be clear on our definition: A black hole is a gravitating entity with mass at the center and surrounded by an event horizon, where the vacuum (aether) is flowing inward at lightspeed. And since the radius of the event horizon is GREATER than the radius of the central mass, the space medium's flow-speed inside the event horizon is necessarily greater than lightspeed. That is, the flow in the region between the central mass and the event horizon has to be greater than lightspeed. The size of the black hole is determined by the quantity of mass and by the diameter of the event horizon (and not by the diameter of the central mass).

Back to their nonexistence. We should all be able to recognize the absurdity of one type of black hole. No one in their right mind believes that singularity-type black holes actually exist. They are merely mathematical fabrications, the fanciful constructs of overzealous astrophysicists, as discussed earlier. Einstein objected to the physical reality of the singularity, pointing out that since superluminal motion is forbidden *Schwarzschild singularities*, as they were called, do not exist in physical reality. Nevertheless, absurdity has its adherents. In 1992 a couple of the experts, seemingly oblivious to the dissonance, reported, "The balance of opinion among the experts is that all matter entering a black hole eventually encounters a singularity of some sort."[1] I guess if you are compensated for believing, you become a believer. (But since when do singularities come in different

---

[1] Paul Davies and John Gribbin, *The Matter Myth* (Simon & Schuster, Touchstone, New York, 1992); p272.

varieties?!)

But what about another type of black holes, the *non*singular kind, the type for which the mass is not confined to a zero-dimensional point? By nonsingular black hole, I mean an extended-mass object surrounded, at some distance, by an event horizon (the boundary of lightspeed inflow). This kind of object cannot exist. And there is a simple reason.

Follow, in your mind, the flow of the space medium for a hypothetical extended-mass black hole. Outside the event horizon the speed of the flow is less than lightspeed, at the horizon the speed of the flow is exactly lightspeed, and below the concave side of the horizon the speed of the flow is *greater than lightspeed* —all according to the definition. The space medium reaching the surface of the black hole's mass would be traveling faster than the speed of light! And that is the problem. Just as you can never have an object moving through the vacuum (aether) at lightspeed, you can never have the vacuum/aether moving onto (or into) a mass body at lightspeed. Both violate a key tenet of Special Relativity.

The experts, of course, know all this. No science-oriented person disputes these facts. Why then did astrophysicists, so enthusiastically believe in black holes, including supermassive ones?

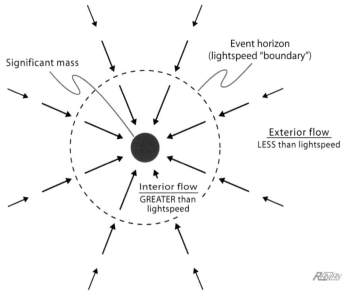

Extended-mass black holes cannot exist. Here is the reason: By definition of an "in space" event horizon, the speed of the space medium on the interior side is greater than lightspeed. This means the space medium would encounter the surface of the stationary mass with a rate above the lightspeed limit —an encounter in blatant violation of Special Relativity. The configuration shown can never be. (Cross-section view)

Fourth Interlude. Key for Decoding the Universe

Belief in black holes rests upon a combination of two factors: The first is that gravitational collapse *does* occur. But this, in itself, does not justify invoking the black-hole concept; for that, another element is needed. The second factor was the complete lack of a mechanism to get rid of the inevitable excess mass. Once you accept the reality of the initial gravitational collapse and you then (with significant additional mass) extrapolate the collapse process unconditionally, you end up with a supermassive black hole.

Let me explain.

The first factor is that gravitational collapse *does* occur. As we saw, this just leads to a Terminal star (end state neutron body). No need to invoke a black hole. It is worth recalling that many scientists of the last century argued against the actuality of total collapse. The favored idea was to use a property of spin. Possibly, rotation would keep the total neutron mass subcritical by having the centrifugal force expel the excess and thereby prevent collapse. American astrophysicists Philip Morrison and Kenneth Brecher *"deny[denied] that the collapse or congregation of stars need proceed so far as to create a black hole. Spinning a massive star might stave off complete collapse for a long time, and new 'anti-gravity' effects may be discovered that prevent the eventual collapse into a black hole."*[2] Theoretical physicist Michio Kaku has a modern version, in which the centrifugal force prevents a rotating ring of neutron mass from collapsing. And don't forget Stanley A. Eddington, he believed "various accidents" might intervene, or some hidden law of physics would preclude a final crunch. The uncompromising fact is that gravity is relentless; it just keeps on grabbing anything within reach; it eventually overpowers any real or imagined mechanism that might oppose final collapse.

Turning to the second factor. This can best be described as a trap that physicists have created for themselves and especially for their astrophysics colleagues. Standard physics does not allow mass to die —to be removed from the universe. Theorists have no mechanism for mass disposal; they have become trapped by a strict adherence to the First Law of thermodynamics. Energy, whether in the guise of mass or some other form, cannot be created or destroyed. As a collapsed structure accrues more and more mass, its belly grows. No material can escape, not even light. Their law of conservation of matter dictates that whatever goes in can never come out (Hawking radiation is irrelevant, too insignificant, just too far-fetched). Mass can be converted to energy and vice versa. But that's it. Material is permanently trapped. No escape. And like a cancer, it grows.

> "A black hole is the nightmare legacy of Einstein's theory of relativity. It is a gaping wound in the fabric of spacetime, an

---

[2] Nigel Calder, *Einstein's Universe* (The Viking Press, New York, N.Y., 1979); p26.

unfillable hole that gets bigger and bigger as it swallows matter."[3]

Thus, the established belief system demanded the hypothesis of black holes —including supermassive ones. Scientists of the 20th century had failed to devise some way of getting rid of the excess mass. They had completely overlooked the law of *Mass Extinction by aether deprivation*.

The take-away message is that total collapse happens, supermassive regions exist, but holes in the Universe are illusory.

Turning to the second half of the opening quote, what does it mean to *understand the ultimate laws of the Universe*? It means you are then able to specify the universe's spatial structure, to predict the patterns of matter distribution, and to specify its temporal nature —a finite origin in the past or perpetual existence. An understanding of the true laws of the Universe means having a scientific theory that explains how the Universe works. Above all, it means the theory's predictions correspond to the observable evidence.

And it is precisely this kind of correspondence that we find in the 21st-century cosmology theory, the DSSU —as will be abundantly revealed in Chapter 7.

Selected issues raised during peer review

The subject matter of the previous chapter, and of the original research Paper, is mass annihilation. Although the focus is on the Terminal annihilated of matter, the Reviewer wanted to know about the opposite. Never mind what is actually in the paper, he wanted a write-up on *matter creation*! How does one explain the emergence of matter? The question automatically leads into the even larger concern about mass/energy conservation.

Here was my response:

> The balance between mass extinction and mass formation/creation is an important point. The details surrounding the relationship between aether and the creation of matter are profound and yet remarkably simple. It turns out that the mechanism for the emergence of matter in the DSSU does *not* require matter particles to be produced (brought into existence) directly from aether.
>
> This mechanism is presented in a separate article entitle "*Law of Cosmic-Scale Conservation of Energy*." For the present, the issue of Conservation is appropriately addressed as follows:

---

[3] Charles Seife, *Decoding the Universe* (Viking Penguin, New York, N. Y., 2006); p225.

Fourth Interlude. Key for Decoding the Universe 155

> "Under the DSSU paradigm, there is a unique way of assuring compliance to the rules. It is recognized that the end-state structures are but components of a much larger system. And within that larger system, there is no violation of the conservation law (and also, no violation of the entropy rule). A more detailed discussion of how energy conservation is achieved and how natural processes manage to maintain entropy stability is published elsewhere."

There was something else, even farther removed from the main topic, that was on the mind of the Reviewer. It concerned gravity waves. He asked, "What is the place of gravitational waves in the aether[-permeated] universe? Do gravitational waves interact with the aether medium?"

To which I responded:

Gravitational waves are not discussed for the simple reason that they do not in any way influence the subject material —*Mass Extinction by Aether Deprivation*— of this Paper

However, this topic is discussed in the referenced article: *The Nature of Gravity –How one factor unifies gravity's convergent, divergent, vortex, and wave effects*, International Journal of Astrophysics and Space Science (2018).

The Reviewer questioned "the mathematical infrastructure of aether."

My response:

I should point out that the mathematical infrastructure of aether is not important to the main theme of the Paper. What is of greatest relevance is that matter (mass and energy) requires a sustaining kinematic and dynamic flow of aether. Without such flow, mass cannot exist. The equations for the dynamic flow are detailed in previously published articles. See references ...

What is interesting is that DSSU aether is compatible with Einstein's aether regarding two major characteristics: Both types are massless and both are dynamic. However, the crucial difference is that Einstein's aether is a *continuum* while DSSU's consists of *discrete units*.

The continuum-versus-discrete difference is extremely important to the mechanism sustaining the existence of matter. Think of it this way: Saying that mass absorbs the "continuum" makes very little sense. But having mass absorb, or feed on, discrete units quantizes the sustaining process —making for a far more intelligible and commonsensical approach.

Stephen Hawking on black holes

A proper understanding of black holes still does not exist. So said the University of Cambridge physicist Stephen Hawking (1942-2018), one of the creators of modern black-hole theory.

Hawking told the journal Nature. Quantum theory "enables energy and information to escape from a black hole." However, a full explanation of the process, the physicist admitted, *would require a theory that successfully merges gravity with the other fundamental forces of nature*. But that is a goal that has eluded physicists for nearly a century. "The correct treatment," Hawking said, "remains a mystery."[4]

So, how black holes work is still a big question mark in conventional cosmology. No surprise —we already know black holes do not exist in the natural world. Most revealing is the false assumption about forces, about the unification of forces. The reality is that gravity is not a true force. It simply cannot be "merged" with the electromagnetic force.

✠ ✠ ✠

---

[4] https://www.nature.com/articles/nature.2014.14583 (accessed 2022-1-16)

Law of Centrifugal Effect Negation

5. Centrifugal Effect Curtailment and Negation

The basic centrifugal effect

The centrifugal effect is a force-like tendency —peculiar to circular motion— that is equal but opposite to the centripetal force striving to keep a particle or object on its curved path. It can be described as the tendency of mass to "pull" away from the center of rotation. For example, a stone attached to a string and whirling around in a horizontal circular path produces a centrifugal force-effect that is exactly balanced by the tension in the string. The tension, as the inward-directed *centripetal pseudo-force*, is balanced by outward-directed *centrifugal pseudo-force*.

A simple expression for the centrifugal effect: If an object, having mass m, is located at some distance r from the axis of the circular motion, then the centrifugal pseudo-force is

$$\frac{(\text{mass}) \times (\text{speed})^2}{(\text{radius})}, \text{ or simply } \frac{m \cdot v^2}{r}.$$

Note that the speed is understood to be tangential to the circular motion. In vector terminology, we would say that the velocity **v** is perpendicular to the radius vector **r**. The term v^2/r serves as the centrifugal acceleration. It is clear from the mathematical ratio (numerator versus denominator) that the effect intensifies with an increase in the speed of rotation and diminishes with an increase in the radial distance.

With these proportionalities (direct, for mass and speed, and inverse, for distance) in mind, and since this chapter explores the very limits of the centrifugal effect, we want to employ a gravitating object that is as small as possible and spin quite rapidly. And since the effect is also proportional to the intensity of gravity, the object should be capable of becoming both extraordinarily massive and dense. In the following exploration, the mass, the density, and the spin speed will be adjusted, and the consequences noted. The size of the structure, however, will not change. In order to make this topic as easy to understand as possible, the discussion is confined, for the most part, to just one physical size. The basic gravitating object to be used to explore the concepts will be a spherical mass with a radius of 10 kilometers. Moreover, naturally occurring centrifugal bulging will be ignored.

There are two factors that limit the centrifugal effect. Both are important. The first is obvious. Spin the stone on a string too fast and the string will break. Any mass structure that spins too fast will tear itself apart resulting in fragments flying off tangentially (within the plane of rotation). The second factor is one imposed by special relativity. No chunk of mass, no particle of mass, can travel through the space medium with a speed equal to, or greater than, the speed of light in vacuum.

Let us examine the first factor in more detail:

The state of rotation at which a structure loses its ability to hold onto its equatorial mass occurs when the centrifugal effect just balances the gravity effect. This situation can be determined by equating the *magnitudes* of the two opposing effects on a test mass (located at the equator) as follows:

(Centrifugal effect) = (Gravity effect);

(mass)×(centrifugal acceleration) = (mass)×(gravitational acceleration);

$$m_{test}\left(\frac{v_{equator}^2}{R}\right) = m_{test}\left(\frac{GM}{R^2}\right).$$

This simplifies to $v_{equator}^2 = \frac{GM}{R}$;

where $v_{equator}$ is the critical speed of rotation at zero latitude; gravitational constant[1] G; M is the structure's mass; R is its spherical radius.

Since G and R are constants, the equation here represents a simple function between the critical rotation speed and the corresponding total mass.

---

[1] Value of G depends on the units system being used. Most often, it is expressed as 6.67×10^{-11} N·m^2/kg^2.

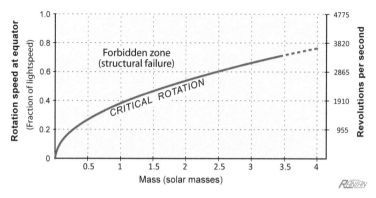

Figure 5-1. Centrifugal effect increases with rotation. But there is a defined limit to the maximum allowable rotation rate (in accordance with the equation in the text). The curve traces the critical combinations of mass and spin for various structures, *all having a radius of ten kilometers*. The graph represents the combinations for which the gravity and centrifugal effects are in balance. Above the "critical rotation" boundary, structures lose their gravitational cohesion.

The balance between the two effects can be represented by a graph relating the tangential speed at the equator (as the graph's vertical coordinate) and the mass (as the horizontal coordinate) as shown in **Figure 5-1**. Remember, the discussion is confined to stars of radius 10 kilometers. For the ones having more than one solar mass, think of them as extremely dense neutron stars.

For the situations falling along the curve, the centrifugal effect will be maximal. If the rotation speed lies above the "critical rotation" curve, the structure simply flies apart.

Another way to express the first limiting factor is to say that, for any given structure to maintain its integrity, the centrifugal effect must never exceed the gravity effect.

Turning to the second factor —the special relativity restriction. Under the traditional view, the rotation speed can never attain the full speed of light. However, there are two ways to define the rotation speed; there are different reference frames to choose from. A proper understanding of the "relativity restriction" depends on how one answers the straightforward question, *rotation speed relative to what?*

How special relativity can restrict rotation

When gravity is treated as a force, the special relativity restriction simply means that the rotation speed at the equator of the rotating mass body must be less than the speed of light. Such is the basic conventional view. But when we include the causal mechanism of gravity —something conspicuously missing in both Newtonian and Einsteinian gravity theories— the situation is considerably different.

Consider a neutron star having a mass value of 2.5 Suns. The radius, for our discussion, continues to be 10 kilometers. According to the DSSU theory of gravity, the speed of inflow of aether at the surface, and perpendicular to the surface, is 86 percent of the speed-of-light constant c. The value was obtained from the basic aether inflow expression detailed in Appendix A.

Before adding rotation to this structure, it helps to be clear about what the mass at the structure's surface "experiences." Think of a block embedded in the surface. This chunk of mass is subjected to an aether headwind of 0.86 c (or about 258,000 kilometers per second, and is safely below the lightspeed limit). This continuous inflow is necessary to sustain the interior mass. As long as the structure is stationary, the surface and the embedded block experience this perpendicular aether inflow, as shown in **Figure 5-2a**.

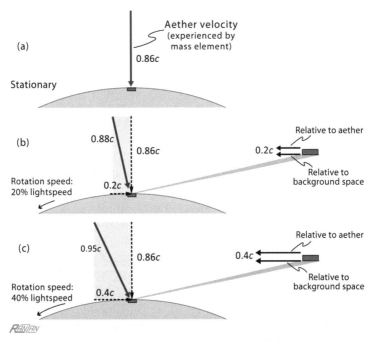

Figure 5-2. An equatorial chunk of mass embedded in a neutron star "experiences" aether inflow depending on the star's state of motion (as indicated). The star's, total mass is 2.5 Suns, and its radius is 10 kilometers. The aether-flow vectors are relative to the mass's reference frame (*left-hand column*, which gives sectional views through the equatorial plane). When the structure is stationary, the flow is perpendicular (part a). When the structure is rotating, the mass experiences the same vertical inflow, but now there is also a tangential flow due to the through-aether motion (parts b & c). The right-hand column compares two distinct velocities of the embedded mass as *viewed from the perspective of background space*: the tangential velocity with respect to aether and the velocity with respect to background space.

Now when rotation is added, the perpendicular component does not change. It does not change because the quantity of total mass has not changed, nor has the distribution (we continue to assume structural sphericity). And so, the perpendicular component remains the same, because mass and size have remained the same. Rotation, however, introduces a tangential aether-flow component, as indicated in **Figure 5-2b**. Consequently there is a change in the headwind experienced by the surface mass element. If the rotation speed at the equator is, say, two-fifths lightspeed, then the Pythagorean Theorem tells us that the new headwind there (at the equator) must be 88 percent lightspeed.

If the rotation speed is further increased say to four-fifths lightspeed, then the equatorial mass will have a relative-to-aether motion of 95 percent lightspeed. See **Figure 5-2c**.

The diagram clearly shows the motion of aether as it encounters the surface of the rotating sphere. But what about the motion of the chunk of mass itself? —In particular, its *tangential* motion?... This can be referenced in two ways. One is with respect to the aether, the other is with respect to background space. This dual referencing is the consequence of having an aether that permeates all space. The block's tangential motion *through* aether is equal in magnitude (and oppositely directed) to that of the aether's tangential-velocity component. The block's motion through background space is self-evident.

And let's remind ourselves so as to leave no doubt as to the meaning of "background space"; it means space in the sense of *an empty nothingness container*. It has no properties whatsoever, other than its three dimensions.

For the rotation speeds shown in **Figure 5-2**, the aether-referenced and background-referenced speeds are equal to each other. The two are shown as such in the figure's right-hand column, but this is not always the case, as will be seen in a moment.

What is important is that the surface element's motion through aether is subject to the special relativity restriction, while motion with respect to the background is not.

Let us continue to increase the spin rate, still using the 2.5-solar-mass-10-kilometer-radius neutron star. When the rotation speed clocks at 51.2 percent of lightspeed, the chunk of equatorial mass finally encounters the special relativity limit. At this rate of rotation, the mass experiences the ultimate aethereal headwind. See **Figure 5-3a**.

From the perspective of the equatorial surface mass, it is moving through the space medium at the nominal speed of light. It simply cannot travel faster (see the vector rectangles in **Figure 5-3**, left-hand column). Mass, of course, cannot propagate at lightspeed. What actually happens with the onset of the ultimate speed is that the affected mass (a thin surface layer) converts to pure energy; the process was described in Chapter 3. The assertion is that the surface really does attain lightspeed with respect to the aether medium (as illustrated).

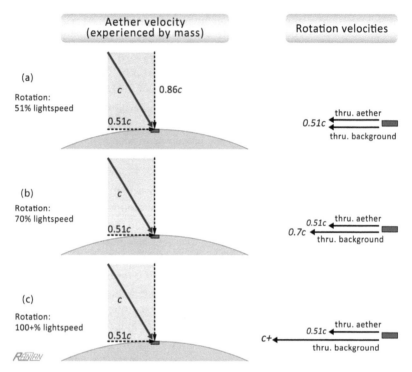

Figure 5-3. Equatorial mass of the same neutron star (having rotation as indicated) "experiences" maximal aether headwind. Part (a) shows the threshold situation —when mass confronts the special relativity limit. This is when the rotation rate pits mass against the uncompromising speed-through-aether restriction. Mass experiences a strict limit, which means *the rotational speed with respect to aether* simply cannot be increased. However, the rotational speed with respect to background space is NOT subject to such restriction (Part b). It could, theoretically, exceed the speed of light (Part c)!

Centrifugal effect curtailment

Here is the lesson of **Figure 5-3**: If the structure does not fly apart at the rotation speed of $0.51c$, then it will not, and cannot, fly apart at even higher speeds! For the mass element to centrifugally "fly free" and be flung away, it would have to travel into the aether flow with a speed greater than the lightspeed constant. And that is quite impossible.

The broader lesson of the special-relativity restriction on rotation: When the flow of aether, relative to a mass structure's surface, attains lightspeed, then the centrifugal effect for that structure will have reached its limit. Then, any increase in rate of rotation will have no effect. In this way the centrifugal effect encounters attenuation. Moreover, this principle is not restricted to the spherical shape; it applies to any contiguous oblate structure. Recall, when a structure manifests the centrifugal effect, the equatorial regions bulge out and the body takes on the shape of a flattened sphere.

For a graphical representation of centrifugal effect attenuation, see **Figure 5-4**. The diagram applies to any compact structure with sufficient total mass and sufficient compactness (density) to be able to reach its centrifugal limit. Instead of going into detailed calculations and difficult analyses of the inevitable structural deformations, let us just say that the spinning body has enough mass and gravitational intensity to avoid tearing itself apart and, instead, is able to actually reach its attenuation level.

Follow the curve. As the spin increases, so does the centrifugal effect. Once the special-relativity limit kicks in, the centrifugal effect cannot increase further. Expressed in terms of aether: Once the tangential motion *through aether* can no longer be increased, the rotational pseudo-force simultaneously attains a limit. However, the rotation itself (with respect to background space) has no theoretical limit. The limiting of rotational velocity *through aether* does not in any way restrict the rate of rotation.

A good question to ask at this point: *How is it possible to keep the speed through aether constant (as when it attains the aforementioned limit) yet still allow for higher rates of rotation of the structure?* ... The answer lies with the process of aether drag —conventionally called *frame dragging* by eschewers

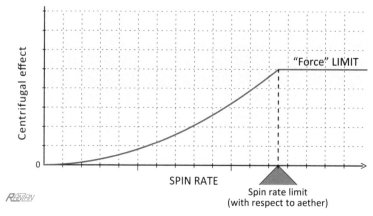

Figure 5-4. Graphic demonstration of the *centrifugal effect curtailment principle*. The centrifugal effect is sketched as a function of the spin rate. It may be said that every sufficiently massive and compact star has a rotational velocity limit, its own identifiable characteristic. Think of high-mass, dwarf stars, consisting of extremely dense degenerate matter. When the spin rate brings equatorial mass face to face with the aether-referenced speed limit, as described in the text, the centrifugal effect, at that point, stops increasing. The pseudo force, having reached its limit, follows a flat line. However, these limits do not in any way restrict the rate of rotation with respect to background nothingness-space. Hence, the horizontal extension has no theoretical limit.

of the aether concept. Aether is dragged along with the rotating structure. It is drawn vortex-like around and around while flowing into the neutron star. Aether drag may be quantified; it is the difference between the velocity through aether and the velocity with respect to background space (or synonymously, with respect to the surrounding distant universe). Frame dragging will be revisited later in association with total centrifugal negation.

The physics rule that emerges from the foregoing discussion on the implications of special relativity may now be stated.

Principle of centrifugal effect curtailment: In astrophysics, if and when a gravitating structure's speed of rotation attains its *aether-referenced speed limit* —a limit imposed by special relativity— it becomes subject to centrifugal effect curtailment. The Principle only applies if the structure is sufficiently massive and compact to actually be capable of reaching said speed limit.

As a general rule, the centrifugal effect is determined by the *tangential* velocity THROUGH the space medium (aether).

Total cancellation of centrifugal effect

The "effect curtailment" described above did not reduce the pseudo force; it only imposed a limit. However, nature does have a way of completely canceling the centrifugal effect.

The explanation leading to the complete negation of the centrifugal tendency requires an understanding of the fundamental difference between absolute rotation and mere relative rotation. This means gaining an appreciation of the fact that absolute circular motion leads necessarily to centrifugal effects, while purely relative motion does not. That sounds simple enough, but it very much depends on how absolute rotation is defined. Absolute motion still has to be relative to something else. We may be convinced that some object possesses absolute rotation, but the inquisitive mind still wonders, *relative to what is it rotating?*

A little bit of history may be helpful here.

Absolute versus relative rotation

In a remarkably simple demonstration, Isaac Newton (1642-1727) convincingly showed that the centrifugal pseudo force is the result of absolute motion, and not just relative motion. His famous bucket experiment is performed as follows: A bucket of water is tied to the end of a length of rope which is secured to an overhead support. The hanging bucket is turned round and round many times, causing the rope to become tightly twisted. While the bucket is held patiently in place, the disturbed water is allowed to come to rest. The experiment begins when the bucket is released and allowed to freely rotate as the rope unwinds. Initially, the contained water remains stationary while the bucket rotates around it; the water displays no centrifugal effects (**Figure 5-5a**). As the rotation speed quickly increases, the water develops a concaved surface, thereby revealing the presence of a centrifugal effect (**Figure 5-5b**). With the rope now markedly untwisted, the turning bucket is grabbed and forcefully brought to a halt. But the contents is still rotating, still manifesting the centrifugal effect of surface concaveness (**Figure 5-5c**). Notice, when the water is rotating relative to the bucket in Part (a) there was no centrifugal effect. Yet when rotating relative to the immediate surroundings, as in Parts (b) and (c), the Effect was obviously present. The centrifugal effect signals a special kind of rotation.

An even simpler pair of experiments reveals the absolute-versus-purely-relative difference. Twirl a ball on the end of a string and watch the ball defy the Earth's gravitational pull as it flies in circular orbits above the experimenter's head. A very real absolute effect is being produced. Compare this with apparent relative circular motion. Spin yourself around amidst a stand of trees; watch the trees circle round and round; no matter how fast you turn

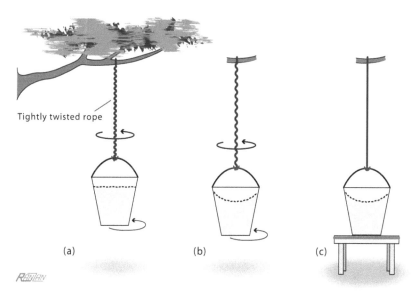

Figure 5-5. Newton's water-bucket experiment. Part (a) shows the situation just a brief moment after the bucket is released and allowed to spin in response to the torque present in the rope. At this stage, the water remains stationary while the bucket rotates around it; the water displays no centrifugal effects. In (b), the water displays a centrifugal effect (while rotating relative to the surrounding world), indicating the presence of absolute rotating motion. The surface of the water becomes concave, as it rises up the side of the bucket. In (c), the bucket is brought to a sudden stop. But the contents is still rotating, still manifesting the centrifugal effect of surface concaveness. Newton concluded that the centrifugal force is the result of absolute motion, not relative motion.

in-place and the trees rotate around you, they do not bend away from you, not in the slightest (even though they rotate, as an aggregate, relative to your reference frame). The ball has motion relative to the experimenter's raised hand; the forest is rotating relative to the observer. One manifests a centrifugal effect; the other does not. Evidently, one undergoes absolute motion; the other is limited to only *relative* motion.

Newton, in connection with the bucket experiment, concluded: "The effects which distinguish absolute from relative motion are centrifugal forces, or those forces in circular motion which produce a tendency of recession from the axis. For in a circular motion which is purely relative no such forces exist, but in a true and absolute circular motion they do exist, and are greater or less according to the quantity of the absolute motion."[2]

If it is only absolute motion that produces centrifugal effects, what is it about absolute motion that makes it different? Does it have something to do with the surroundings? Look at the summary of the observations of Newton's water-bucket experiment in the following table.

Table 5-1. Rotating water-bucket experiment.

Figure reference	Water rotating relative to bucket	Water rotating relative to background	Centrifugal effect (signature of absolute motion)
Part (a)	Yes	No	No
Part (b)	No	Yes	Yes
Part (c)	Yes	Yes	Yes

The centrifugal effect only manifests when there is rotation relative to the background environment. Clearly, there is something special about the surrounding world that defines absolute motion and imparts centrifugal effects.

Identifying the effect-inducing "background"

The problem boils down to identifying, specifically, the background to which absolute rotation has an underlying connection (or to which it is relative, so to speak).

One influential scientist was convinced he had the answer to the "background" question —Austrian physicist and philosopher Ernst Mach (1838-1916). The centrifugal force, he believed, is produced only if the rotation is relative to the surrounding universe, in his words "relative to the fixed stars."

"For me only relative motion exists ... When a body rotates relative to the fixed stars centrifugal forces are produced; when it rotates relative to some

---

[2] As in E. R. Harrison, *Cosmology, the Science of the Universe* (Cambridge University Press, Cambridge, UK, 1981); p181.

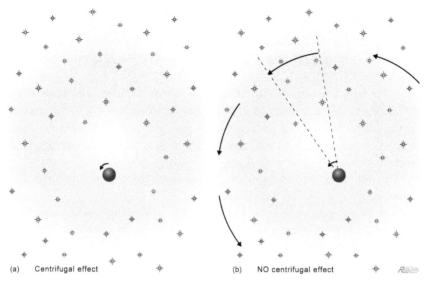

(a) Centrifugal effect   (b) NO centrifugal effect

Figure 5-6. Mach's rotation hypothesis. (a) Body rotates relative to the background universe, "relative to the distant stars." According to Ernst Mach, this is the cause of the inertial centrifugal effect. (b) Mach's premise logically means that if the background universe were to rotate in synch with the test body, then there would be no centrifugal effect.

different body and not relative to the fixed stars, no centrifugal forces are produced."[3] In other words, for Mach, *absolute motion* is only meaningful in the sense of being relative to the universe as a whole (as in **Figure 5-6a**).

Now, if Ernst Mach is correct (and it turns out that Mach was partially right but for the wrong reason), then the following argument must be true: Assume a body rotates; it can be a large or small body; it does not matter. If the universe were to rotate around the body, *at the same rate* about the same axis, then there would be no centrifugal effect. There would be no rotation "relative to the fixed stars"; and, therefore, there would be no tendency of recession from the rotation axis. The rotation would be quite undetectable. **Figure 5-6b** shows the universe rotating in synchronization with a representative spherical body.

The modern view, however, interjects a crucial element; one that changes the way matter relates to the rest of the universe. It is not the distant stars that are important but, rather, the space medium present between the stars and between all bodies. A body's entire "sensory" connection with the surrounding universe is by way of the universal space medium —the medium that empowers gravity. Pause and think about this. The entire universe need not

---

[3] E. Mach, *The Science of Mechanics* (Open Court, LaSalle, Illinois, 1942).

rotate for the argument to remain valid. Only the body's immediate local universe, the surrounding aether, needs to rotate, in order to produce the same negating effect. Say, the body is enveloped by an aether vortex. The rotating body, then, will "believe" itself to be stationary. It will manifest no centrifugal effects.

Something scientists of the 20th century (not to mention also those of the 19th century) failed to recognize is that the absolute motion —that special motion that Newton had deduced to be the key factor in the centrifugal effect— was none other than motion with respect to the ethereal universal medium.

The all-important *background* that defines absolute motion and that facilitates the centrifugal effect is aether.

The distant stars, the far off galaxies, the surrounding universe, all are irrelevant. Only the circular motion *with respect to aether* matters. When the relationship between the rotating body and the inflowing aether meets a simple limiting condition, the Effect entirely vanishes.

Negation of the centrifugal effect

The phenomenon of *Effect cancellation* is contingent upon a surprisingly simple condition, a necessary and sufficient state. The essential condition arises when a gravitationally collapsing body possesses sufficient mass to transform itself into a *Terminal star* (an end-state neutron star). If a body has collapsed but lacks the total mass needed, then it must first acquire the necessary additional mass before transforming into the Terminal state. As we learned in previous chapters, this kind of star is unique. Its density is the ultimate in all of nature and, hence, is often called a *Superneutron star*. Being a defined "Terminal" structure, it cannot be altered in any way other than its rotation. Such an object is truly in an end state of existence.

The Terminal star's defining feature —the very feature that guarantees the negation of the centrifugal effect— is that its aether inflow at its surface is equal to the speed of light.

When the surface inflow is the same as the speed of light, lateral motion through aether becomes impossible. The uncompromising special-relativity rule precludes it. Such a body cannot rotate "through" the aether medium. It can, however, rotate with respect to background container-space; but in order to do this, it must drag the aether as in a vortex. In effect —looking at this from the perspective of background space— the aether spirals its way into the mass body.

There are two ways to argue the negation phenomenon:
• The special relativity argument. Since lateral motion through aether is precluded, it follows that no centrifugal effect can manifest. (Think of this as the *Principle of centrifugal effect curtailment* taken to its extreme.)

FIVE• Centrifugal Effect Curtailment and Negation   169

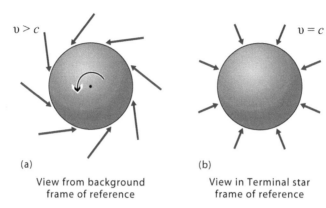

(a) View from background frame of reference

(b) View in Terminal star frame of reference

Figure 5-7. Rotating Terminal star examined from two perspectives. In (a), the view is with respect to background Euclidean space (space as an empty container) and shows significant aether dragging. The velocity magnitude of the flow must necessarily be greater than lightspeed. In (b), the view is in the rotating frame of the rotating star; and regardless of the spin rate, the flow speed at the surface is always the speed of light and the flow direction is always perpendicular to the surface. From the perspective of the Terminal star, the situation is equivalent to having the external world rotating in exact harmony WITH the star —essentially, and remarkably, the star senses NO rotation whatsoever!

• The ultimate-drag argument. Since, in the reference frame of the Terminal star, the aether inflow over the entire surface is perpendicular and equal to the speed of light; the situation is categorically equivalent to having the surrounding universe rotating in synchronization with the Terminal star. See **Figure 5-7**. Again, it follows that no centrifugal effect can manifest.

With respect to background space, the Terminal star may have significant circular motion (**Figure 5-7a**). But from the perspective of the star's own reference frame, there is no rotation (**Figure 5-7b**) —and no rotational effects of any kind. The aether that is streaming onto its surface is streaming from a direction that is *perpendicular to the surface*. The space medium inflow is perpendicular to the surface exactly as it would be if the body were *not* rotating.

The next diagram illustrates the situation from the perspective of a purely imaginary observer on the surface of the spinning Terminal star. He "sees" an object, comoving with aether, falling perpendicular to the surface. The corresponding vector diagrams are instructive on two counts: One, the "perpendicular component" vector at the surface is constant and must remain constant; while the other two vectors can vary in relationship to the rate of rotation. Two, the planted observer in this thought experiment has the ability to measure the motion of the aether (via the comoving object) but has no way of determining the rate of rotation. For him, the body is not rotating at all. We, as the distant observers, can see, or deduce, the reality of the situation: With respect to the background Euclidean space, the comoving falling object (comoving with the aether) has a velocity magnitude that is considerably greater than lightspeed; however, its comoving velocity with respect to the Terminal star has a magnitude of 300,000 kilometers per second and a direction perpendicular to the surface.

In fact, everything impacting the surface —radiation or particulate matter, whether comoving or not— will "appear" to the imaginary surface observer to have a perpendicular trajectory. More specifically, *only what is arriving from directly overhead would be visible*. The phenomenon is known as relativistic aberration[4], a focusing of radiation in the direction of motion, that is, the direction of the motion of the observer relative to the aether flow (which, for the Terminal star, is necessarily perpendicular).

Under less extreme circumstances, it would be possible to observe radiation coming from all directions; it would just be most intense in the direction of motion (that of either the observer or the source). However, when applied to the ultimate situation of the Terminal star, *relativistic aberration* means that the surface observer would see only what aligns with his direction of motion (motion with respect to aether, motion equal to lightspeed). And this direction is perpendicular to the surface as shown in **Figures 5-7** and **5-8**. Everything else is invisible.

Relativistic aberration is also known as the headlight effect. For a pictorial explanation, see the Scientific American issue of May 1982; for the mathematical proof[5] see the 1966 book by John Archibald Wheeler, Spacetime Physics.

In summing up the negation effect —a long overlooked aspect of physics— we may state as follows:

Principle of centrifugal effect negation. As a consequence of special relativity considerations, any critical-state structure (a gravitating structure for which the surface inflow of the space medium, popularly called aether, equals the speed of light) is categorically precluded from manifesting a centrifugal effect. It logically follows that such a body has no theoretical limit to its rate of rotation.

The Principle applies specifically to end-state neutron stars (Terminal stars), these being the only *stable* contiguous structures existing in the universe

---

[4] Roger D. Blandford, Mitchell C. Begelman and Martin J. Rees, *Cosmic Jets*. As in, *The Universe of Galaxies, Readings from Scientific American,* edited by Paul W. Hodge (W.H. Freeman and Co., New York, 1984); p83.

[5] E. F. Taylor and J. A. Wheeler, *Spacetime Physics* (W. H. Freeman, San Francisco, 1966); p69.

FIVE • Centrifugal Effect Curtailment and Negation   171

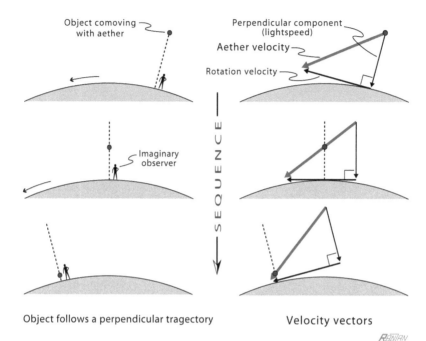

Object follows a perpendicular tragectory   Velocity vectors

Figure 5-8. Imaginary observer on a spinning Terminal star "sees" an object (comoving with aether) falling perpendicular to the surface. The right-hand side shows the corresponding vector diagrams. Three things to note: (i) The "perpendicular component" vector at the surface is, by definition, constant; the other two vectors can vary in relationship to the rate of rotation. (ii) The observer in this thought experiment has the ability to measure the motion of the aether (via the comoving object) but has no way of determining the rate of rotation. For him, the star is not rotating. (iii) Because of the phenomenon of relativistic aberration, or headlight effect, visibility is strictly limited to a perpendicular line of sight. (Viewed from background frame of reference, looking down the axis of rotation.)

that have the requisite surface inflow.

No rotation limit

As discussed earlier, the tangential velocity component (through aether) restricts the rate of rotation. Too much of it and the system flies apart. However, the Terminal star does away with this control; there simply is no tangential velocity component —at least not with respect to aether. The structure is perfectly free to spin with respect to background space (space as an empty vessel); but not with respect to the universal space medium, as the special-relativity-lightspeed restriction must always be respected.

It accomplishes this seemingly magical motion by dragging the aether medium.

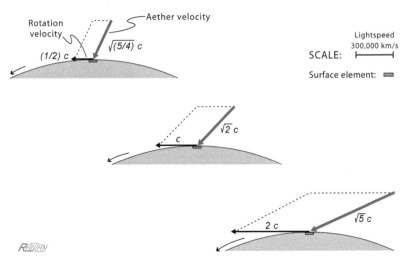

Figure 5-9. Terminal stars have no rotation speed limit. It is because (as explained in the text) there is no centrifugal force, no speed limit applicable to bulk aether with respect to background space, and no limit to aether-frame drag. The aether flow and rotation velocities (shown in these three extreme examples) are with respect to the background Euclidean, or nothingness, space. Note, the aether-drag velocity is the same as the rotation vector. Also note, the heights of the vector parallelograms are all the same; they all equal lightspeed, c.

Note again, there is no law of physics that limits the speed of aether's motion —after all, it is not moving through *anything*, rather, it is merely moving through *nothingness* background space. It follows that there is no limit to the speed with which aether can be dragged. It further follows that there is no theoretical limit to the rate of spin that a Terminal star may possess or acquire.

Thus, a spinning Terminal star can wrap itself with dragged aether to any degree or any speed whatsoever (whatever is compatible with the angular momentum the star possesses). There need be no violation of Einstein's relativity restriction. To see how this works, in terms of the velocity vectors, see **Figure 5-9**. (And keep in mind, the cause of the spin is of no concern here.)

In the broader context, once a star becomes enveloped by a critical-state boundary, its rotation can have no theoretical limit.

Summary discussion

Summary presented by way of two thought experiments

The centrifugal phenomenon associated with rotating bodies requires motion *through* aether —tangentially, or laterally. The importance of this tangential motion is summarized in **Figure 5-10**. The focus is on what a test block of mass "experiences" while resting on the surface of a rotating solid body whose axis of rotation is perpendicular to the page. The two thought experiments entertained in the drawing produce opposite results. One set of conditions leads to an *increase* in the centrifugal effect; another set of conditions leads to its *decrease*. (Except for its rotation, the gravitating structure is considered to be at rest with respect to the surrounding space medium.)

Because of the rotation, the test block experiences the inflowing aether passing through itself at an angle. This aether vector (thick solid arrow), as viewed in the frame of the rotating body, has two components. One is perpendicular to the surface, the other is lateral to the surface.

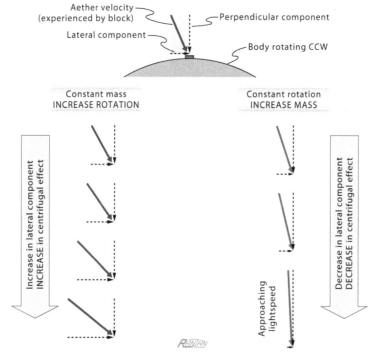

Figure 5-10. Key element in the cause of centrifugation is the presence of a lateral-motion component vector. Left-hand column: With an increase in rotation, there is a corresponding increase in the lateral component; and, clearly, the centrifugal pseudo-force must increase. Right-hand column: However, with an increase in mass (i.e., an intensification of gravity and all else remaining unchanged), there is a *decrease* in the lateral component; for logical consistency, the centrifugal effect must decrease. Note, the view (and the vector analysis) is within the frame of the rotating body. Rotation is counterclockwise.

There are two factors that affect the magnitude and orientation of these velocity vectors: rotation speed and gravitational intensity.
- The left-hand column of the figure pictorializes what happens as the rotation increases, step by step —*while the mass of the planet remains constant.*
- The right-hand column shows what happens as the mass of the structure increases, step by step —*while the rotation rate remains constant.* Picture in your mind the increase in the gravitating mass as a transformation from planet to dwarf star and then to neutron star. Or even simpler, just think of an increase in the intensity of the body's gravity.

Notice the intuitive nature of the left-hand-column sequence: As the rotation increases the tangential velocity component also increases; no one would dispute the fact that the centrifugal effect will increase accordingly. The aether vector (with respect to the test mass) can never become excessive (that is, can never approach lightspeed). A centrifugal "explosion" would occur should an innate limit be exceeded.

The right-hand column reveals a very different effect: As the mass increases, as gravity intensifies, the lateral velocity component diminishes. As the aether flow that the block experiences approaches the velocity of light, the lateral velocity component approaches zero. The centrifugal force progressively decreases; and in the limit, the centrifugal effect ceases to exist.

Overlooked Principle

General Principle governing the centrifugal effect: When the speed of rotation with respect to background space (nothingness space) exceeds the speed of rotation with respect to the universal space medium (aether), the centrifugal effect can then no longer increase. That is the lesson of the vector pairs shown in **Figure 5-3**. The centrifugal effect is always determined by *the rotation speed with respect to aether*. There are two situations for which the effect is zero. This happens for the trivial case when there is practically no rotation; and it happens for end-state neutron stars, regardless of rotation.

The curtailment and negation of the centrifugal effect applies only to structures that are sufficiently massive and compact.

Complete cancellation/negation of the centrifugal effect: Once a lightspeed boundary forms on the surface of an astronomical body, such body totally loses its ability to sense its own rotation. The reason is straightforward. In order to sense rotation, *there must be a tangential motion through aether*; but when the aether inflow attains lightspeed, such motion through the aether becomes impossible (as explained with **Figure 5-7**). In the absence of the lateral aether component, no structure can manifest any centrifugal effects.

Implications

Altering the gravitational signature

An important implication of centrifugal curtailment and negation has to do with the change in the potency of gravity.

Here briefly is how the overlooked Principle affects a body's gravitational signature:

- Any reduction or elimination of the centrifugal effect permits significant increases in rotation.
- This in turn intensifies the vorticular aether motion —conventionally interpreted as frame dragging.
- Which, in turn, increases the stress on the aether, leading to an increase in its self-dissipation.
- And according to the aether theory of gravity, any increase in aether self-dissipation manifests as an amplification-of-gravity effect —an increase in the efficacy of basic Newtonian gravity.

The profound implication is that spinning Terminal stars are far more potent gravitationally than their actual mass content would indicate. In fact, it is not the mass itself that produces the amplification but rather the surrounding zone of secondary gravity (commonly called the *gravitational field*). When *that* zone is stressed, it generates additional gravity. How this amplification/intensification of gravity comes about is explained in greater detail in Chapter 8.

Hidden Matter or Hidden Principle?

The fundamental law governing rotation within the field of astrophysics was not recognized by 20th-century scientists. It was an oversight that severely handicapped their understanding of spiral galaxies. Here is a brief look at several of the influential and distinguished experts and how they arrived at a very strange conclusion.

James Peebles and colleague Jeremiah Ostriker had sought to analyze the rotational stability of disk galaxies using computer modeling. When they ran the program, however, their simulated galaxy went haywire. "To our surprise, the disk went wildly unstable," said Ostriker. "The stars' orbits went from being nearly circular to being very eccentric." Some of the stars even became detached from the disk and flew off into space. Peebles and Ostriker, noting that such chaos does not occur in real galaxies, concluded that some invisible reservoir of mass must be present and provided the additional gravity needed to prevent the disk structure from flying apart.[6] For these eminent astrophysicists, the puzzle was not some hidden principle but simply some invisible mass —some *mysterious dark matter.*

As if to compound the mystery, Peebles eventually received a Nobel physics prize (for 2019), in large part, for this theoretical discovery —more properly, for a speculation of something lacking any real evidence, something that in fact was not needed.

Back in the 1970s and early 1980s, Vera Rubin (1928-2016) and her colleagues at the Carnegie Institute in Washington, D.C., conducted a detailed mapping of the orbital velocities of stars in a large sample of spiral galaxies. They made a seemingly reasonable assumption. The assumption was that the stars, dust, and gas were all traveling *through* the space medium —as they were orbiting about the galactic center— in accordance with Kepler and Newton's celestial mechanics. But that's not what they found. To their surprise and bafflement, redshift measurements showed that all the observable stuff was orbiting much too fast. It was as if *the centrifugal effect was somehow weakened!* Gravity was somehow stronger than theory predicts! Instead of considering that here was the evidence of a new fundamental law of nature, these scientists joined the consensus view and invoked the presence of massive amounts of invisible matter. More mass, theoretically at least, solves the problem; but additional mass has never been found. Nothing ever showed up… even though the search has been going on for over half a century. And it continues, as this is being written. Still nothing.

Realize, the quantity is by no means trivial. According to galaxy expert Vera Rubin, "In a spiral galaxy, the ratio of dark-to-light matter is about a factor of 10."[7] In order to make 20th-century gravity work for spirals, theorists required a mass-correction multiplier of about 10. It was an embarrassing disjuncture between theory and observation! To her credit, Rubin did recognize a need to rethink the nature of gravity completely. She has said, "I suspect we won't know what dark matter is until we know what gravity is."[8]

It seems no one identified Newton's problematic premise. Twentieth-century scientists failed to recognize how a basic premise of Newtonian gravity limits its applicability: Newton's law, in practice, assumes objects (orbiting bodies and rotating bodies) are moving *through* the space medium

---

[6] *Computers and the Cosmos*, Editor Jason McManus (Time-Life Books, Alexandria, Virginia, 1988); p97.

[7] Vera C. Rubin, *A Century of Galaxy Spectroscopy*, The Astrophysical Journal, **451**:419-428 (1995); p427.

[8] Paul Halpern, *Collider, The Search for the World's Smallest Particles* (John Wiley & Sons Inc., Hoboken, New Jersey, 2010); p190.

(vacuum, quantum foam, aether, etc.). Remember, Newtonian gravity is built around an absolute-type space, it treats space as something absolute. But in spiral galaxies, stars are, in large part, moving WITH the space medium! See why conventional gravity fails?

It was a fateful oversight —one that led directly to the false belief in dark matter.

The unavoidable conclusion. The failure of 20th-century scientists to recognize the centrifugal mechanism at a fundamental level —the deep connection between matter in motion and the universal space medium— led to the flawed modeling of spiral galaxies with the inclusion of wholly unnecessary *dark matter*.

The distinction between motion THROUGH aether and motion WITH aether is critically important for understanding the spin of compact stars, the rotation of large-scale structures, and the centrifugal effect.

☥ ☥ ☥

Fifth Interlude

Strange Historical Perspectives. Strange Reviews

Historical perspectives on the nature of rotation

Mach's strange universe

According to Ernst Mach (1838-1916), the inertia of a material object — the object's resistance against being accelerated— is not an intrinsic property of matter, but a measure of its interaction with all the rest of the universe. In his view, matter only has inertia because of the presence of other matter in the universe. So when a body rotates, its inertia produces centrifugal forces, but these forces appear only because the body rotates "relative to the fixed stars," as Mach worded it. If those fixed stars were to suddenly disappear, the inertia and the centrifugal effects of the rotating body would disappear with them.[1] In Mach's universe with only a single star and no inertia, you might think that rotation would have no limit. But no, in the absence of any other stars and objects in the universe, rotation loses all meaning. If rotation cannot be referenced to something, then it cannot exist.

This conception of inertia, which became known as Mach's principle, had a deep influence on Albert Einstein and is said to have been his original motivation for constructing the general relativity theory. Physicist Fritjof Capra, in his book The Tao of Physics, has pointed out that due to the considerable mathematical complexity of Einstein's theory, the experts were not able to agree whether it actually incorporates Mach's principle or not. Most physicists believed, however, that it should be incorporated in one way or another, into a complete theory of gravity.

Now here is where Mach's universe reveals itself to be very strange indeed. As reported in Scientific American, *"Mach inferred that the amount of inertia a body experiences is proportional to the total amount of matter in the universe. An infinite universe would cause infinite inertia. Nothing could ever*

---

[1] Fritjof Capra, *The Tao of Physics, An Exploration of the Parallels Between Modern Physics and Eastern Mysticism*, 4th Ed. (Shambhala Publications, Boston, 2000); p209.

move."[2] In an infinite universe nothing would move, nothing would rotate! But here's the problem for Mach's idea: The real Universe *really is* infinite; and things do move. So, Mach's strange universe fails.

As we now know, Mach was wrong —wrong because he ignored the universal medium. What made his oversight such a serious setback for 20th-century astrophysics and cosmology was that too many theorists (including Einstein) went along with his highly-abstract speculation. The search to find the true nature of space, as being permeated by aether, was not in vogue. No, the big thing at the time was to relativize physics.

Gödel's strange universe

The view of another famous thinker is worth noting. Although a great mathematician, Kurt Gödel (1906-1978) was unaware that a rotating universe is *not* subject to the centrifugal effect!

Gödel devised a rather strange model of the universe (based on general relativity) and presented it as a gift to Einstein on his 70th birthday. The universe Gödel described to his skeptical friend had unique properties. One of which is relevant to the present discussion. It rotated, which Gödel believed provided centrifugal force that would prevent gravity from crunching together all the matter in the cosmos, creating the stability Einstein demanded of any cosmic model.[3]

It is said that Gödel pored over catalogs of galaxies, looking for clues that his theory might be true. Astronomers, of course, have found no evidence that the universe is rotating.[4]

One wonders if Einstein or anyone else asked Gödel the simple question, *rotation of the universe relative to what*? In what possible framework (reference frame) is his universe performing its absolute motion?

Einstein on rotation

Here is another version of the "relative to what" problem.

It might seem at first that Einstein had fallen into a similar trap. Back in 1967, Scientific American carried an article by Gerald Feinberg (physics professor at Colombia University, at the time) in which there was this comment about Einstein's view on rotation:

> "*Einstein pointed out that if the earth were alone in the universe and Newton's laws were still valid, it would be possible for an observer on the earth to determine whether or not the earth was*

---

[2] Jean-Pierre Luminet, Glenn D. Starkman and Jeffrey R. Weeks, *Is Space Finite?* Scientific American April 1999, p92.

[3] Tim Folger, *Gödel's Strange Universe*, Scientific American, September 2015, p71.

[4] Ibid., p72.

rotating by measuring the flattening of the poles. This conclusion seems counterintuitive, since one is inclined to ask: Rotating relative to what?"[5]

But in this case, since Einstein believed in the existence of some kind of aether, one cannot say he was entirely wrong. The presence of a space medium is, after all, essential for the manifestation of centrifugal effects. The ambiguity may be expressed this way: Although Einstein believed in the reality of aether, he never incorporated it into his ideas.

Strange world of science journals
Reviews and reactions

The **International Journal of Astronomy and Astrophysics** took four weeks to decide that the Paper *Centrifugal Effect Negation* contained nothing new, no fresh insights. Their Referee, who already understood everything, commented as follows:

> This paper is not a scientific work but a collection of ideas that do not work. We do not need to explain things in a different way that we already understand. Sometimes a new way of looking at things can have a value but this is clearly based on unfounded and incorrect assumptions.
>
> General relativity must be used since special relativity does not apply. There is no aether of the kind used here through which anything moves.
>
> The author does not seem to be aware of gravitational redshift and the change in energy of radiation as it enters and leaves clusters of galaxies. Nothing has been overlooked and we teach it to our students.

The article was rejected even though the Reviewer failed to specify any fault in the physics, nor point to any flaw in the logic of the arguments presented! This is a clear example of a reviewer using Old Physics! The Reviewer is sadly unaware that general relativity theory breaks down for totally collapsed structures; unaware of the several experiments confirming the detection of aether. Moreover, the Reviewer raises issues such as the "gravitational redshift" and "galaxy clusters," which have no relevance to the topic discussed in the Paper. (This is so typical: express objection to, and find fault with, things that are not part of the presentation.) Evidently, and sadly, Old Astrophysics still pervades the indoctrination institutions; and this educator-indoctrinator is proud to admit, "and we teach it to our students."

Here is what the Reviewer might find worthwhile to think about:

---
[5] Gerald Feinberg, *Ordinary Matter*, Scientific American, May 1967, p134.

Why would I trouble myself to build a theory on "a collection of ideas that do not work"?

Why would I want to explain things in a different way the things that "we already understand"?

How does this Reviewer know that my "assumptions are unfounded and incorrect"?

Isn't the Reviewer curious about these questions? What kind of an uninquisitive scientist is this, who, nevertheless, bothers to review original research?

International Journal of Astrophysics and Space Science (IJASS): The editors of this publication took only a few hours to arrive at a decision.

> Following careful consideration by the journal's editorial board, we think the subject of your manuscript is not fit for our journal. We regret to inform you that we are unable to accept your submission.
>
> We hope to see more of your work in the future.
> ...

Hmmm, sounds like they actually had a board meeting and carefully thought about what to do. And then came to the conclusion that a new insight with important ramifications for astrophysics (not to mention cosmology) is a "subject ... not fit for our journal."

Foolish me, I had believed their posted claim. *IJASS is an international scientific journal dedicated to the publication of original research covering the entire range of astronomy,* **astrophysics***, and* ***astrophysical cosmology****. This includes both observational and* ***theoretical research****.*

The **American Journal of Astronomy and Astrophysics**, instead of initiating a review of the manuscript, accused me of making simultaneous submissions!

With the next submission, I lost several more precious weeks. This time with a publication dedicated to physics —but, as it turned out, not interested in an overlooked aspect of physics. **Physics International** journal rejected the Article outright on the advice of their reviewers.

> **1st Reviewer:** This article is not suitable with the journal where it has been submitted. The themes between the article and journal are different.
>
> **2nd Reviewer:** I don't know how to put this more delicately but I do not think that this Paper is suitable for publication. The idea of

an inertial force has been studied before. See D. W. Sciama, The Physical Foundations of General Relativity (1969).

Progress in Physics is an on-line journal with an undeclared qualifier. Its devotion to "progress" is confined to orthodox physics. Progress beyond the established narratives is not welcome.

It seems they could not find anything wrong with the Paper's physics. So, instead of attributing the Rejection Decision on scientific issues, they came up with stock excuses:

> Your submitted manuscript "Centrifugal Effect Negation" has been reviewed. Unfortunately it has been found to not be suitable for publication in our journal as it does not meet the requirements of our journal.
>
> Your paper is too long for our journal. ...
>
> Please note that we do not engage in discussion of submitted manuscripts.
>
> Sincerely yours,
>
> Pierre Millette, Subject Editor (2021-2-27)

Next in line was the journal of **Advances in Astronomy**. The Editor politely stated, "I regret to inform that your manuscript has been rejected for publication in Advances in Astronomy for the following reason: [see below] ... Thank you for your submission, and please do consider submitting again in the future."

> Dear Dr. Ranzan,
> Having read your manuscript, I found that it is more an overview of current theories. You deal with many topics, and the manuscript seems to be out of a clear focus. Please note that we only consider novel manuscripts with significant scientific merit. In consequence, my firm decision is rejection.
>
> Thanks in any case for considering Advances in Astronomy to submit your paper.
>
> Prof. Dr. Josep M. Trigo-Rodríguez
> Editor in Chief,
> Institute of Space Sciences (CSIC-IEEC), Barcelona, Catalonia
> (2021-7-9)

The following expresses my thoughts after reading the Editor's rejection email:

Not focused?! Too many topics?! ... On the contrary: The Article deals with the very specific topic of the Centrifugal Effect as it affects compact objects of extreme gravity. And very specifically shows how and why the Centrifugal Effect can actually become completely impotent. Yet the Reviewer (the Editor in Chief of Advances in Astronomy) says "You deal with many topics, and the manuscript seems to [have no] clear focus."

Surely the Editor is aware that there is no mechanism known to conventional science by which Centrifugal Effect negation occurs; yet he suggests that the manuscript "is more an overview of current theories," lacks novelty, and has no "significant scientific merit"! WHAT?! ... "In consequence, my firm decision is rejection."

Thank you Professor Dr. Josep M. Trigo-Rodríguez of the Institute of Space Sciences in Barcelona, for your brilliant perceptive assessment.

There you have it, another fake review. What can one say?

In the end, a publisher was found and the Article became part of the scientific literature for physics. Shortly thereafter an independent thinker sent me this:

> *I read your Paper, Centrifugal Effect Negation. Another brilliant novel work.* –Australian engineer Mac Rynkiewicz
> (2021 November 10)

Invoking the centrifugal effect to support a mythical cosmology

Every so often someone will come up with the bright idea of using universe-wide rotation as a convenient explanation for the purported universe expansion. The scheme is straightforward. The centrifugal effect associated with cosmos rotation causes, supposedly, an outward dispersion of galaxies and related structures.

Not long ago, I was asked to review a paper of this type. The cosmology being proposed required mega rotation. In the words of the author, *"The universe ... is a single giant vortex."* My reaction was to confront the author with the same crucial question discussed earlier.

Quoting from my report:

> The Universe is necessarily the ultimate reference frame. If the claim is that it rotates as "a single giant vortex" then the question is *"With respect to what is it rotating? Is it rotating clockwise or counterclockwise?"* ... These are invalid questions with no logical answers. In other words, the author's premise is meaningless!
>
> The following argument, as presented, is demonstrably wrong. It seems the author associates the vortex idea with the concept of

Fifth Interlude. Strange Historical Perspectives. Strange Reviews

spin (advanced in one part of the paper). Then in a later section, it is postulated that vortices rotating in opposite directions attract and those in the same direction repel. Presumably then, like spins repel and unlike spins attract. This would mean that if an up-spin electron encounters a down-spin electron, then the two should attract each other!! ... A clear contradiction of the reality.

Glaring omission. Although the central theme of the article is the identification of particles and forces and structures with the concept of vortices, there is no mention of the originators and pioneering developers of the idea. No mention of Hermann von Helmholtz, Maxwell, and Kelvin, in connection with small-scale vortices. And on the grand scale, no mention of **René Descartes'** ***whirlpool universe*** in which the whole cosmos was a system of interlocking vortices or "tourbillons."

(End of excerpt)

Returning to the main point, the author seemed unaware of the physical and metaphysical ramifications of his cosmic scale assumption —physically impossible to verify and philosophically meaningless. He had over-extrapolated a feature of rotation in an effort to explain a nonexisting condition. He had trapped himself in a common pitfall as part of an amateurish attempt to explain the popular 20th-century myth of universal cosmic expansion.

✠ ✠ ✠

> *Law of Cosmic-Scale*
> *Conservation of Energy*

6. Energy Conservation of the Universe

Two questions

As an introduction to this chapter's topic, I want to show the need for a clear fundamental definition of energy.

Consider the energy of motion —kinetic energy. There is something peculiar about it. It can be shown that this energy may be acquired without any interaction, without any transfer of energy. A situation of this sort commonly arises in certain regions of the universe.

Always surrounded by a web of galaxy clusters, the typical cosmic Void is about 300 million lightyears across. The space medium of the Void is expanding —expanding because this is its nature when subject to tension. (The tension is caused by the gravitational "pull" of galaxy clusters located on opposite sides of the Void.) Within the Void, the *vacuum*, or quantum foam, or aether (our preferred term), is substantially free of contractile-type gravity fields[1] and, consequently, is allowed to freely expand. This property of expansion is a foundational property of all modern cosmology.

Within such a Void we drop two test objects, say, a pair of galaxies. See **Figure 6-1**. Placed sufficiently far apart and with the center point of expansion between them the test galaxies will drift apart as they comove with the flowing aether medium. Moreover, as they move apart, they will do so with an accelerating speed. Remarkably, without any physical interaction, the galaxies

---

[1] Contractile-type gravity is produced by mass bodies and defines a region where the *contraction* of the space medium occurs.

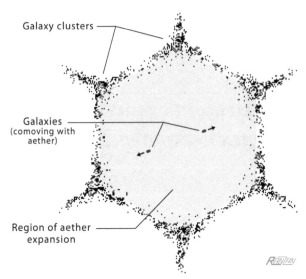

Figure 6-1. Mass acquires kinetic energy within a typical cosmic Void, but without any transfer of energy taking place. Voids, like the one shown here in a schematic cross section, are typically 300+ million lightyears across and surrounded by a web of galaxy clusters. The aether in the Void continuously expands, but the Void itself does NOT. This is because the galaxy clusters compensatingly contract aether.

are "pushed" apart. Without any transfer of energy, the galaxies gain enormous kinetic energy. Each galaxy acquires significant energy of motion — significant when referenced to the distant galaxy clusters and particularly when referenced to the destination clusters into which each will eventually merge.[2]

The question is *Where did the energy come from*?

Here is the problem: Without a fundamental definition of energy, there is really no way to explain the source of the energy and the veritable cause of the motion. Moreover, there is a deeper aspect to the question. There is the issue of energy conservation that must be addressed.

We turn next to another situation, one that is even more remarkable, one that leads to the second question. When exceedingly massive stars collapse to the Terminal state (our familiar *end-state neutron star*); or when a sufficient quantity of degenerate mass accretes; or when previously collapsed stars collide; then, a staggering quantitative loss of mass will occur. Nature, as we learned in Chapter 4, has a way of *totally* annihilating mass without requiring any input energy and without any surviving energy whatsoever. It is a

---

[2] C. Ranzan, *Ellipticity, Its Origin and Progression in Comoving Galaxies*, American Journal of Astronomy and Astrophysics, Vol.3, No.2, pp.12-25 (2015). (Doi: 10.11648/j.ajaa.20150302.11)

mechanism of the total vanishment of mass.[3]

The question now is *Where did the mass energy vanish to?*

For an explanation, once again, a truly fundamental definition of energy is needed.

Two scenarios, two questions. *Where did the energy come from?* and *Where did the energy go?* The challenge is to provide not only explanations for the simple example of kinetic energy acquisition and the strange disappearance of mass energy but also to make them fit into a functional conservation-of-energy framework.

For these and other energy-related questions that persistently arise in 21st-century cosmology, what is needed is a deep understanding of the connection between energy and aether. Understanding begins with a basic definition.

Energy defined at the most fundamental level

The 20th-century concept of energy was not well understood, and certainly not at a deep level. The following description, taken from a contemporary textbook *Foundations of Space Science*, summarized the situation.

> "Energy is a very useful concept that appears in one way or another in almost every significant theory. But *it is impossible to say exactly what 'energy' is*. It is recognized by the action or changes it produces. Every definition of energy ends up saying that it is what causes something to happen. An object at rest does not move unless energy is spent to accomplish the motion. Energy is the cause of the motion but there is no answer to the question of *what causes energy.*"[4]

There is another problem: the lack of a unifying framework.

> "The concept of energy is used in the same way [as Newton's Law of Gravitation] except that it appears in so many forms that *no single theory or law can describe how it operates.*"[5]

The experts of the time were baffled by energy in terms of what it is, what causes it, and how it operates. However, in the early part of our 21st century, the fundamental mechanism of energy was discovered. It was recognized and refined in conjunction with the development of DSSU cosmology, in which it is a key element.

Foundational definition of energy: Any quantitative change in the space

---

[3] C. Ranzan, ... *Mass Extinction by Aether Deprivation*, Journal of High Energy Physics, Gravitation and Cosmology Vol.7, No.1, pp.191-209 (2021).
(Doi: https://doi.org/10.4236/jhepgc.2021.71010)

[4] W.J. Krynowsky, W.L. Ramsey, C.R. Phillips, F.M. Watenpaugh, *Foundations of Space Science* (Holt, Rinehart and Winston of Canada Ltd., 1967); p251. Emphasis added.

[5] Ibid. Emphasis added.

medium represents a manifestation of energy. It means any quantitative change in aether (itself defined as a non-material non-energy essence) manifests as one or another form of conventionally recognizable energy. If there is no quantitative change in aether, then there is no manifestation of energy. *Fundamental energy* consists of essentially two processes: One is the emergence of discrete aether units/entities; the other is their vanishment. All forms of recognizable energy fall into one or the other of the two categories.

The non-energy aspect. It must be emphasized that the aether units themselves, contrary to popular expectation, are NOT energy entities. This may come as a surprise. But it is important to realize that only the change in their numbers represents energy. What this means is that (in accordance with the DSSU definitions of energy and of aether) the discrete fluctuations of the vacuum *cannot* be called quanta of energy. Units of aether, as such, are not the manifestation of quantized energy but, rather, the manifestation of what is termed the ***essence process***. Aether, then, consists of ***essence-process units***.

The aether that fills all space is the essence medium of the physical universe. Bulk aether is normally dynamic (it can expand and contract). However, when in a static state (no volume change, not expanding and not contracting), aether is in its non-energy ontological state.

Let's look at some examples of how the energy definition applies to the various forms of energy found in nature.

- The energy of mass, as well as radiation, at the most fundamental level, is identified with the excitation and absorption-annihilation of aether units. Mass and radiation possess energy simply because they absorb and annihilate aether.
- The energy of gravity fields, that is, the energy of the *contractile-gravity* regions that surround planets and stars, is caused by the self-dissipation (vanishment) of aether units.
- The energy of Lambda, that is, the energy of the Void portion of cosmic gravity cells, is caused by the *addition* of aether units. "Lambda" is just a technical term for the antigravity effect.
- The energy of electromagnetic fields involves a complex patterned excitation of aether accompanied by the absorption-annihilation of aether.
- Thermal, and innate kinetic, energy is another form of aether excitation accompanied by essence destruction.
- In fact, anything that places stress on the aether medium tends to manifest as some kind of energy. Two notable examples: Strong gravity waves carry energy because they produce an associated compressive stress on aether; and there is the stress of extreme rotation (as found around gravitationally collapsed objects and in spiral galaxies).

Energy classification

We need to delve into a bit of history pertaining to the classification of subatomic particles; their original sorting into positive and negative forms of energy; and show, specifically, why all matter can be deemed to represent *negative* energy.

The mechanical energy of a particle, any particle, may consist of several components: rest energy, kinetic energy, and momentum energy. They can be arranged into a right-angled triangle, as was done back in Chapter 3 (Figure 3-1). But this time we will add some symbols. See **Figure 6-2**. But don't panic. If you are not into math and find symbolic functions as impenetrable as thick fog, do not let the equations here cause discouragement. There will not be any calculations. We are not at all concerned with the functionality of any equation —only with the form and the simple underlying meaning.

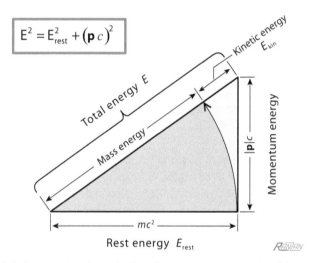

Figure 6-2. Energy triangle embodies the Pythagorean relationship among the mechanical energy components that may be associated with bodies or particles; and most relevantly to elementary particles —be they mass or massless, be they stationary or moving at the speed of light.

Originally, in the history of the development of particle physics, one of the solutions for the energy-of-particles equation was a representation of negative energy and the other solution was for positive energy. But, in time, it was decided that *all* matter should be considered as *positive* energy.

It all started in the 1920s with Paul Dirac's search for a solution to the state of the electron and the relativistic energy-momentum relation

$$E = E_{rest} + E_{kin} = \sqrt{(mc^2)^2 + (\mathbf{p}c)^2} \ .$$

In the equation and in the diagram, E is the total energy, E_{rest} is the rest energy[6], E_{kin} is the kinetic energy, m is the mass, **p** is the momentum, and of course, c is the lightspeed constant. The relationship qualifies as being "relativistic" because the particles it describes may be moving at any speed, even at lightspeed.[7]

Because of the Pythagorean form of the energy equation (with E being a squared parameter, as expressed in **Figure 6-2**), it follows that there are two solutions for E^2 —one positive and one negative.

And so the equation for energy E, allowed for two solutions,[8]

$$E = +\sqrt{m^2c^4 + \mathbf{p}^2c^2} \quad \text{and} \quad E = -\sqrt{m^2c^4 + \mathbf{p}^2c^2}.$$

"The positive root is associated with particle states, and the negative root with antiparticle states."[9] as quoted from a textbook on the subject. In the Paul Dirac version of the energy equation, applicable to one-half spin objects such as electrons, there are, as we will see, four independent solutions.[10]

Although originally, during the 1920s and 1930s, the two solutions shown above were interpreted as representing positive and negative energy, respectively; the view changed during the 1940s. Henceforth, physicists no longer claimed that matter and antimatter represent opposite forms of energy, one positive and the other negative. Ever since then, *both particles and antiparticles have been considered to be the same form of energy* —strictly positive energy.

The same "form of energy," yes indeed; but why positive?

The question is *What determined the sign assignment?* —Why make the energy positive instead of negative? The standard sign designation is not imposed by the above energy equation! It turns out to be an arbitrary choice; it is merely an assumption; and it could just as easily have been the reverse. A similar free choice holds for the particle-versus-antiparticle designation. It is in principle completely arbitrary which one you call the "particle" and which the "antiparticle" The positron could have been designated as the *particle* and the electron then serve as the *antiparticle*. For us, this presents a great relabeling opportunity, one that we will exploit in a moment.

Theorists, long ago (1940s, Feynman and Stückelberg) decided to place particles and antiparticles on an equal footing —both were deemed to

---

[6] Rest energy (and mass energy), expressed here as E_{rest}, is more commonly symbolized by E_o.
[7] L. B. Okun, *The Concept of Mass*, Physics Today, Vol.**42**, No.6, 31 (1989).
  (Doi: http://dx.doi.org/10.1063/1.881171)
[8] D. J. Griffiths, *Introduction to Elementary Particles* (John Wiley & Sons, New York, 1987); p18.
[9] Ibid., p219.
[10] Ibid., p217.

represent positive energy. Physicist David Griffiths, in *Introduction to Elementary Particles*, describes how Richard Feynman and Ernst Stückelberg provided a way around the intractable problem of infinite energy radiation predicted with the negative energy solution. "In the Feynman-Stuckelberg formulation the negative energy solutions are reexpressed as *positive*-energy states of a *different particle* (the positron); the electron and positron [as a particle and antiparticle pair] appear on an equal footing ..."[11]

Incidentally, the reason why the "negative solution" was not considered to represent negative energy was mainly mathematical. If the positive solution $+\sqrt{m^2c^4 + \mathbf{p}^2c^2}$ is taken as positive energy and $-\sqrt{m^2c^4 + \mathbf{p}^2c^2}$ is taken as negative energy, it would mean, given the natural tendency of every system to evolve in the direction of lower energy, that the electron, for instance, would "runaway" to increasingly negative states. According to the mathematical interpretation, the electron in this process would *radiate an infinite amount of energy* (presumably as a consequence of the acceleration during its orbital motion).[12] Nevertheless, Paul Dirac's early view was that electrons could have positive and negative energy states.

As stated earlier, the Paul Dirac version of the relativistic energy-momentum equation allows for four independent solutions. Why four? — because the two expressions in the above paragraph have two solutions each. Notice that the momentum parameter $\mathbf{p}$ can be mathematically plus or minus. Regardless of which sign is inserted, it makes no difference to the outcome. Square a negative and you get a positive. This means $(-\mathbf{p})^2$ is equal to $(+\mathbf{p})^2$, and the value of the expression does not change. But what it does is allow the two opposite signs to represent *two different states*, two opposite SPIN directions. One up, one down.

For the record, here is the basic definition of the Dirac equation: A relativistic wave equation for an electron in an electromagnetic field, in which the wave function has four components corresponding to four internal states specified by a two-valued spin coordinate and an energy coordinate which can have a positive or negative value.[13] The Dirac equation provides a description of elementary spin ½ particles, such as electrons, consistent with both the principles of quantum mechanics and the theory of special relativity.

The importance of the Dirac equation is that it allows for two classes of objects: particles and antiparticles. Furthermore, each of these may have two spin states (spin up and spin down). The 20th-century interpretation for spin ½

---

[11] D. J. Griffiths, *Introduction to Elementary Particles* (John Wiley & Sons, Canada, 1987); p21.

[12] Ibid., p18.

[13] McGraw-Hill Dictionary of Scientific and Technical Terms, 3rd Ed. Sybil P. Parker, Editor in Chief (McGraw-Hill Book Co., New York, 1983); p464.

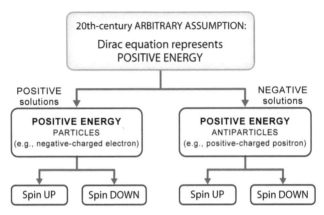

Figure 6-3. Twentieth-century interpretation of the Dirac energy-state equation. The initial assumption was that the equation represents *positive* energy. The equation has two sets of solutions. The positive solution is associated with *particle* states. The negative solutions, as physicist David Griffiths made quite clear, "... we now interpret the 'negative energy' solutions as representing *anti*particles with *positive* energy."

particles is charted in **Figure 6-3**. Categorically, both positive and negative solutions represent POSITIVE energy . Quoting from the textbook by David Griffiths and retaining his emphasis: "... we now interpret the 'negative energy' solutions as representing *anti*particles with *positive* energy."[14]

The physical interpretation of the Dirac equation, "while providing a wealth of information that is accurately confirmed by experiments, nevertheless, introduces a new physical paradigm that appears at first difficult to interpret and even paradoxical. *Some of these issues of interpretation must be regarded as open questions.*"[15]

The door to a reinterpretation is open.

Theorists had made the assumption that the Dirac equation represents *positive* energy. However, they could just as easily have declared that the Dirac equation represents *negative* energy! But, of course, they did not; which is unfortunate —unfortunate because it placed matter-energy in opposition to gravitational energy. In effect, it delayed the recognition of the connectedness between the two —the one was wrongly believed to be positive and the other was correctly believed to be negative.

We will now take advantage of the relabeling opportunity. We reverse the historical assumption. In the physical interpretation that accords with DSSU theory, matter is deemed to represent *negative* energy . (The negative solutions

---

[14] D. J. Griffiths, *Introduction to Elementary Particles* (John Wiley & Sons, Canada, 1987); p217.

[15] *Dirac Equation*, Wikipedia, http://en.wikipedia.org/wiki/Dirac_Equation (accessed 2012/4/13)

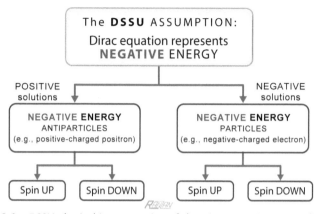

Figure 6-4. DSSU physical interpretation of the Dirac equation. Here the initial assumption is that the equation represents *negative* energy. Both sets of solutions also represent *negative* energy. The positive solution is associated with *antiparticle* states. The negative solution is associated with *particle* states. Contrast this with the previous flowchart (**Figure 6-3**).

will represent particles instead of antiparticles; this is a reversal of the 20th-century system. But this choice is more of a suggestion than of necessity, the choice is arbitrary.) There appears to be nothing preventing the implementation of the DSSU interpretation as presented in **Figure 6-4**.

With the revised interpretation, the electron and its twin, the positron, can be classified as negative energy. Since the Dirac formulation applies to all spin ½ quantum objects, they can all be classified as negative energy.

Thus, all fermions (particles with the spin ½ property) represent negative energy.

But it is possible to go further. As will be shown next, *all* matter can be deemed to represent negative energy.

How the photon unifies all fundamental Negative Energy

The photonic theory of particles (including both mass and radiation particles) appears to have originated with James H. Jeans early in the 20th century. He may well have been the first to realize the fundamental nature of mass. Mass is essentially a state of photon confinement. Particles of mass are just various configurations of self-looping waves of light. He wrote, back in 1931:

> "The tendency of modern physics is to resolve the whole material universe into waves, and nothing but waves. These waves are of two kinds: bottled-up waves, which we call matter, and unbottled waves, which we call radiation or light. The process of annihilation of matter is merely that of unbottling imprisoned wave-energy and setting it free to travel through space. These concepts reduce the

whole universe to a world of radiation, potential or existent, and it no longer seems surprising that the fundamental particles of which matter is built should exhibit many of the properties of waves."[16]

Modern champions of the photonic theory of particles include physicists John G. Williamson and M. B. van der Mark. They have presented compelling evidence that the electron is composed of a confined photon.[17,18] It has been clearly demonstrated that when a photon doubles back on itself, confines itself into a one-wavelength double loop, the photon configuration (i) acquires mass, (ii) generates an electric field, a *concentric* electric field, and (iii) generates a magnetic dipole. This is most remarkable. A self-orbiting and suitably-polarized photon provides an exemplary model of the electron and its antiparticle (**Figure 6-5**). And by predicting differing configuration patterns, the theory may be extended to all particles of mass. Furthermore, it turns out that the strong-force particles (gluons) are not needed; they can simply be replaced by the principle of *loop completion*.

Even the neutrino, the other ubiquitous radiation particle, fits nicely into the photonic particle theory, as will be explained in a moment.

Understand the simplicity of what is happening during the self-looping of the one-wavelength photon in **Figure 6-5**. It is a single electromagnetic wave; defined, in textbooks, as a propagating oscillation of coupled electric and magnetic fields. While following its helical path, the photon is twisting itself in such a way so that its electric field vectors are always radially aligned (directed inward for the electron, outward for the antielectron). This "twisting" is what the circular polarization, mentioned in the drawing, is all about. As for the magnetic vectors, remarkably, they all remain *axially* directed and are responsible for the presence of a magnetic dipole —a property exploited in *magnetic resonance imaging* technology.

Now to be consistent, the photon itself also has to be connected to Negative energy. The rationalization can be summarized as follows: The electron (and the positron) represents negative energy (per **Figure 6-4**), and by extension, its electric and magnetic fields are also so designated. The electron consists solely of a trapped photon. Therefore, photons must also be classified as manifesting Negative energy. Moreover, since all mass particles are configurations of photons, then it must be that all mass represents Negative energy. But we don't stop here. We can extend the designation to *all* matter.

---

[16] James Hopwood Jeans, *Chapter: Matter and Radiation*, The Mysterious Universe (Cambridge University Press, 1931); p69.

[17] J. G. Williamson and M. B. van der Mark, "Is the electron a photon with toroidal topology?" *Annales de la Fondation Louis de Broglie*, Vol.**22**, No.2, 133 (1997).
(Posted at: https://www.researchgate.net/publication/273418514)

[18] J. G. Williamson, *On the nature of the electron and other particles*, The Cybernetics Society 40th Annual Conference (2008) in London. (www.cybsoc.org/cybcon2008prog.htm)

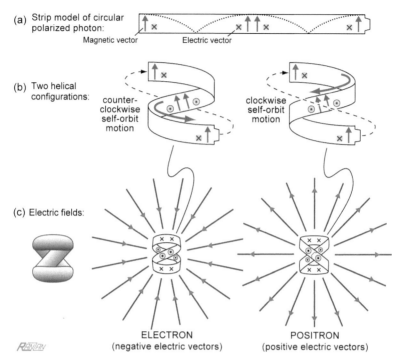

Figure 6-5. The strip-model of a circularly polarized photon in (a) can be arranged in two ways as shown in (b). The helicity and rotation-sense of the confinement configuration determine the direction of the electric field vectors, and of the type of mass particle. (c) By convention, an inwardly directed pattern is associated with the negative charge of the electron, and the outwardly directed pattern with the positive charge of the antielectron (or positron). (The "strip" represents a single electromagnetic wavelength. The symbol ⊙ is for the point-end of electric vectors and ⊗ is for the tail-end. In color version, electric vectors are shown in blue, magnetic vectors in green.)

Accordingly, to this classification scheme, we add the neutrino, a particle with no mass, just pure energy. Second only to the photon itself, the neutrino is the most abundant energy particle in the universe. It turns out that the neutrino is a neutralized packet of electromagnetic energy —essentially two equal-wavelength photons "locked" together while maintaining a phase offset of 180 degrees (π radians). All the electric and magnetic vectors interfere destructively, as shown in **Figure 6-6**. In other words, neutrinos are cleverly concealed double photons.[19] Neutrinos, then, must also be classified as Negative energy.

---

[19] C. Ranzan, *Natural Mechanism for the Generation and Emission of Extreme Energy Particles*, Physics Essays Vol.**31**, No.3, pp.358-376 (2018). (Doi: 10.4006/0836-1398-31.3.358)

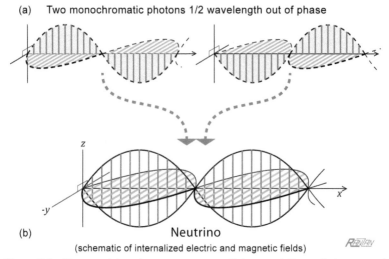

Figure 6-6. The neutrino is an energy particle consisting of two equal-wavelength intertwined photons that are 180° out of phase. Their electric and magnetic fields, rather than being projected to the outside world, are entirely internalized —hence the ghostly behavior of this massless particle. (In quantum mechanics, the neutrino is often referred to as a superposition of electromagnetic waves.)

The premise whereby mass exists as "bottled" radiation appears to be wholly justified and overcomes some serious problems in the standard model of particles. By unifying all mass and radiation particles (including the neutrino), it greatly simplifies our understanding of basic matter. Therefore, we recognize the *primacy of the photon* and adopt it into the framework of energy unification and energy conservation.

Putting the pieces together: mass and antimass, leptons and baryons, light and other forms of electromagnetic radiation, and neutrinos; all are either free-radiation particles or confined-radiation particles.[20] All are manifestations of Negative energy. All are brought about by the Universe's only fundamental energy particle —the photon.

Categorically, the photon is the unifier of all Negative energy. It manifests this energy **directly** in the form of mass and radiation particles —by virtue of the fact that all matter consists entirely of photons. See **Table 6-1**.

The photon is also responsible for Negative energy **indirectly**. In this guise, the photon generates the energy of gravitation. This is by virtue of the fact that all mass possesses a gravity field. Simply put, yet having profound

---

[20] Obviously under this system, based on the *primacy of the photon*, quarks are not required. They become merely mathematical artifacts.

consequences, the negative energy of mass produces the negative energy of gravity. The photon's indirect influence extends to any energy component involving an *indirect* loss of aether. For example, coulomb energy involves charged particles, in the *direct energy* sense or manifestation; as well as a surrounding electromagnetic field in which there occurs an additional *indirect* loss of aether —an indirect manifestation of Negative energy.

Table 6-1. The photon, a negative energy particle, unifies all "Negative" energy —directly via matter and indirectly via fields (regions of stress-induced aether loss).

Photon is responsible for ...	
Negative Energy DIRECTLY	Negative Energy INDIRECTLY
• Mass • Antimass • Radiation, including neutrinos • Coulomb energy • Electromagnetism • Thermal energy	• Gravity • Self-dissipation of aether (associated with gravity) • Electromagnetic field energy • Any other stress-induced loss of aether

As a general conclusion, the photon, directly and indirectly, is the key component of all fundamental Negative energy. Essentially, it (the photon) is the energy of the universe's physical realm.

We next consider another unifier —the unifier of the physical *and* sub-physical realms.

Aether unifies Positive and Negative Energy

The photon, whether propagating in the free state or in the confined state, owes its existence to the excitation-absorption-annihilation of aether. Photon propagation is a process of aether excitation accompanied by aether destruction. Aether excitation is followed immediately by aether vanishment. The process applies to any form, or guise, of electromagnetic radiation. Essentially, the photon is a consumer of aether and a major, albeit indirect, cause of the self-dissipation of aether. Since aether self-dissipation involves a loss of aether it is, therefore, classified as a form of negative energy (and entered as such in **Table 6-1**). Stated more broadly, any stress-induced loss of aether represents negative energy. This aspect relates to gravity and is explained in more detail in Chapter 8.

But for the present let me clarify this point. Whenever the space medium, aether, is excessively stressed —as during its bulk converging flow into bodies or its vorticular flow around rotating bodies— there occurs a disappearance of aether. The cause is related to one or more of the stresses of compression, shear, or excitation. The loss is what is important. If there is no aether loss, then there simply is no manifestation of fundamental negative energy.

Aether is fundamentally responsible for ALL Negative energy.

Turning to the connection between aether and Positive energy: In terms of modern cosmology, Positive energy is the expansion of the essence medium. It is the emergence or quantitative growth of aether. This axiomatic emergence of aether (non-ponderable non-energy aether) is Nature's only manifestation of Positive energy. Physicists recognize the existence of this energy and generally call it Lambda or *vacuum energy*, but are uncertain, even baffled, about the cause. As mentioned earlier, Lambda is the energy of the Void portion of cosmic-scale gravity cells and is actually the process by which new aether comes into being.

Positive energy may be described as a fount process on large and small scales. When referring to large regions of aether (or aether as a bulk fluid) then the Positive energy manifests as wholesale expansion. But at the submicroscopic level, the energy is described as the spontaneous emergence of new aether units.

The unification here is obvious. By supporting the photon's unique mode of existence, on the one hand, and by the *sui generis* fount process, on the other; *aether is the unifier of all energy* —Positive and Negative, as summarized in **Table 6-2**. It should be emphasized, this fount process is extraordinary indeed, for it is Nature's only, yes *only*, Primary Cause process. Long unrecognized as such, it is Nature's *causeless* process. The chart lists all the manifestations of Negative energy in the right-hand column and the sole manifestation of Positive energy in the left-hand column. A word about gravity's position: Because mass/matter has been categorized as Negative energy and mass/matter produces the gravity effect, it is only logical that gravity is also classified as Negative energy. Also note, *contractile* gravity has been specified in order to distinguish it from *divergent* gravity, the effect illustrated earlier in **Figure 6-1**. Whatever the process, aether is the

Table 6-2. Aether is the fundamental unifier of all forms of energy — Positive and Negative.

Aether is responsible for ...	
Positive Energy	Negative Energy
• Space medium expansion (Lambda) The emergence of new aether units	• Mass and antimass • Radiation (photons and neutrinos) • Coulomb energy • Electromagnetism • Electromagnetic field energy • Thermal energy • Contractile gravity (i.e., normal gravity) • Gravity amplification caused by extreme rotation • Any other stress-induced loss of aether
Defining feature: Aether SOURCE	Defining feature: Aether SINK

fundamental unifier.

Now let us briefly examine where the experts of the previous century went wrong. According to the generally accepted way of accounting for the energy content of the universe, the energy associated with gravitation was considered to be negative and all the other kinds were considered to be positive, as listed in **Table 6-3**. You can find it all in 20th-century physics textbooks. The historical classification of gravitation as a form of Negative energy was a good choice. It logically served as the opposite to Lambda (Positive energy). But how did mass and radiation end up on the Positive side of the Table? As was explained earlier, it was the arbitrary choice made in connection with the solution to the Dirac equation during the historical development of particle physics.

Table 6-3. Conventional classification of various forms of energy into positive and negative categories.

Energy Balance Sheet Twentieth-century physics	
Positive energy	Negative energy
• Lambda (as dark energy) • Lambda (as Einstein's cosmological constant) • Vacuum energy (per quantum and string theories) • Mass, radiation, neutrinos • Coulomb energy • Electromagnetism • Electromagnetic field energy • Thermal energy	• Contractile gravity
Problems: Various	Problem: Causal mechanism is missing

See the problem here? With the 20th-century view, there is a serious disconnect between mass/matter and gravity. The energy of matter and the energy of gravity are intimately connected at the fundamental level (both consume aether) and yet they were placed in opposition to each other — separated into opposite categories. (Based on the new understanding of the energy process, one can see that there is no compelling reason for assigning a *Positive* qualifier to most of the various forms of non-gravitational energy.)

This "disconnect" is the main reason that scientists of the previous century were unable to resolve the question of energy conservation on the grandest scale.

Cosmic-scale energy conservation

We have our unifiers. One is from the physical realm and one from the sub-physical realm. When we bring them together, we end up with a remarkably natural system.

The fundamental system of energy conservation within the Universe rests upon two pairs of self-balancing processes.

Energy conservation on the material level of existence

All energy of the *material realm*, as we saw, is tied to the photon. Because of this unifying feature, the establishment of a law of energy conservation for the entire physical domain is greatly simplified. All that is required is the existence of a balancing mechanism, a harmony of opposites so to speak, between photon energy loss and photon energy gain. Ideally, it should be a perpetual and steady-state set of processes.

It turns out, such opposing processes do exist. See the Photon Energy Gain-Loss table (**Table 6-4**). The energy gain is brought about by the *Blueshifting process*[21] occurring within the surface of Terminal stars (end-state neutron stars), as explained in Chapter 2. The process shortens the wavelength of trapped photons and neutrinos; resulting in some staggering amounts of energy gains.

Meanwhile, the energy loss is brought about by two separate and distinct processes. One is familiarly called the cosmic redshift; but technically known as the *velocity-differential redshifting process* [22,23]. We covered this in Chapter 1. The other is the process discussed in Chapter 4, *mass extinction via aether deprivation*. This unique process brings about the utter vanishment of photons existing in the confined state (or as James Jeans referred to it, "in the bottled state") and subjected to the *deprivation* condition that is present only in the core of Terminal stars.

It is important to note that the three processes listed in the Photon Gain-Loss table do not represent any sort of transfer of energy in the conventional sense. The energy gain via the Blueshifting process is not the result of some interaction with some other form of physical energy. The energy losses via the redshifting process and the mass-extinction process do not result in the manifestation of some other form of physical energy.

Everything else beyond the three processes stated in the Photon Gain-Loss table (**Table 6-4**), every other activity, every other process, is classified as an energy-conserving transformation or conventional-law-obeying interaction. Among the "other activities" should be mentioned the conversion of radiation

---

[21] C. Ranzan, ... *Energy Generation via Velocity Differential Blueshift*, Physics Essays Vol.**33**, No.3, pp.289-298 (Sept 2020). (Doi: http://dx.doi.org/10.4006/0836-1398-33.3.289)

[22] C. Ranzan, *Cosmic Redshift in the Nonexpanding Cellular Universe: Velocity-Differential Theory of Cosmic Redshift*, American Journal of Astronomy & Astrophysics Vol.**2**, No.5 pp.47-60 (2014). (Doi: http://dx.doi.org/10.11648/j.ajaa.20140205.11)

[23] C. Ranzan, ... *The Velocity Differential Propagation of Light*, Physics Essays Vol.**33**, No.2, pp.163-174 (2020). (Doi: http://dx.doi.org/10.4006/0836-1398-33.2.163)

to mass; and the reverse conversion of mass to photons and to the energy of neutrinos. Also worth mentioning is this familiar aspect of particle physics; since photons do carry momentum, photonic energy can be transformed into kinetic/momentum energy.

There is something else to note. Turning briefly to one of the listed "energy loss" items in the Photon Energy table, the *mass extinction process via aether deprivation* is, undoubtedly, a revolutionary process. It involves the photon, in the bottled form of mass, but it does not appear in **Table 6-1**, which charts the photon's connection to Negative energy, nor does it appear in **Table 6-2**, which charts the aether's connection to energy.

So, why does the *mass extinction process* not appear in the Photon functionality chart (**Table 6-1**)? ... Because it is a process of the removal, in the sense of extinguishment, of photons —something for which the photon is NOT responsible. It is not the expeller, but rather, the expellee. It is the victim of what is happening. One cannot say, as called for in the table's title, the "Photon is responsible for ..." this mass-removal process.

As well, the mass extinction process does not appear in the Aether functionality chart (**Table 6-2**). The reason? Because it is a process that happens as a consequence of the ABSENCE of aether; it is not because of some action *of* aether. One cannot say, as in the table's title, the "Aether is responsible for ..." this process.

Table 6-4. Photon energy gain-loss balance sheet. Photon energy is the energy that exists in the form of any and all physical particles —radiation, neutrinos, subatomic elements (baryons, leptons, and mesons). The way that energy conservation functions in the cosmic-scale system is not by a strict unbendable law; rather, it is a self-balancing mechanism that is perpetually at work. (Redshifting refers to both cosmic and gravitational kinds.)

Photon Energy of the Universe	
(The quantized level of energy, the energy of the material realm)	
ENERGY **GAIN**	ENERGY **LOSS**
• Blueshifting process (Chap. 2) (occurs only within surface of Terminal stars; aka end-state neutron stars)	• Redshifting process (Chap. 1) (occurs when traversing large and small gravity wells) • Matter extinction process via *aether deprivation* (Chap. 4)
Everything else, every other activity, every other process, is classified as an energy-conserving transformation or interaction.	

And yet, as the Photon Gain-Loss **Table 6-4** makes clear, the *mass extinction process* fits perfectly into the balancing scheme for photonic energy. Although, it is not a form of photonic energy, this process, nevertheless, reduces such energy (photonic energy). So, the question is: How does it meet the requirements of the fundamental energy definition? Consider a small concentric core within a Terminal star. Prior to its extinction, this mass core is an ongoing consumer of aether. Post extinction, there is no longer this consumption of aether. (The lost mass of the core is, of course, instantly replaced by other mass falling inward.) Thus for the larger system, there is a sudden quantitative change, a reduction, in aether consumption; and so, the fundamental definition is satisfied. Loss of mass (by extinction/annihilation) is a loss of energy simply because there occurs a reduction in aether consumption.

From the perspective of process physics, we may sum up the cosmic-scale energy-conservation mechanism applicable to the material level of existence this way: The conventional energy conservation principle is, in a profound way, superseded by *the harmony of gain and loss processes*. Through a balanced self-regulation, the energy of the macro-scale physical realm is maintained in perpetuity —while intimately linked to the micro realm of quantized electromagnetic energy.

Energy conservation on the sub-material level of existence

At the sub-material level of existence, energy conservation means the maintenance of a balance between aether gain and aether loss. And this automatically means the preservation of a balance between the Positive energy process and the two Negative energy processes.

As established earlier, all energy is tied to the aether whether the energy is related to the material world (of mass and radiation) or the nonmaterial world (Lambda expansion, essence medium emergence). Aether ties the two "worlds" together, yet aether itself is solely a sub-physical medium —it is an element, an essence, that exists entirely in the nonmaterial realm. It acts as the ultimate unifier between the two realms. An energy conservation law for this unifying essence fluid is simply a matter of specifying a self-balancing mechanism (another harmony of opposites) between aether gain and aether loss (**Table 6-5**). Conservation, here, amounts to simply maintaining a quantitative equilibrium between aether emergence and aether vanishment.

Let us be quite clear what is being conserved. Since aether is continuously being lost, it may be misleading to speak of the conservation of aether. It is not the aether itself that stays the same, it is, after all, constantly being replaced. Rather, the overall *quantity* stays the same. What is conserved —or more properly, maintained— is the *balance* between aether emergence and consumption. The balance sustains a more or less fixed overall quantity. But this balance is not a local thing. It only operates on the cosmic scale —the scale of cosmic gravity cells.

The chart, opposite page, shows the gain-and-loss balance sheet for aether —the fundamental sub-physical component of energy. Aether gain comes about via an axiomatic emergence process. As mentioned earlier, this gain operates as the Universe's Primary process. That is to say, it proceeds without prior cause. The profound implication is that this formation/emergence process makes the entire energy mechanism a *perpetual* system.

Every other energy-manifesting process operates on the basis of aether consumption and, hence, falls on the "loss" side of the chart. This includes the enigmatic process of excitation-absorption-annihilation of aether by matter, as required to sustain the very existence of such matter. And it includes the most prodigious loss modality of all, the *self-dissipation* of aether —the process that maintains the efficacy of contractile gravitational fields. Moreover, the loss side encompasses any other stress-induced annihilation of aether, notably the shear stress of significant rotation.

Table 6-5. Balance sheet for fundamental energy pertaining to the sub-material realm of existence. One aether-gain process, the source-energy process, is balanced by the aether loss from two consumptive processes. The conservation manifesting in this system, operating as it does on a cosmic-scale, is not a strict unbendable law; rather, it is a self-balancing mechanism that is at work, forever tending towards a harmony of opposites —between aether emergence and aether vanishment. .

Fundamental Level of Energy (The sub-material realm)	
Positive side AETHER GAIN	Negative side AETHER LOSS
• **Formation / Emergence** of *aether** (colloquially, the expansion of the vacuum or space medium) * As defined by DSSU theory	• **Excitation-Absorption-Annihilation** of aether by matter (as required to sustain the existence of such matter). This is the ontological nature of the physical realm. • **Self-Dissipation** of aether —the process that occurs as a consequence of the stresses suffered in conjunction with contractile gravity and rotation —in accordance with the DSSU aether theory of gravity.
Definition: **Fundamental process of energy:** is defined as any quantitative change in the number of fundamental fluctuators (the discrete units of nonmaterial aether).	

Summary and implications

We can now answer the question from the opening discussion, *Where did the kinetic energy of the diverging galaxies come from?* It came from the sub-physical realm in the guise of Positive fundamental energy. It was the Positive energy process in action. New aether formed/emerged in the mostly-empty region between the two galaxies causing the galaxies, along with the aether in which they are embedded, to move apart.

It is worth noting that there are two fundamentally different ways by which Nature generates the energy of motion:

- The non-interaction way, as illustrated in **Figure 6-1**. It begins purely as comotion —a motion with the bulk flow of aether. Such derived kinetic energy, however, really only manifests when the co-conveyance ends, as eventually it must, when the galaxy reaches the interface between a neighboring Void.
- The interaction way found in standard physics. For example, charged particles interacting with an electromagnetic field; the release of chemical or nuclear energy; and of course, the transfer of the momentum energy from one particle or object to another.

The balancing of opposite processes

Because of the intimate relationship between the Universe's essence medium (the sub-physical fluid) on the one hand and the Universe's various manifestations of physical-realm energy on the other, the conservation of energy on the cosmic scale requires two pairs of balancing mechanisms.

As shown schematically in **Figure 6-7**, there is one self-balancing system for the physical realm and another for the sub-physical realm.

In terms of the quantities involved in the balancing:

(Photonic energy gain via 1 process) ≈ (Photonic energy loss via 2 processes)
(Aether emergence via 1 process) ≈ (Aether vanishment via 2 processes)

These relationships apply to any large region of the universe, namely to the autonomous cosmic gravity domains of our cellular universe.

In each case there is a balance between a fount process and more than one negation-or-consumption process.

The founts of photonic energy are the Terminal stars, as detailed in Chapter 2. And the "annihilators" are the countering processes of cosmic redshifting (and gravitational redshifting) and matter extinction by aether deprivation, as detailed in Chapter 1 and Chapter 4, respectively.

The fount of aether is its self-emergence. The negation processes are all the manifestations of Negative energy listed in **Table 6-2**, all destroyers of aether. But these countering components fall into two main categories — "excitation-annihilation process" and "self-dissipation process" (**Figure 6-7**).

The law of energy conservation of the Universe

I don't think the essential ideas of cosmic conservation can be brought together in a more useful summary than what is presented in **Figure 6-7**. Consider it as a concise graphic representation of the *Law of energy conservation of the Universe*.

A universe of only two fundamental particle species

A remarkable implication of the cosmic-scale energy-balancing system is that all processes and all energy manifestations depend entirely on the

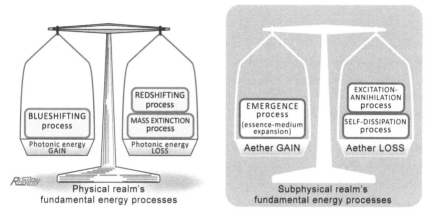

Figure 6-7. Energy conservation in our Universe is the preservation of a balance between fundamental processes. (a) For the physical realm, there is a systematic equilibrium between the Blueshifting process and the two processes of redshifting and matter Extinction. (b) For the sub-physical realm, there is a systematic equilibrium between the aether emergence/expansion process and the consumption processes of excitation-annihilation and self-dissipation.

existence of only two fundamental particles. One particle —the photon— rules the physical realm. And one particle —the subquantum aether oscillator— rules the sub-physical realm.

Aether, dynamic aether, exists. Its ontological nature is axiomatic.

Photons exist because aether exists. That is to say, aether sustains the existence of photons and (by the photonic theory of particle physics) the existence of *all* corporeal entities.

Energy exists as processes of aether. Energy also exists as processes of photons.

Thus, the Universe exists as a two-level system inhabited by only two kinds of most-fundamental particles. One of them, as described earlier, acts as the *ultimate* unifier. Aether underpins all mass and energy of the physical realm AND manifests the dynamics (the energy of emergence/expansion and vanishment/contraction) of the sub-physical realm.

From the thermodynamics perspective, the Universe exists as a perpetual balance of energy between Positive and Negative fundamental energy processes.

Positive process. Nature has only one fundamental-level Positive energy process —the self-emergence of aether. This process takes place within the cosmic Voids.

Negative processes. Any activity in nature that consumes aether directly or indirectly defines this category. Examples include: (1) Photonic phenomena such as mass, radiation and electromagnetic effects are forms of energy that manifest, at the most fundamental level, as an excitation-absorption-annihilation process —a process whereby flickering units of essence (the nonmaterial aether) literally vanish. (2) The energy of a contractile gravitational field is the result of the stress induced self-vanishment of aether; the stress arises from the convergent inflow caused by the presence of the central mass acting as an aether sink. (3) Any other stress-induced loss of aether, such as the vorticular shear stress caused by significant rotation.

In our Universe of only two basic particle types, both enter into the definition of fundamental-level energy. While the Positive process is exclusively the domain of aether units, the Negative energy is the domain of both —carriers of electromagnetic energy *and* discrete aether oscillators.

Evaluating validity

The Conservation mechanism has been shown to be an all-encompassing highly-successful theory. But is it true in the sense of actually reflecting reality? —Is it valid? That is what will be explored in Chapter 7.

But for now let me say, this mechanism in conjunction with the other overlooked laws has, for the first time in history, made it possible to explain in unambiguous terms an extensive list of diverse astrophysical phenomena and observed structural features. Most notably, it has led to the resolution of what had long been the most perplexing pattern of galaxy clusters in observational cosmology. The law governing the cosmic-scale conservation of energy is an essential element for explaining and fully understanding the famous Abel-85 system of galaxy clusters.[24]

✠ ✠ ✠

---

[24] C. Ranzan, *DSSU Validated by Redshift Theory and Structural Evidence*, Physics Essays Vol.**28**, No.4, pp.455-473 (2015). (Doi: http://dx.doi.org/10.4006/0836-1398-28.4.455)

Sixth Interlude

Mytho-Science "Wisdom"

The previous chapter contains the first book-published version of the discovery of the *Cosmic Law of Energy Conservation*. What it details is an unprecedented development in modern cosmology. Utilizing a powerful definition of the fundamental process of energy, DSSU theory has solved one of the deepest mysteries of our Cosmos. Highlights include:
- The new understanding of the **conservation of energy** mechanism, pertaining to the universe system, overturns the long-held belief that the energy conservation law does not apply to the Universe.
- This conservation mechanism makes the DSSU the first true scientific theory of the Universe —one that complies with the paramount law of nature.
- Remarkably, no radical physics was required; rather, the *conservation of energy* mechanism emerged as a logical extension of the highly-successful *unified theory of gravity* based on DSSU aether.
- The mechanism is pleasingly straightforward —everything fits together in a self-evident way.
- Nature's energy secret has been revealed —its hand in how cosmic-scale energy conservation is maintained within the physical realm *and* within the sub-physical realm; and in how physical existence is entirely dependent on the aether medium.
- A bonus highlight: What emerged from the theory and the evidence was the profound conclusion that *we live in a two-particle Universe*. It was as unexpected as it was unavoidable: **All existence is based on only two elementary particles.**

Reactions to the original research paper from several journals

Since the concept of the conservation of energy on the cosmic scale contradicts textbook dogma and conflicts with professional articles of faith, several journal editors declined to publish. However, as on so many previous occasions, they reported NO flaws in the logic or the physics.

The people at the Journal of High Energy Physics, Gravitation and Cosmology decided to stick with the Old Physics (for this term's precise meaning see the Glossary). They felt that "After careful consideration," their energy-and-cosmology Journal "might not be an appropriate publication" for my energy-and-cosmology article.

Their anonymous Reviewer gave two reasons for rejecting the paper.

- "There is no aether."
  The Reviewer's argument, stripped of the irrelevant wordage, was that there is no aether because Wikipedia says so, and included the link (https://en.wikipedia.org/wiki/Luminiferous_aether) so that I could see for myself. (Reviewer also sent along pages of irrelevant third party material.)
- "No conservation of energy in cosmology."
  The justification here was similar. There is no energy conservation for the cosmos because a blog posting at www.preposterousuniverse.com[1] says so. The blog argues against conservation. (Reviewer again provided pages of irrelevant third party views.)

Maybe my consistency, my sticking to the same ideas, was a third reason: Quoting the Reviewer's concluding remark, "The paper uses two erroneous ideas repeatedly. Hence it is not a scientific document."

Transported to another time and place, this reviewer, given the chance, would have told Nicolaus Copernicus, *We know the Sun orbits the Earth because Ptolemy's monograph, the* Almagest, *says so. You must know, that's the way it has been for almost 15 centuries. You have used this erroneous idea of Heliocentricity repeatedly throughout the presentation in* your Revolutions of the Heavenly Orbs. *Yes, repeatedly. Hence it is not a scientific document.*

Another journal still stuck in the previous century is the International Journal of Astrophysics and Space Science. They too, "Following careful consideration," felt that their Journal would not be appropriate for my astrophysics research article. There was no review report, just the comment, "we think the subject of your manuscript is not fit for our journal."

My reality-based astrophysics was simply not fit for their myth-based astroscience.

---

[1] Full address: https://www.preposterousuniverse.com/blog/2010/02/22/energy-is-not-conserved/

Sixth Interlude. Mytho-Science "Wisdom"

Space Science International promotes itself as "an international peer reviewed journal dedicated for publication of research and review articles in all of the various science fields that are *concerned with the study of the Universe,* ..." Surely they would be interested in a logically-structured model of cosmic-scale conservation of energy. Not so. Same story. Their editorial board decided that my well-presented astrophysics discovery, "does not quite fall under the journal's scope." No, its scope is just too narrow. No review report. No other comments.

With the next publishing attempt, it was my misfortune to encounter still another editor clueless about the history of the aether theory! Another gatekeeper guarding the corpses of old failed hypotheses. This one was in charge of Progress in Physics —just one more journal promoting itself as *encompassing all aspects of physics* and yet is not interested in a well-reasoned reexamination of key fundamentals! ... No wonder mainstream Physics (and Cosmology) finds itself in crisis mode. The lack of a meaningful response makes the situation most tedious. But on the positive side, I did experience something new.

First time ever! After having published over 60 research papers, all related to the universal space medium in one way or another, a physicist has essentially stated that the space medium does not exist. The paper entitled "… Cosmic-Scale Conservation of Energy," when submitted to the journal *Progress in Physics,* clearly stated (in red font, no less) that the universal space medium is being called *a nonmaterial aether.* Yet the editor Dr Pierre Millette, by stating "The aether, material or not, is a long-discredited nineteenth century concept that is not required," is disavowing the very existence of a space medium.

My manuscript emphasized the connection, "call it what you like, vacuum, quantum foam, general-relativity fluid, cosmic fabric, etc.; while, herein, it is being called *aether.*"

See why there is a crisis in the old pre-DSSU physics and cosmology?

It is hard not to come to the conclusion that only rarely are science journals willing to risk publishing meaningful new discoveries.

Maybe human nature is a factor. I suppose there are more than a few scientists who, after having spent many years doing research without ever having experienced the excitement of scientific discovery or the feeling that comes from developing a successful theory or originating a new understanding, would gain a certain degree of perverse satisfaction in disparaging and trashing someone else's efforts, especially successful ones.

Just a passing thought.

Mytho-science non-conservation views

It's almost a given that if one believes in an expanding universe, then energy conservation has to be precluded.

An expanding universe may well be in violation of energy conservation, according to Yale University professor Robert K. Adair who had this to say about the issue: "If our universe is expanding from a beginning 15 billion years ago, a conclusion supported by [the old 20th-century interpretation of] astronomical evidence, do we not have an absolute time; is the universe not changing with time; and is energy really conserved in our universe? Perhaps not!" In reference to expansion constructs, "Some models of the origin and character of the universe contain implicitly a violation of the conservation of energy."[2]

Most interestingly, Professor Adair further points out that ultimately it depends on how energy (or its manifestation as a force) is defined.

Cosmologist Edward Harrison, a believer in the universe-wide expansion, stated in his popular textbook this: "The conclusion, whether we like it or not, is obvious: Energy in the universe is not conserved."

> "The conservation-of-energy principle serves us well in all sciences except cosmology. In regions that do not partake of the expansion of the universe, that are dense compared to the average density of the universe, we can trace the cascade and interplay of energy in its multitudinous forms and claim that it is conserved. But in the universe as a whole it is not conserved. The total energy decreases in an expanding universe and increases in a collapsing universe. To the questions where the energy goes in an expanding universe and where it comes from in a collapsing universe the answer is – nowhere, because in this one case energy is not conserved."[3]

Harrison had presented the typical 20th-century view. However, since the real Universe is definitely not expanding, we definitely must have a cosmic conservation-of-energy principle.

(Make what you will of this: Edward Harrison, in another cosmology book he published many years later in 2003, made no mention of cosmic non-conservation of energy. The book's title was Masks of the Universe, Changing Ideas on the Nature of the Cosmos, second edition.)

---

[2] Robert K. Adair, *The Great Design* (Oxford University Press, New York, 1987); p27.

[3] Edward R. Harrison, *Cosmology, the Science of the Universe* (Cambridge University Press, Cambridge, UK, 1981); p276.

Sixth Interlude. Mytho-Science "Wisdom"

Arthur Stanley Eddington, one of the early pioneers in cosmology and a firm believer in the non-conservation of energy, expressed the situation as "The Running-Down of the Universe." He had used that expression as the title of Chapter IV in his popular book The Nature of the Physical World written in 1929.

Another version of the running down of the Universe, came from another expert who had overlooked a few things. Quoting from a book by mathematical physicist Paul Davies: "essentially all physical processes that we observe in the Universe are finite and nonrenewable. ... The supply of material for new stars is limited."[4] Thus, based on what "we observe," energy is not conserved; rather, it is being used up. But there is a caveat Davies failed to mention, and it's a big one, observations *do* need to be properly interpreted.

So, what happens when all the energy has been used up? ... American Professor Grant L. Voth includes just that contingency in his *Myths of Cosmic Destruction* lecture. He says this about the expanding universe: "Even the Big Bang theory includes a return to eternal unity and stasis once all the energy has been used up."[5] The assumption is that the cosmic order originally arose by differentiation out of a simple state of things, conceived as a primitive confusion in which everything was fused together.

Of course, when all the energy has been used up, the universe will be in a very bad entropy state. What does that mean? It depends on the myth you are entertaining. It may mean sinking back into primordial chaos of some sort. Professor Voth addresses this issue in a rhetorical way as he wonders: "What keeps us from sinking back into primordial chaos?"[6] The answer is *nothing*. Unless maybe, just maybe, there is some conservation law that prevents sinking into oblivion.

So much for mytho-science based on the expanding-universe myth; but what about the real world?
Our Universe can never run out of energy.

---

[4] Paul Davies & John Gribbin, *The Matter Myth* (Simon & Schuster, New York, 1992); p123.
[5] Grant L. Voth, Lecture 11, Myths of Cosmic Destruction, in *Myth in Human History* (The Great Courses, Chantilly, Virginia, 2010); Guidebook, p35.
[6] Ibid., Lecture 2; Guidebook, p7.

It can never run out of energy because Terminal stars are forever spewing out regenerated energy —old energy amplified by the Blueshifting process.

Moreover, its ongoing energy conservation is performed with two self-adjusting mechanisms. Cosmic conservation of energy is dynamically achieved through (i) the balance of the process of aether formation/emergence (positive Lambda) on the one hand and the process of aether disappearance (via excitation-annihilation and self-dissipation) on the other; and (ii) the balance of radiation energy gain by the Blueshifting mechanism on the one hand and the loss by the redshifting mechanism and the Matter extinction process on the other. (See **Figure 6-7**.)

Another good thing about the real world is the existence of a knowledge-gaining system called science —with a remarkable tendency to self-correct.

> *The time might come when an idea that has seemed as firm as steel must be given up —even laws of conservation ... It is the glory of science that it occasionally corrects itself, however reluctantly. No other variety of human intellectual endeavor seems to have quite the same built-in machinery with which to do so.* –Isaac Asimov (1991)

✠ ✠ ✠

Law of Cosmic Cellular Structure

7. Law of Cosmic Cellular Structure

> *[I]n the final analysis, only true definitive predictions can justify the promotion of a theory from being viewed as one of many plausible hypotheses to being recognized as the best available approximation of how nature actually works.*
> –Robert Oldershaw[1]

A combination of simple processes predict cosmic scale structural patterns

The first half of this chapter presents the four key determiners of cosmic-scale structure. These determiners include, first and foremost, a completely self-evident assumption, and three indispensable mechanisms (based on most of the overlooked laws discussed in this book). Our single assumption serves as the foundation onto which we overlay the three important mechanisms of physics that 20th-century scientists overlooked —namely (i) *matter regeneration* associated with the Blueshifting process, the energy escape mechanism, and the subsequent formation of mass particles; and (ii) the *cosmic-scale conservation of energy* resulting from the balance between *matter gain* by the foregoing *matter regeneration* and the *matter loss* due to Extinction by aether deprivation; and (iii) gravity based on aether. The

---

[1] R.L. Oldershaw, *The New Physics: Physical or Mathematical Science?* American Journal of Physics Vol.**56**, No.12 (1988). (Doi: https://doi.org/10.1119/1.15749)

assembled system is then used to make a number of model-specific predictions.

The second half of the chapter compares the predictions with the actual observable evidence. The theory-anticipated patterns and the actually observed evidence are compared and shown to be in remarkable agreement. The match is unequivocal and irrefutable. The profound conclusion is that the patterns of the Universe's cosmic-scale structure are not phenomenological but are inherent. Cosmic structure exists by virtue of a perpetual self-sustained mechanism —a timeless steady state system.

First Determiner of cosmic-scale structure: the empirical proposition, EXPANSION

The most fundamental determiner of cosmic structure is the expansion of the space medium. As Leonard Susskind, a physics professor and expert on the cosmic landscape, has stated, regarding the expansion and the growth of the space medium; *it seems rock solid, no cosmologist questions it.*[2] The highly respected cosmologist Edward Harrison wrote, "The expanding space paradigm lies at the heart of modern cosmology."[3]   In cosmology "expansion" is a big thing; it's what the substrate does. Whatever one chooses to call it —the vacuum, the quantum foam, the cosmic fabric, or the aether— it is the emergence/expansion of such substrate that sustains vast near-empty regions. It is the emergence of new vacuum (what herein will continue to be called *aether*) that sustains the existence of cosmic Voids. (The specific nature of the aether will be reviewed later.)

The existence of great voids is a fact and is based on observational evidence. However, the mechanism underlying their existence is, at this stage of the discussion, an assumption. Well actually, it's stronger than a mere assumption. Because of the obvious evidence, "expansion" is really an *empirical proposition*. Based on this proposition that the universal space medium, aether, has the innate ability to expand, we should expect a landscape of Voids. There is something else we should expect. But we do need to be careful of a potential pitfall.

It is critically important that the expansion proposition be limited in scope. The assumption must not be extrapolated to whole-universe expansion, as was done in the last century. The claim is *not* that aether emergence/growth sustains the expansion of the Universe. The definitive reason for refraining

---

[2] L. Susskind, *The Cosmic Landscape, String Theory and the Illusion of Intelligent Design* (Back Bay Books, New York, 2006); p309.

[3] E.R. Harrison, *Masks of the Universe, Changing Ideas on the Nature of the Cosmos*, 2nd Ed. (Cambridge University Press, Cambridge, UK, 2003); p207.

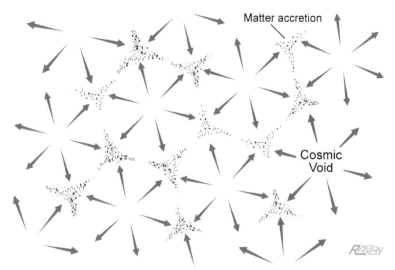

Figure 7-1. Cosmic Voids are sustained by aether emergence/expansion as shown in the schematic cross-section view. Whatever mass there is in such a universe "landscape" must logically accrete at the interfaces between the Voids.

from this sort of speculation was explained in Chapter 1 and will be revisited in a moment.

We logically expect the near-empty Voids of any cosmic pattern to be surrounded by clustered material, conveyed there by the pattern of expansions. But there is a serious potential problem here. It is not with the arrangement of the Voids, as presented in **Figure 7-1**, but rather, with the mass aggregation. It is clearly evident that all the galaxies, stars, dust and gas —all the material stuff of the universe— tend to aggregate within the pockets between the Voids. Over time, the pockets of aggregations, the galaxy clusters, would, *in the absence of some countering mechanism*, undergo gravitational collapse. Ultimately, everything would vanish into an end-state object, what is popularly called a black hole. There would be one such object at each "pocket" or aether-sink location.

Having everything vanish into blackness certainly makes for a bleak future; but that's not the worst part of the problem. Total gravitational collapse is plausible and readily understandable and does not require hypothetical black holes. No, the deeper problem is the question of where all the matter came from in the first place. Where indeed? In the absence of some form of continuous matter formation, the unavoidable implication is that the universe must have had a beginning! There is no more philosophically-unsound concept than this. It makes the problem intractable; it immeasurably deepens the bafflement. Yet that is exactly the course the scientist of the 20th century followed. They adopted and embraced the view of *universe genesis*, a mathematically-derived grand event, imagined as a real event despite lacking a proper physics foundation. The result? The legacy has not been one of enlightenment. Rather, what has been passed on to the 21^{St} century are new and embarrassing levels of bafflement as to how the universe operates. In short, the 21^{St} century inherited a problem-plagued cosmology.

And it all began with the misinterpretation of the cosmic redshift —which, in the 1920s and 1930s, was assumed to be caused almost entirely by space-medium expansion. The EXPANSION assumption was then unscientific extrapolated to the expansion of the entire universe.

First Determiner must not be extrapolated, the Universe is steady state by default

The assumption that the aether space medium expands is a good one. The evidence supports it.

The assumption that the universe itself expands was rejected as a thoroughly misguided 20th-century extrapolation. There is absolutely no evidence. There is no so-called Hubble expansion; no systematic recession of distant galaxies and distant clusters. Scientists had simply overlooked the true cause of the cosmic redshift —the true cause being the *velocity differential redshift* mechanism. (The *velocity differential* refers to the motion of the aether as it affects the wavelength of the propagating light waves.) The proof was presented in Chapter 1, and was first proposed in 2014 in the published article *Cosmic Redshift in the Nonexpanding Cellular Universe* [4]. The proof is impeccable.

In addition to the lack of evidence, the assumption of uninhibited expansion has a troubling philosophical aspect. Given that the Universe is infinite, how can it expand to become more infinite?! Or, why would it bother to become more infinite?! What does that even mean?

Needless to say, there are numerous other problems, a discussion of which would take too long and divert attention from the main point. But do be aware that the Big Bang experts (like physicist Sean M. Carroll) readily admit, their extrapolated expansion and accelerating scenario leads to a Preposterous universe.

Also, as mentioned above, there is the unresolvable problem of the origin of the Universe —some unimaginable all-encompassing beginning that issues from a theoretical backward-in-time extrapolation of the "Preposterous" version. The problem is as serious as it could possibly be —a complete breakdown of physics.

---

[4] C. Ranzan, *Cosmic Redshift in the Nonexpanding Cellular Universe: Velocity-Differential Theory of Cosmic Redshift*, American Journal of Astronomy & Astrophysics Vol.2, No.5, pp.47-60 (2014). (Doi: http://dx.doi.org/10.11648/j.ajaa.20140205.11)

Thus, by not making this kind of an unscientific assumption (not to mention, philosophically untenable) one is left with the default position. This is a position that requires no beginning and no assumption about the Universe, other than asserting that *The Universe IS*. Period. This is the default position.

The default position is that the universe, the real Cosmos, is *a steady state Universe*.

By rejecting the "Preposterous" assumption of whole-world expansion and simply recognizing the universe's steady-state status, it stands to reason that just as the Voids always remain as they are, so too, the galaxy clusters remain coherently in their proper places. While Voids are sustained by the expansion process; galaxy clusters have their own sustaining mechanism —a mechanism of ongoing matter regeneration.

Second Determiner: matter regeneration

It is here, with the mechanism for the regeneration of mass and energy particles, that a couple of overlooked laws of physics function as essential components:

- The Blueshifting process. It affects photons and neutrinos and occurs within the surface of Terminal neutron stars.[5]
- The escape mechanism. This is the means by which photons and neutrinos are able to break free of what is otherwise an impenetrable barrier enveloping Terminal stars.[6]
- The transformation of the escaped radiation (the photons and neutrinos emitted by the Terminal stars) into mass particles. This is nothing more than a standard physics phenomenon —involving basic high-energy collisions. It represents the last step of the matter regeneration mechanism.

Matter Regeneration, the Blueshifting process

This is the process detailed in Chapter 2. It is exceedingly important and worthy of a quick review.

The Blueshifting process takes place within the surface of the most extreme type of contiguous object found in the universe —the Terminal neutron star.

Because a Terminal star contains mass and energy, it absorbs and consumes aether (as a fundamental aspect of the aether theory of gravity).

---

[5] C. Ranzan, ... *Energy Generation via Velocity Differential Blueshift*, Physics Essays Vol.**33**, No.3, pp.289-298 (2020). (Doi: http://dx.doi.org/10.4006/0836-1398-33.3.289)

[6] C. Ranzan, *Natural Mechanism for the Generation and Emission of Extreme Energy Particles*, Physics Essays Vol.**31**, No.3, pp.358-376 (2018). (Doi: http://dx.doi.org/10.4006/0836-1398-31.3.358)

Because it is the most extreme type of contiguous body, the inflow of aether *at its surface* is correspondingly the most extreme. The inflow is equal to the speed of light, nominally 300,000 kilometers per second.

Because its surface is a pure energy surface, the aether inflowing at lightspeed there (at the surface) is *not* a violation of special relativity. This situation is safely within the bounds of basic physics. As was shown in **Figure 2-6** (Chapter 2), the energy surface consists of photons and neutrinos all propagating radially outward, while aether is flowing radially inward. The speeds are equal; the directions are opposite; the energy particles, therefore, remain "stationary." They remain trapped within the Terminal star's energy layer.

Why are only photons and neutrinos trapped (within this layer) and not other particles? It is simply because they are the only particles known to travel at the speed of light. Only they can propagate in-place against the lightspeed inflowing aether.

The remarkable thing is that while trapped in the energy layer the photons and neutrinos continuously gain energy. In other words, while propagating in the surface layer, these particles undergo a gradual, but relentless, wavelength contraction. The proof of this is surprisingly simple.

Consider a representative photon trapped in the surface, as shown in **Figure 7-2**. In accordance with the aether theory of gravity, the inflow speed (of aether) varies as indicated by the graph. Obviously there exists an inflow velocity difference at the photon's location. The photon "experiences" a velocity difference between its two ends. When analyzed, it is found that the front and back ends are actually moving closer together.

Given that the photon always travels at speed c with respect to the aether medium, the following must be true.

$$\begin{pmatrix} \text{Relative velocity} \\ \text{between photon ends} \end{pmatrix} = \begin{pmatrix} \text{velocity} \\ \text{of front end} \end{pmatrix} - \begin{pmatrix} \text{velocity} \\ \text{of back end} \end{pmatrix},$$

$$= (c + v_1) - (c + v_2),$$

$$= c + v_1 - c - v_2,$$

$$= (v_1 - v_2) < 0.$$

Note that aether flows v_1 and v_2 are both negative (negative because aether flow is in the direction of decreasing distance to the center of gravity). But because v_1 is more negative than v_2, the bracketed expression must be negative —indicating a converging situation. Front and back ends of each energy particle are very, very slowly moving closer together.

Consequently, any surface-embedded photons and neutrinos will undergo energy amplification —they slowly gain energy.

In fact, they can gain energy far greater than anything that can be produced

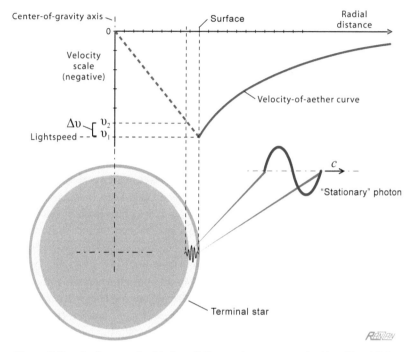

Figure 7-2. Surface embedded radiation gains energy via the Blueshifting process. The representative surface photon, by virtue of propagating in-place within a zone of decelerating aether, is subjected to continuous wavelength contraction. (The front and back ends of the photon "experience" a flow differential Δυ.) Essentially and in accordance with the *law of velocity differential propagation*, it slowly gains energy.

in the laboratory, even during the most violent collisions of elementary particles in the largest manmade accelerators.

If nature is to harness this energy generating process, there needs to be a way for this recharged radiation to escape the Terminal-state stars.

Matter Regeneration, the escape mechanism

Neutron stars, including Terminal stars, are known to possess powerful magnetic fields, which actually originated with the individual neutrons[7]. With rapid rotation of these stars, the lines of force of their magnetic fields become twisted into narrow beams —the lines of force become collimated. Significant

---

[7] An individual **neutron** possesses what is called a **magnetic moment**. This means each neutron has a magnetic field with positive and negative poles at opposite ends. Any collection of magnetic fields tends to align along a common axis. It follows that the Terminal star is like a gigantic nucleus with all the tiny magnetic fields lined up, with all the individual magnetic moments contributing to the star's total field.

rotation is almost always present at some stage in the gravitational collapse (or gravitational aggregation) leading to the Terminal structure. In the course of collapse, as the structure's diameter shrinks, the speed of rotation increases. The increase in rotation speed is in accordance with the law of conservation of angular momentum. Terminal stars may, therefore, be pictured as having very tight magnetic beams (one at each pole) with truly enormous energy densities.

Now remember what was said about mass and energy. They both require a flow of aether to sustain their existence. It follows that, when there is a greater mass density, more aether must flow into the same volume; when there is a greater energy density, more aether must be consumed. The collimated magnetic field represents an extraordinarily high energy density and is, therefore, a prodigious consumer of aether. And this voracious extraction of aether is the key —it precludes the formation of a lightspeed boundary at the star's poles, at the bases of the tightly-bundled magnetic force lines (**Figure 7-3**).

In other words, a Terminal star's magnetic field provides the channels through which those surface-trapped photons and neutrinos are able to escape. Once they reach the opening, they escape. They shoot out at lightspeed; and be reminded, this speed is not with respect to the surface but, rather, with respect to the inflowing aether.

It should be emphasized that this energy dispersal mechanism operates continuously. There is no theoretical limit to the amount of energy that Terminal stars can expel into the greater cosmos. As we learned before, Terminal stars cannot collapse further, can never become smaller, and can never become bigger —regardless of mass input. So they continue beaming energy ... on and on, until, sooner or later, each is absorbed into, or is meshed with, some other Terminal star.

Next we examine what happens to this limitless energy flow.

Matter Regeneration, the formation of mass particles

Although the supply of ejecta is limitless, we need something more for producing mass. If the emitted particles themselves are of low energy, their transformation into mass would be impossible or negligible, at best. Fortunately, Terminal stars produce and eject the most energetic particles to be found in the universe. No other process even comes close —no collision, no other stellar phenomenon, and no manmade particle accelerator can approach the extremes generated by Terminal stars. For example, neutrinos having energy in the peta-electron-volt range (that is, 10^{15} eV) have been detected by the IceCube Neutrino Observatory located at the South Pole.[8] The journal

---

[8] Francis Halzen, *Neutrinos at the Ends of the Earth*, Scientific American, October 2015, p61. (Doi: https://doi.org/10.1038/scientificamerican1015-58)

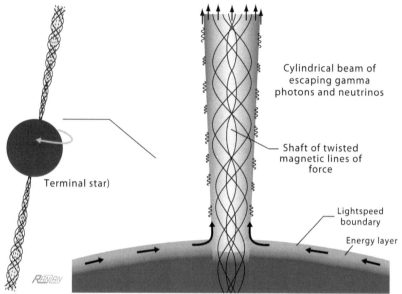

Figure 7-3. Escape mechanism for releasing photons and neutrinos. Shown in cross section, radiation escapes from the Terminal star by means of the opening in the lightspeed boundary. The openings originate with the neutron star's magnetic field and are maintained in the open state by the collimated nature of the magnetic field and the intense particle beam itself, both of which are voracious absorbers of aether. Both diminish the radial inflow of aether. Within the surface energy layer, there is considerable lateral pressure (due to the ultra-extreme density present) that drives the radiation particles toward the edge of the portals from which they emerge as a ring of radiation. The streaming of escaping energy is a continuous phenomenon —sustained by the ongoing energy-generating Blueshifting process. (Thickness of the energy layer is greatly exaggerated.)

New Scientist (2016, April 30-May 6) reported, "Neutrinos captured at the South Pole carry more energy than we [the experts] can explain." To give another example, this one from 2017 September: "The neutrino that triggered everything last September in IceCube was at approximately 300 tera electron volts (3×10^{14} eV) —nearly a factor of 300 million times more energetic than the neutrinos that come prolifically from the fusion production in our sun." So said Darren Grant of the physics department at the University of Alberta and a spokesman at the IceCube Observatory.

Terminal stars emit a vast quantity of ultra-energetic photons; oftentimes in spurts. Such ejecta are observable and recognized by astronomers as gamma-ray bursts.

This is all Nature needs to forge mass —high energy light. It is entirely within the realm of standard physics that gamma and ultra-gamma photons can, and do, generate mass particles. Light can interact with light to produce mass. Light can interact with preexisting mass to produce additional mass. To borrow a line from physicist Frank Wilczek, "You start with massless particles and you get mass." The ability to create matter from light is amongst the most striking predictions of quantum electrodynamics. Experimental signatures of this have been reported in the scattering of ultra-relativistic electron beams with laser beams, intense laser–plasma interactions and laser-driven solid target scattering.[9]

Neutrinos can be just as potent. The mass generating ability of some of the ejected neutrinos can be truly astonishing. A neutrino in the peta-electron-volt category has the equivalent mass energy of about one million resting protons. This means a PeV neutrino has the potential to "create" a million protons. Think about it; one such neutrino is capable of producing about 1,000,000 hydrogen atoms. And not to be overlooked is the inexorable fact that the neutrino will, in due course and time, strike some preexisting mass particle and transfer its energy —a large portion of which will go into the formation of new mass particles.

Two other scenarios worth mentioning: An escaped photon or neutrino may, by chance, encounter a similar one coming from the opposite direction and result in a mass-producing collision. Also, escaping particles, while propagating along the escape shaft (**Figure 7-3**), may interact with the Terminal star's intense magnetic field and generate mass that way.

Typically though, expelled particles propagate through the universe until, sooner or later, near or far, they collide and interact with whatever lies in their path. The result (provided the incident particle has sufficient energy) is the production of new mass particles. Naturally, this mass regeneration/formation occurs predominately where matter is already present. The reasonable expectation is that more mass is generated in galaxies and galaxy clusters than in less dense regions.

Third Determiner: matter gain is balanced by matter loss

Now pause and reflect on the picture developed up to this point. The potential problem of all galaxy clusters collapsing —of each cluster collapsing into some sort of single compact structure— has been solved, solved by the matter regeneration mechanism described above. But the picture is still incomplete. As things stand, we now have the opposite problem. The Second Determiner, our formation/creation process is perpetual —it literally has no limit. No quantity limit, no time duration limit. So, what prevents the universe from

---

[9] O.J. Pike, F. Mackenroth, E.G. Hill, and S.J. Rose, *A Photon–Photon Collider in a Vacuum Hohlraum*, Nature Photonics Vol.**8**, pp.434-436 (2014).
(Doi: http://doi.org/10.1038/nphoton.2014.95)

becoming buried in mass? It is at this point that the mass-loss process enters the picture.

The process, covered in Chapter 4, is called *mass extinction by aether deprivation*, and is remarkably easy to explain.

Recall, once again, the ontological fact about mass and energy: They require a continuous flow of aether to sustain their existence. It follows that when, for whatever reason, the available flow is insufficient, the vital flow is somehow restricted, then the unavoidable result is the vanishment of the so-deprived mass.

Follow the simple logic:
- The greater the quantity of mass in a structure, the more aether must flow into that structure. It is straightforward. The more massive the star is, the more aether it needs to consume.
- Consider a spherical body. Assume, as a fairly accurate approximation, it has uniform matter density. The demand for aether then depends directly on the sphere's *volume*.
- The availability of aether, however, depends directly on the sphere's *area*. The aether has to flow through the limited surface area; there is no other way to reach the mass.
- Now since the volume is proportional to the *cube* of the radius, while the area is proportional to the *square* of the radius, the demand for aether can exceed the availability.

And this is exactly what happens when too much mass comes together. When that happens, aether deprivation will occur.[10] It will occur, as should be self-evident, at the core of the structure (**Figure 7-4**).

Mass extinction by this process is a common occurrence. It generally happens when a multiple solar-mass star (or the star's core) suddenly changes in density, namely, when it collapses to the neutron-density state. Whenever a structure, or a mass aggregation, collapses and the result is a Terminal star, there will occur a significant loss of mass. The extinction process also takes place when mass falls onto any existing Terminal star.

Let me emphasize, there are only three scenarios under which the process of mass extinction by aether deprivation arises. One, is when a supermassive star suffers a core collapse, usually as part of a terminal collapse to the end state. Two, is when a dead star (a non-thermal compact dwarf object) accumulates sufficient additional mass which then triggers total gravitational collapse to the Terminal state. Three, is when large or small mass objects collide with an already existing Terminal star.

As expected, Terminal stars are found wherever normal stars exist. But

---

[10] C. Ranzan, ... *Mass Extinction by Aether Deprivation*, Journal of High Energy Physics, Gravitation and Cosmology Vol.7, No.1, pp.191-209 (2021).
(Doi: https://doi.org/10.4236/jhepgc.2021.71010)

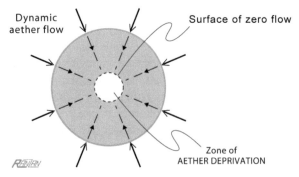

Figure 7-4. Mass extinction by the process of aether deprivation. The quantity of mass, represented in the cross-section schematic, exceeds the amount sustainable by the aether able to flow through the surface. Shown is the situation of aether being totally consumed by the time it reaches the "surface of zero flow." The result is a core region of **aether deprivation** —a place where mass instantly vanishes. What remains is a nothingness hole in the universe that instantly closes. This sequence, simplified here, generally occurs when a multiple solar-mass star (or a star's core) suddenly changes in density, viz., when it collapses to the neutron-density state.

they are most abundant within the supermassive cores of galaxies. This means most of the mass loss occurs at those cores and, by extension, at the center of galaxy clusters.

With the resolution of the excessive mass situation, an amazing interconnection has emerged. It has just been shown that mass formation (brought about by energy generation) is effectively countered by mass Extinction. But in this resolution, we find something deeper, something profound. Terminal stars, it turns out, are creators AND destroyers. They are *the energy generators for mass formation as well as the uncompromising destroyers of mass*. Moreover, the two are self-balancing on the cosmic scale. They are the key to the maintenance of the Universe's material flow as a steady state system.

Terminal stars, collectively, act as the beating heart of the Cosmos.

Thus, we have our Third Determiner, a remarkable balance between matter gain and matter loss, between too much mass and too little, between too much energy and too little. It is a balance performed through a converging feedback between too much and too little matter within the cosmic cellular environment.

Fourth Determiner: gravitation based on aether

Placing gravity last among the Determiners should not be interpreted as a reflection of its true importance. Gravity is a vital component in achieving the energy balance covered in Chapter 6 and the more narrowly-focused gain-loss balance discussed above.

Allow me to briefly explain why it had to be an aether-based gravity and a particular type of aether.

Gravitation is the most important force-like effect governing the universe. The question that confronted 20th-century theorists was *Which gravity?* ... There were, categorically, only three models to choose from: the force type (Newtonian gravity), the geometric class (Einstein's curved continuum), and the aether-based gravity (almost always built on the density variation of a material aether). The twentieth century came to an end and with the advent of the new millennium came a new theory of gravity. It was uniquely based on a *nonmaterial aether.* In due course, it developed into the successful *DSSU aether theory of gravity* which stands as the only known model that unifies all five of gravity's aspects and agrees with observations.[11] Success depends entirely on its unique non-mass, non-energy aether. It is this unified gravity that served as the backdrop for the detailed explanation of the matter regeneration mechanism.

How important is the Fourth Determiner of cosmic structure? Without aether-based type of gravity, there would be no Terminal-state stars — meaning no mass-Extinction cores, no energy amplifying surface, no energy ejection, and no gain-loss equilibrium of mass. Never forget, Terminal stars underpin the Universe's entire energy-and-matter regenerating system.

The detailed workings of DSSU aether gravity are not needed here (but may be found in Chapter 8). For the most part, we only need to remember this critically important aspect of the theory: The simple fact is that all particles absorb/consume aether, and thereby sustain a continuous inflow of more aether.

With all the determiners of cosmic structure in place, it becomes possible to predict the distribution of matter, to predict the very shape of cosmic structure. And if everything described so far is rooted in reality, if those long-overlooked factors and processes are valid determiners, then their combination as a coherent system should lead to meaningful predictions —the various features and phenomena that should be observable.

Prediction of a natural cosmic pattern

A conviction of philosophers across the ages, and of scientists of the modern age, has been that knowledge of the proper laws of physics would enable mankind to predict the universe as it is known to be, and to elucidate its key

---

[11] C. Ranzan, *The Nature of Gravity –How one factor unifies gravity's convergent, divergent, vortex, and wave effects*, International Journal of Astrophysics and Space Science Vol.**6**, No.5, pp.73-92 (2018). (Doi: http://dx.doi.org/10.11648/j.ijass.20180605.11)

features. This cherished belief was the inspiration that brought forth a great variety of cosmology theories. Everything depended on finding those *proper laws*. The Ancients may be forgiven for having failed in this endeavor. But not the Moderns. They, unfortunately, turned the 20th century into an embarrassing fiasco. These are strong words, but they do not in any way exaggerate the situation. Just consider the evidence. Several of the proper laws of physics were not found, several others were just plain wrong. Consider the extreme disconnect between prediction and reality: The Moderns adopted laws that predicted a single-cell expanding universe; the reality turned out to be a multi-cellular nonexpanding Cosmos. They (most cosmologists) embraced laws that predicted a chaotic cosmic "explosion"; the reality turned out to be an ordered steady state system.

The universe is known to be cellular; and is observed to be cellular. And now, with the proper laws of physics in place, we want to spell out the predictions and examine the specific nature of the "ordered cellular structure."

The four Determiners of cosmic structure are all interconnected. They work together as an integrated system —a system that presents a picture with two dominant effects:

- *Expansion effect of the steady state Voids* —the manifestation of the First Determiner.
- *Contractile gravity effect of the galaxy clusters* —the manifestation of the Fourth Determiner. It is a never-ending effect; all because the galaxy clusters themselves are maintained in steady state permanency by means of the Second and Third Determiners.

When these effects are combined on a 2-dimensional plan, they predict a hexagonal pattern as in **Figure 7-1**.

When brought together into a 3-dimensional arrangement, the expansion and contraction effects take us into the full-dimensional real world: When the Voids as *fonts of aether* are combined with galaxy clusters as the *sinks of aether*, the prediction is a close-packed array of rhombic dodecahedra. For the detailed proof, see the article *Large-Scale Cell Structure of the Dynamic Steady State Universe*[12]. An isolated schematic cell is shown in part (a) of **Figure 7-5**; and a representative close-packed array of such cells is presented in part (b). Notice that the prediction of the rhombic dodecahedral structure automatically includes the prediction of two distinct cluster sizes. The smaller class is associated with the four-armed nodes (or Minor nodes) and the larger class is associated with the eight-armed nodes (or Major nodes). Another interesting characteristic is predicted; there are 8 small clusters and 6 large

---

[12] C. Ranzan, *Large-Scale Cell Structure of the DSSU*, American Journal of Astronomy & Astrophysics Vol.**4**, No.6, pp.65-77 (2016). (Doi: https://doi.org/10.11648/j.ajaa.20160406.11)

SEVEN • Law of Cosmic Cellular Structure

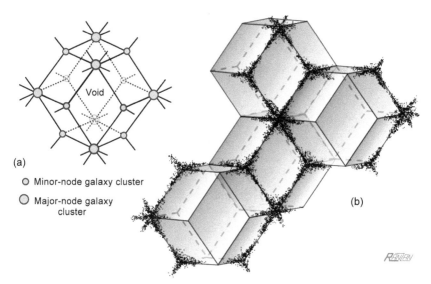

○ Minor-node galaxy cluster
◎ Major-node galaxy cluster

Figure 7-5. Predicted cosmic structure. The basic unit is the rhombic dodecahedral cosmic cell (a). Notice the natural way by which the model incorporates two sizes of galaxy clusters. In part (b) a schematic of a close-packed array of rhombic dodecahedra is matched-up with a partial network of associated galaxy clusters.

clusters surrounding each Void.

Note an important distinction. The prediction is *not* that galaxy clusters will form at the cosmic polyhedron's nodes. It is important to keep in mind that the Determiners are the components of a true steady-state system; therefore the prediction must be that galaxy clusters have always existed at the nodes and are forever being sustained there. In a nutshell, what takes place is *not* the formation of galaxy clusters but rather the sustainment of existing clusters —the *sustainment* of cosmic-scale structure. The concept is of paramount importance. Take a pencil and underline "true steady-state system" and "*sustainment* of cosmic-scale structure."

(Individual galaxies, with one class being exempt, are not sustained. They are forever being replaced. The exemption applies to each cluster-dominating galaxy, or *nodal Supergalaxy*.)

The rhombic dodecahedron is one of only three possible space-filling shapes (shapes that can be close-packed with no gaps in between) useful for explaining an ordered universe. The other two are the cube and the truncated octahedron. There are no others; unless, that is, twinned shapes are employed, such as a regular tetrahedron along with its twin the regular octahedron. Of the three candidates, the rhombic dodecahedron is the only one with non-identical nodes.

A structure unfamiliar (even peculiar) to many people, it is not one of the

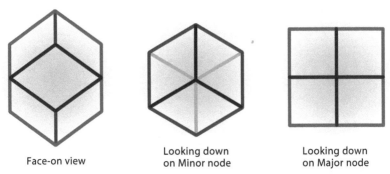

| Face-on view | Looking down on Minor node | Looking down on Major node |

Figure 7-6. Stick model of a rhombic dodecahedron viewed along three different axes of symmetry reveals two hexagonal profiles and an unexpected perfect square.

well-known five Platonic shapes and possibly for this reason has generally been ignored by researchers. Moreover, when viewed in profile it is almost unrecognizable. Depending on the orientation, the rhombic dodecahedron's profile can appear as a hexagon or as a square (**Figure 7-6**). The unusualness of the shape is responsible for some remarkable patterns that may be found within any extended dense packing.

An awareness of the peculiarities of this rhombic structure will be quite useful for interpreting and understanding the evidence that we will look at next.

As a recap, consider the striking contrast. Under the 20th-century view, the observational evidence of cellular structure was merely a random phenomenological condition, while under the 21st-century view —of a dynamic steady state universe— it reflects a natural inherent state. Cellularity is rooted in the theory. The void-cluster network is predicted.

Structure prediction matches observational evidence

> *Immanuel Kant thought deeply about the relation between theory and observation. He recognized that a scientific cosmology must be based on observation, but he also recognized a less obvious and less widely appreciated truth —that **only observation guided by theoretical insight is likely to uncover the deep regularities underlying phenomena**.* –David Layzer[13]

---

[13] D. Layzer, *Constructing the Universe,* Scientific American Library (W H Freeman & Co., New York, 1984); p114. David Layzer was Professor of Astronomy at Harvard University.

Evidence that the Universe *is* cellularly structured is abundant and incontrovertible

Astronomers began suspecting that the universe is cellular around the middle of the 20th century. Early evidence came from redshift studies of galaxies conducted by Gérard de Vaucouleurs. Building on those data, Estonian astronomer Jaan Einasto, beginning in the 1970s and continuing for some decades, reported on galaxy surveys that revealed astonishing structures. Galaxies were distributed in the form of long filaments and giant walls. He spoke of the distribution of galaxies being a honeycomb with huge voids which contained practically no galaxies.

By the 1980s astronomers were routinely reporting on the structural details of superclusters —real super aggregations consist of multiple intertwined filaments of galaxies. Naturally, *great voids* were always part of the description.

The general consensus settled on the largest structures in the universe being membrane-like distributions of galaxies enclosing enormous voids, and as physicist Evan Harris Walker pictures it, "as if the galaxies were the molecules in the foam on some giant mug of beer."[14]

On the basis of overwhelming evidence, cosmologists now recognize the architecture of the Universe as being a vast interconnected system of filamentary structures and great voids —a Veronoi honeycomb.[15]

Galaxy clusters (the nodal structures) come in two sizes

Just as was predicted with the close-packed dodecahedron (**Figure 7-5**), clusters fall into two categorical sizes. Back in 2002, following a study of 79 distant clusters of galaxies (redshift range 0.1 < z < 1), astronomers Naomi Ota and Kazuhisa Mitsuda announced the "discovery of two classes of cluster Size." They presented a histogram, based on the measured core size of each cluster, revealing a distinct double-peaked distribution. The pattern cannot be explained by any selection bias or instrument effects. They, therefore, concluded that it reflects the real nature of the clusters.

The evidence was clear "the histogram of the core radius shows two peaks at 60 and 220 kpc." In other words, cluster core diameters could be classed as either 390,000 lightyears or 1,430,000 lightyears. The enormity of the difference made this a significant discovery. However, Naomi Ota and Kazuhisa Mitsuda were baffled by the underlying cause. They failed to apprehend "through which physical processes such discrete cluster structures

---

[14] E.H. Walker, *The Physics of Consciousness* (Perseus Books, Cambridge, Mass., 2000); p313.

[15] J.R. Gott, *The Cosmic Web, Mysterious Architecture of the Universe* (Princeton University Press, New Jersey, 2016); p67. (https://doi.org/10.1515/9781400873289)

are formed."[16]

Supergiant elliptical galaxies

One of the predictions (but not mentioned earlier) is that there will always be at least one supergiant galaxy at each node of the dodecahedral structure. This means that such a galaxy should always be present at the center of a nodal galaxy cluster.

Astronomers classify these giants as *ellipticals* and further identify them as cD galaxies[17] and have long confirmed the prediction. At the center of any significant cluster there is always to be found a giant dominant galaxy. Importantly, they are nonrotating or have negligible rotation and "tend to outshine the next brightest cluster galaxies by as much as a factor of two."[18] The best known examples include M87 at the center of the Virgo cluster (our nearest nodal cluster), ESO444-46 in the Shapley Cluster (also identified as A3558 cluster), and NGC4874 at the heart of the Coma Cluster.

In general, the structure of the cosmos is revealed in the distribution of masses and the motion of those masses. However, the lack of motion can also reveal the nature of grand structure. Nodal cD galaxies, located as they are at the Universe's stationary points, are examples of bodies practically motionless. Each rules a cluster, while forever a captive of a cosmic node and forever absorbing its surrounding subordinate galaxies.

Right-angled walls of galaxies

The eminent astronomer Anthony P. Fairall (1943-2008) repeatedly reported finding walls of galaxies having right-angled bends. He and his colleagues emphasized the usefulness of such a feature as a critical test for any theoretical model. They had immediately realized that if the 90-degree bends are real, as it was believed they are, then any theory based on randomness would be untenable.

But the right-angle feature is precisely what the dodecahedral shape predicts, as is clearly evident in the right-hand image of **Figure 7-6**. Theory agrees with observation.

In the year 2000, a Mexican professor of astronomy, Renée C. Kraan-Korteweg, published a study focusing on the galaxies behind the Milky Way in the direction of the so-called Great Attractor. The professor pointed out that only in recent years have astronomers developed the techniques to peer

---

[16] Naomi Ota and Kazuhisa Mitsuda, *X-ray Study of Seventy-nine Distant Clusters of Galaxies: Discovery of Two Classes of Cluster Size*, The Astrophysical Journal **567**(1) (2002). (Doi: http://dx.doi.org/10.1086/339852)

[17] The "c" stands for *supergiant*; the "D" stands for *diffuse outer envelope*.

[18] G.O. Abell, *Exploration of the Universe, 4th Edition* (Saunders College Publishing, New York, NY, 1982); p613.

through the Milky Way's dense disk, the gas-and-dust region called the Zone of Avoidance, and uncover the formerly hidden distribution of galaxies.

One of the startling discoveries was an unmistakable right-angled wall of galaxies directly behind the Zone of Avoidance (between Galactic longitudes 200° and 350°). But lacking a theory or any kind of reasonable explanation for ninety-degree bends, the professor chose to describe the pattern of galaxies as a *thin wave* —a "continuity of ... thin filamentary sine-wave-like structure that dominates the whole southern sky and crosses the Galactic Equator twice."[19] The structure is, by cosmic standards, relatively near to us, as evident by its quite large angular measurements, being $\Delta 150°$ (longitude) by $\Delta 75°$ (latitude). The redshift distance, as Kraan-Korteweg's figure 14 indicates, is between 500 and 3500 kilometers per second, which places the structure at a distance of not more than 190 million lightyears.

Wall-like structures

Certain redshift surveys, what are called *pencil beam redshift studies*, in which data is collected from within a very narrow and very extended cone, have revealed remarkable regularity in the distribution of galaxies. A notable example is that of R. Broadhurst and colleagues, which involved a deep-space ($z < 0.5$) survey aligned with the polar axis of the Galaxy and extending out to distances of more than 1 gigaparsec (3.26 billion lightyears) in two opposite directions. The data revealed a succession of wall-like galaxy concentrations at fairly regular intervals —with an average spacing of about 130 megaparsec (420 million lightyears). More than 13 such evenly-spaced "walls" of galaxies were found.[20]

Motivated be these unexpected observations in the distribution of galaxies (redshift $z < 0.5$) a team of physicists undertook a statistical analysis "by comparing the data with models in which galaxies reside on the surfaces of bubbles or sheets." The analysis indeed found that the structure was arranged into a regular cellular pattern . "It is striking that the standard deviation of the observed distribution is close to the most likely deviation from periodicity expected from a regular cellular structure."[21]

They found that the closest fit to the data was obtained with "a close-packed face-centered cubic lattice" of cosmic voids. Their statistical analysis

---

[19] R.C. Kraan-Korteweg, *Galaxies Behind the Milky Way and the Great Attractor*. In: D. Page and J.G. Hirsch, editors, *From the Sun to the Great Attractor*, Vol.556, pp.301-344 (Springer, Berlin, 2000). (https://doi.org/10.1007/3-540-45371-7_8) (http://arxiv.org/abs/astro-ph/0006199)

[20] R. Broadhurst, R.S. Ellis, D.C. Koo, and A.S. Szalay, *Large-scale distribution of galaxies at the Galactic poles*, Nature Vol.**343**, pp.726-728 (1990). (Doi: https://doi.org/10.1038/343726a0)

[21] H. Kurki-Suonio, G.J. Mathews, and G.M. Fuller, *Deviation from periodicity in the large-scale distribution of galaxies*, Astrophysical Journal, **356**(1 Part 2) (1990). (Doi: http://dx.doi.org/10.1086/185738)

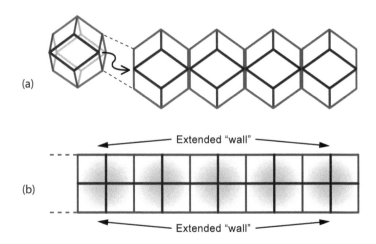

Figure 7-7. Wall-like structures can be explained with close-packed rhombic dodecahedra. Part (a) shows a row of such cells assembled so that rhombic face joins rhombic face; any face can be used, as long as the assembly remains linear. When viewed from above, as in Part (b), the "walls" are unmistakable as is their parallel nature. The parallel aspect accounts for the reported periodicity.

revealed that cosmic voids are distributed ("packed together") in space according to the *close-packed face-centered cubic arrangement*. In terms of the density of the packing this is identical to the *hexagonal closest packing* of spheres[22,23] —which is exactly the configuration of the space-filling packing of rhombic dodecahedra. This is exactly what DSSU theory predicts!

It should be noted that the **face-centered cubic close packing** of spheres actually corresponds to a space-filling arrangement of a combination of rhombic dodecahedra *and* rhombic-trapezoidal dodecahedra. This differs from the *hexagonal closest packing* only in that it reduces somewhat the symmetry of the cosmic tessellation.

The clearest explanation of wall-like structures (as well as their periodicity) can be found in a simple face-to-face assembly of rhombic dodecahedra. **Figure 7-7** shows cells lined-up in a row so that they are all joined together via rhombic face to rhombic face. Be aware that an idealized rhombic dodecahedron has all its faces identical; thus, any face can be used, as long as the assembly follows a linear axis. The "walls" are most evident when the face-on row (a) is viewed from above —a view illustrated as row (b) in **Figure 7-7**.

---

[22] B.H. Mahan, *University Chemistry* 2nd Ed. (Addison-Wesley Publishing Co., 1969); pp.109-110.

[23] D.D. Ebbing and S.D. Gammon, *General Chemistry* 6th Ed. (Houghton Mifflin Co., New York, 1999); pp.476-477.

There is even evidence that the walls of galaxies are parallel. Fairall and his colleagues have stated, there is "the tendency for right-angled intersections and *parallel structures* to occur." "It is remarkable that this second wall [the Fornax Wall] runs virtually parallel (<5°) to the Sculptor Wall."[24]

Ribbon-like structure

Astronomers have reported the tendency of walls to be "ribbon-like" with filamentary "structures interconnect[ed] with one another to form the labyrinth."[25] The plane of the wall structure, in reference to the Cetus Wall, "appears to twist" and "to take on an almost ribbon-like characteristic."[26]

Look at the row of structures of **Figure 7-7a**; note the predicted arrangement. Is there any better way to describe the string of rhombuses shown there as "ribbon-like"? Or as a "twisted" structure? The match between observation and prediction is impressive; the conclusion is clear. The Universe is a zig-zag labyrinth of galaxy distributions.

Great walls of galaxies

Great walls are said to be the largest structures of the Universe. The most famous is known as the Coma Wall (**Figure 7-8**), named after the Coma galaxy cluster located at Right Ascension 13h and first discovered by Margaret. J. Geller and John P. Huchra. Several others even longer have been found, such as the Sloan Great Wall. These structures are chains of Major clusters (located at the larger vertices of the dodecahedra) linked together by filamentary galaxy distributions along those zig-zag rhombic boundaries pictured in **Figure 7-7a**.

Holes in the cosmic structure

Needless to say, galaxies within wall structures are never uniformly distributed. Walls, as commonly observed, have gaps or holes. From another perspective, this is much the same as saying that the great Voids are all

---

[24] A.P. Fairall, W.R. Paverd, and R.P. Ashley, *Visualization of Nearby Large-Scale Structures*, ASP Conference Series, Vol.**67**, pp.21-30 (1994); p23 & p25.
(http://www.adsabs.harvard.edu/full/1994ASPC...67...21F)

[25] A.P. Fairall, D. Turner, M.L. Pretorius, M. Wiehahn, V. McBride, G. de Vaux, P.A. Woudt, *Percolation Properties of Nearby Large-Scale Structures: Every Galaxy has a Neighbour*, (*Nearby Large-Scale Structures and the Zone of Avoidance*, A.P. Fairall and P.A. Woudt, conference proceedings editors) ASP Conference Series, Vol.**329** (2005).
(https://ui.adsabs.harvard.edu/abs/2005ASPC..329..229F)

[26] A.P. Fairall, G.G.C. Palumbo, G. Vettolani, G. Kauffmann, A. Jones, and G. Baiesi-Pillastrini, "Large-scale structure in the Universe: plots from the updated *Catalogue of Radial Velocities of Galaxies* and the *Southern Redshift Catalogue*," Monthly Notices of the Royal Astronomical Society, **247**, pp.21-25 (1990). No Doi.

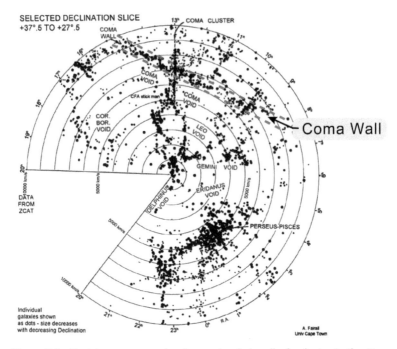

Figure 7-8. Best known example of an extended wall of galaxies is the Coma Wall (also known as the CfA2 Great Wall). When first discovered in the 1980s by M. Geller, J. Huchra, and V. Lapparent it was the largest known structure in the Universe. (Galaxy-map image by kind permission of Estate of Anthony Fairall.)

interconnected.[27] As Anthony Fairall described it for the ASP Conference in 2005, "Were it [physically] possible, one could travel throughout the universe, passing from void to void." The holes in great-wall structures were reportedly one tenth the size of the large voids separating the great walls themselves.[28]

Here, in brief, is the theoretical explanation: The hole is simply the central portion, the least dense region, of a rhombic face or wall (**Figure 7-9**). The direction of motion (in accord with aether gravity theory) is radially away from the rhombus center and towards the boundary edges and nodes —hence, the center is constantly being swept clean, so to speak. As for the size ratio mentioned, it is compatible with the predicted cellular structure.

---

[27] A.P. Fairall, *Large-Scale Structures in the Distribution of Galaxies*, Astrophysics and Space Science, Vol.**230**, Issue 1-2, pp.225-235 (1995). (Doi: http://doi.org/10.1007/BF00658183) (https://ui.adsabs.harvard.edu/abs/1995Ap&SS.230..225F)

[28] A.P. Fairall, W.R. Paverd, and R.P. Ashley, *Visualization of Nearby Large-Scale Structures*, ASP Conference Series, Vol.**67**, pp.21-30 (1994); p23 & p25. (http://www.adsabs.harvard.edu/full/1994ASPC...67...21F)

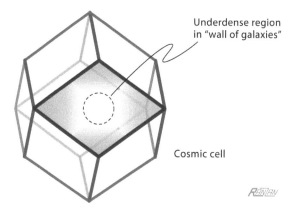

Figure 7-9. Schematic of a cosmic cell showing a rhombic-shaped portion of a "wall of galaxies." The underdense region is the interconnecting opening between Voids. It is the gaping hole in the observed cosmic structure.

Now, if one's line of sight passes through several such holes, something entirely possible since rhombic faces on opposite sides of a Void are more or less parallel to each other, it would seem as if one is looking into a deep cosmic cavity —a hole in the universe, if you will. Probably the most dramatic example of such a configuration is the "WMAP[29] cold spot" located in the southern hemisphere of the celestial sphere in the direction of the constellation Eridanus. It was headlined in Scientific American (August 2016) as **The Emptiest Place in Space** and described as "A pocket of almost nothing [that] tells us something about the cosmos." The authors of the article also called it the "cold spot anomaly." The initial idea proposed by the experts was that the cold spot, this apparent hole in the universe, was some kind of a supervoid; but then more realistically suggested, "If several spherical voids are stacked next to one another in the direction of the cold spot (like a snowman), then the void could more easily explain its presence." That is, a snowman-like triple void could easily explain the presence of the observed cold spot.

The explanation applies to any such cold spots observed. When the line of sight is aligned across opposite parallel faces (and especially when extending through several Voids), because of the sparsity of galaxies, the view will be one of a dark region. It would be like looking into a colossal multi-chambered hole in the cosmos.

---

[29] WMAP is the acronym for *Wilkinson Microwave Anisotropy Probe*, a satellite designed to measure miniscule temperature variations in different directions of the celestial sphere.

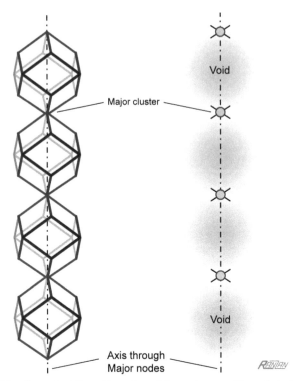

Figure 7-10. Characteristic pattern found in a close-packing of dodecahedral cosmic cells. The line of sight through opposite Major nodes, as shown here schematically, encounters an alternating sequence of Major galaxy clusters and great Voids. (The surrounding units of the packing have been omitted, for the sake of clarity.) The evidence for this kind of pattern can be found in the Abell-85 system of galaxy clusters.

Extraordinary sequences of galaxy clusters

The evidence under this category is rarely discussed by astronomers and theorists. What these unusual sequences reveal is cosmic structure that is far too systematic —much too orderly to be explained by any 20th-century model. The evidence consists of the cluster-void sequences shown schematically in **Figure 7-10** and **Figure 7-11**.

The first pattern we examine is that of periodic galaxy clusters associated with the Major nodes. The Abell-85 sequence, which also includes the background clusters Abell 87 and Abell 89, are undoubtedly the most unusual arrangement of galaxy clusters ever observed. The near-regular spatial periodicity of the clusters is completely inexplicable with any other theory or

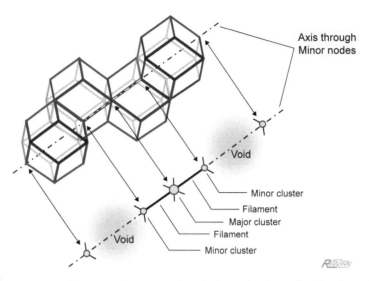

Figure 7-11. Characteristic pattern found along the line of sight through opposite Minor nodes of idealized dodecahedral cosmic cells. The sequence along an extended axis through Minor nodes always has Voids separated by triple clusters. In other words, the sequence alternates between Voids and triple-chain clusters. (Again, the surrounding cells of the close-packing have been omitted to better reveal the pattern. In the perspective view, the end units are closer than the middle pair.) For evidence of this kind of pattern, see Figure 7-8 where the Major node is occupied by the Coma cluster.

hypothesis. No other universe model can explain a cluster-and-void sequence extending for over 10 repetitions![30,31]

Another example of this kind of cluster periodicity is the structure known as **DC1842-63** with three periodic clusters. A published histogram shows them located at 4,500 km/s; 10,500 km/s; and about 16,000 km/s; evidently evenly spaced.[32] (It could be that DC1842-63 is also known as Abell S0301.)

The second sort of pattern involves periodic linear clusters. A completely different sequence arises when the line of sight runs through opposite Minor nodes of the dodecahedral tessellation (**Figure 7-11**). The order in this case is as follows: void, triple cluster, void, triple cluster, and so on. Each triple

---

[30] C. Ranzan, *DSSU Validated by Redshift Theory and Structural Evidence*, Physics Essays Vol.**28**, No.4, pp.455-473 (2015). (Doi: http://dx.doi.org/10.4006/0836-1398-28.4.455)

[31] F. Durret, P. Felenbok, D. Gerbal, J. Guibert, C. Lobo, and E. Slezak, *Redshift and Photometric Survey of the X-Ray Cluster of Galaxies Abell 85*, The Messenger Vol.**84**, pp.20-23 (1996). (http://adsabs.harvard.edu/abs/1996Msngr..84...20D)

[32] E.M. Malumuth, G.A. Kriss, W.V.D. Dixon, H.C. Ferguson, and C. Ritchie, *Dynamics of clusters of galaxies with central dominant galaxies*, Astronomical Journal Vol.**104**, No.2, pp.495-530 (1992). (Doi: https://doi.org/10.1086/116250)

cluster falls along the same axis and consists, first, of a Minor cluster, then a filamentary cluster, a Major cluster, another filamentary cluster, and ends with another Minor cluster. The axis then passes across another Void. The familiar "CFA stick man" is an example of such a void-and-triple-cluster sequence (**Figure 7-8**). In the image, the near Void is part of what is called the Northern Local Supervoid[33] and the Major node is AGC1656 (the Coma cluster). Regarding the far Void, it has not been identified; it does not seem to have been given an official name.

Again, evidence agrees with prediction.

Thoughts on cells non-expanding, order eternal, aether uniqueness, reason triumphant

On the non-expansion of our cosmic cells

With so much talk in science circles of "space expansion" and of general expansion in the context of cosmology, it seems like a good idea to underscore the reason cosmic cells do *not* expand. Since this is often misunderstood, here is a recap of what is probably the most important aspect of the cosmic structural cells. Given that the cell interiors, the Voids, are filled with expanding aether, what prevents Voids from growing larger in size? The explanation is simply that mass absorbs the aether. (This is also true of energy particles and energy fields.) The existence of matter is sustained by the absorption-consumption of this essence fluid. The material surrounding each Void acts as the sink for the expanding aether —thereby limiting the overall expansion and helping to maintain each cosmic cell as a steady-state system.

We had adopted "expansion" as the core assumption —but it was expansion without extrapolation. The quantitative expansion/growth of the universal medium is balanced by an equal quantity of contraction/consumption/self-vanishment. In other words, there is *NO NET expansion* on the largest scale!

This harmony of opposites, as it turned out, was one of the overlooked laws of cosmic-scale physics. More than any other factor, this oversight led to the failure to recognize the steady state nature of cosmic structure.

On discovering eternal natural order

Remarkably, we started with only that one empirical proposition. Not even the *cosmological principle* was adopted as an assumption; and for good reason. Both theory and evidence point to a universe lacking homogeneity and

---

[33] The Northern Local Supervoid is the great underdense region between the "nearby" Virgo cluster and the Coma and Hercules superclusters (both of which are part of the CfA2 Great Wall).

isotropy. Everything else beyond our one proposition was based on good evidence and on four important laws of physics that 20th-century scientists overlooked —namely those detailed in Chapters 1, 2, 4, and 8:
- *The Law of velocity differential propagation of light*. It gave us the definitive cause of the *cosmic redshift*.
- *The Law of Blueshift accrual*. It detailed Nature's fundamental energy amplification process.
- *The Law of mass extinction by aether deprivation*. It provides the remarkable explanation of how Nature annihilates mass. No ordinary destruction process, this is annihilation in the irreversible terminal sense.
- *Law of mass property acquisition and the aether theory of gravity*. This is the great unifier of expansion and contraction phenomena. In a true sense, it is the rule book for the behavior of aether itself. (See Chap. 8.)

Something else that scientists had overlooked was a deep and fundamental contrast between individual bodies and collectives of bodies. Although they understood the mutability of stellar and planetary bodies, they utterly failed to recognize the eternal order of the distribution patterns. They never did discover the natural and perpetual order of cosmic-scale structure.

Also, there was something at a truly fundamental level. Although they understood the importance of the space medium, *the vacuum*, they failed to grasp its ontological nature —its essentialness for natural eternal order.

On the *sine qua non* component

Far the most important component of the Universe is the ubiquitous space medium. We purposely call it *aether*, as a way of acknowledging the contribution of the Ancient Greeks; and because Albert Einstein had emphatically stated that aether exists. But he also had made it quite clear that it was not a *material* medium; it was not like the 19th-century aether.[34] Einstein was correct on both counts; unfortunately, the nonmaterial-medium concept was never exploited. When Physics rightfully rejected the material aether, it wrongfully also discarded the whole aether concept. The result was an entire century of obfuscation of the unique nature of the space medium. In any case, it helps to be clear on meanings and avoid any misunderstanding. Here are three relevant definitions of a much-discredited term.

(1) The original *ether*: In Aristotelian physics, it is the fifth element, the *quintessence*, of which the heavens including the stars are made. In Classical physics, it is the invisible medium that diffuses all space.

---

[34] A. Einstein, *Sidelights on Relativity, Ether and the Theory of Relativity*. Translated by: G. B. Jeffery and W. Perret (Methuen & Co. London, 1922); posted at http://www.gutenberg.org/ebooks/7333; republished unabridged and unaltered (Dover, New York, 1983).

(2) The historic aether: The *material* medium that fills the apparent emptiness of the universe. Invented by René Descartes in France and by Isaac Newton in England; reinvented by others, including James Clerk Maxwell who used it for his electromagnetic theory; but was discredited and discarded by the *young* Einstein.

(3) The DSSU aether: The *subquantum* medium that permeates all space. It is the *nonmaterial essence* of the Universe; it consists of discrete units — fundamental essence oscillators. As a basic space medium, it serves as the propagator of electromagnetic waves. As a space-permeating dynamic medium, it manifests gravitation; its nature is responsible for the several guises of gravity, accomplished without the need to invoke any force-carriers.

Aether was actually detected and verified in at least six separate experiments during the 20th century (in addition to the Michelson and Morley detection in 1887). But since there was no supporting theory, the findings were either ignored or relegated to the obscurity of footnotes. It was a baffling situation. How could a supposedly nonmaterial medium be detectable? But since the space medium proved to be detectable, then it would seem to be material after all! Theorists were trapped in a contradiction. Physical experiments said the space medium is material; while theory said it is nonmaterial. They were trapped by a failure to understand the profound connection between the material and nonmaterial realms of existence. That connection, long-overlooked, is a *subquantum* aether.

The essential nature of the space medium is that it is nonphysical — nonphysical because it exists at the *subquantum* level. Furthermore, and this is what is most misunderstood, it is mechanical —it consists of discrete entities. See **Figure 7-12**.

DSSU aether is the essential component for sustaining all things and all structures, and, in the context of the current chapter, for the patterns of cosmic structure.

Consider this analogy, borrowed from physicist Evan Harris Walker, on the need for a special kind of mechanics. Just as the electromagnetic forces govern the structure of atoms and molecules, the gravitation effects dominate the structure of the large-scale objects and the structure of the entire universe. And in order to explain those gravity effects and to understand how the universe as a whole works, a special kind of mechanics is needed. For 20th-century scientists, "That special mechanics is known as the general theory of relativity."[35] *That*, of course, was Einstein's geometric/mathematical model of gravity. (Remember, the older and wiser Einstein acknowledged aether's existence but did not explicitly employ it.) Although of limited applicability in

---

[35] E. H. Walker, *The Physics of Consciousness* (Perseus Books, Cambridge, Mass., 2000); p313.

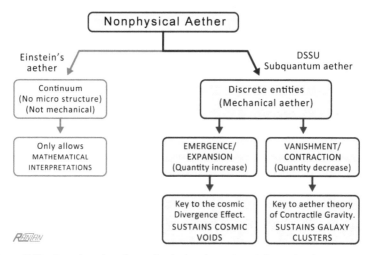

Figure 7-12. Two theories of nonphysical aether. Einstein's version has no micro structure, hence it is not mechanical. DSSU aether consists of discrete units, hence it IS mechanical. Moreover, existing as they do at the subquantum level, these units possess no mass and no energy. Therefore and profoundly, the key mechanical processes of EMERGENCE and VANISHMENT cannot represent a violation of the conventional thermodynamic laws. Thus, a nonmaterial medium underpins cosmic structure.

local reference frames, the theory was, and is, incomplete, by Einstein's own admission, and is quite useless, in that it makes wrong predictions, when applied to the greater universe.

The special kind of mechanics essential to understanding the Universe is *dynamic-aether gravity* —the gravity theory based on the above mechanical aether featured in **Figure 7-12**. The key component of the overlooked laws of physics is the nonmaterial-and-dynamic aether with its unique properties.

Again, there is nothing more fundamental, or so fundamentally important, than DSSU-defined aether.

A triumph of reason

The great discovery of the last century was *cosmic cellularity*. During many past decades, astronomers succeeded in showing some amazing regularities. Meanwhile, theoretical experts struggled to come up with plausible explanations using implausible universe models —hypothetical constructions that had failed long ago and should have been abandoned. Since regularity of structure was not supposed be out there, the concern of contemporary cosmologists focused on the question of actuality. Maybe it's merely an illusion. Let's just hold onto our old theories; besides, it is only a problem *if* the patterns really do exist. Yet the challenge was straightforward.

It was not a matter of "if" but of "how." The question should have been, *In what way is the universe cellular?*... The resolution of this is no more difficult than a high school science project. And the answer could be had by a strategy of elimination. Just examine all the possible suspects that may be behind the "anomalous" patterns, the various configurations of aggregations of galaxies described earlier. The list of candidates is quite short:

- First, consider what the old theories predicted: *Random chaotic spherical cells*, random in size and chaotic in motion. But randomness was just the opposite of the orderly structure evident in the real-world.
- What about cube shaped cosmic cells? Too simplistic and, in a gravity-ruled world, highly unstable.
- What about truncated-octahedra? This was examined by "The Cosmic Web" expert Richard Gott.[36] It is true that truncated-octahedra do close pack in a space-filling arrangement. However, this polyhedron has two obvious problems —it lacks right-angled walls and its nodes are all the same. With all nodes being identical, it becomes difficult to explain galaxy-cluster-size variation.
- Lastly, there is the rhombic dodecahedron —a shape for which all the pieces of evidence fit. There simply are no other cell types —no other symmetrical or semi-symmetrical space dividing cell shapes.

There is only one cell shape, only one tessellation pattern, that can explain all three key structural features described earlier and illustrated in **Figures 7-7, 7-10**, and **7-11** —only the rhombic dodecahedral configuration will work.

But replacing randomness with systematic cellularity comes with consequences. The body of the evidence and its interpretation within the DSSU framework leads to an unavoidable reassessment of a long-cherished belief. The implication is that our Universe is not entirely subject to the *cosmological principle*. A universe inherently cellular is structurally somewhat like a liquid crystal. Both have symmetry properties; both are nonisotropic. The Universe may well be a uniform system of cells; and be described as a homogeneous "packing" of those cells; but it cannot be described as being isotropic.

The resolution of the "how" question is elementary, but it requires a radical rethinking of our worldview. Involved is a revolutionary difference of interpretation of the observational evidence of the void-cluster network. Under the 20th-century view cosmic cellular structure is merely a phenomenological condition, a statistical fluke, while under the 21st-century view —the DSSU worldview— it is an inherent state.

A triumph of reason has led us to the deep understanding of cosmic

---

[36] J.R. Gott, *The Cosmic Web, Mysterious Architecture of the Universe* (Princeton University Press, New Jersey, 2016); p107.

structure. As the DSSU foundational postulate and the overlooked laws of physics were identified and pieced together and applied to the Universe, a point was reached where we could finally grasp the nature of cosmic structure —its diaphanous geometry, its underlying drivers, and its steady state aspects. Theory-based predictions were matched up with astronomical evidence. Theory, predictions, and observations, all were found to fit smoothly together into a reasoned comprehensive vision of reality.

Without philosophical absurdities (no universe genesis); without unscientific extrapolations (no exploding cosmos); without outright paradoxes (no singularity black holes); without physical-law breakdown during extreme gravitational collapse (no infinities); and without fantasy matter (no *invisible dark matter*); the Dynamic Steady State Universe stands *sui generis* as the problem-free cosmology.

A closing thought

When the wise men of their time, long ago, assured the common people that the heavens, the celestial sphere, had a rotational mobility they were wrong.

When the wise men of the 20th century assured the molded masses that the heavens have a radial mobility —an accelerating outward-bound mobility— they too were wrong.

When the wise men of today assure themselves that this time they "succeeded in addressing the ultimate question of creation" that the universe originated in a big bang[37], they are, once again, wrong.

---

[37] Simon Singh, *BIG BANG, The Origin of the Universe* (Fourth Estate, Harper Collins, New York, 2004); p463.

Seventh Interlude

Experts, Obstructers, and Fraudsters

> To the entrenched guardians of orthodoxy —be they in politics, religion or science— logic is an enemy and truth is a menace.
> –variation on a saying

Orthodox experts say orthodox cosmology does not match reality

Physicist Max Tegmark, in his book Our Mathematical Universe: My Quest for the Ultimate Nature of Reality, stated, "The mathematical description of the external reality that theoretical physics has uncovered appears very different from the way we perceive this external reality."[1] That remarkable statement appears under the heading *The Bottom Line*. MIT Professor Tegmark is quite right, the mathematical universe does not match reality. The mathematical description of the external reality that theoretical physics has constructed does not include any of the observable features presented in our previous chapter.

What physicists had been doing was following the advice Paul Dirac dispensed back in 1931: "The most powerful method of advance that can be suggested at present is to employ all the resources of pure mathematics in attempts to perfect and generalize the mathematical formalism that forms the existing basis of theoretical physics, and after each success in this direction, to try to interpret the new mathematical features in terms of physical entities."[2] This led to the belief that the physical world is, literally, a mathematical structure. Professor Tegmark accepts this idea as fact and extends it to the whole universe (as expressed in the title of his book); his "Mathematical

---

[1] M. Tegmark, *Our Mathematical Universe: My Quest for the Ultimate Nature of Reality* (Alfred A. Knopf, New York, 2014); p242.

[2] Ibid., p355.

Universe ... *is* a mathematical structure, and we're simply uncovering this bit by bit." And how is it working out? ... Tegmark says, "The various approximations that constitute our current physics theories are successful because simple mathematical structures can provide good approximations of certain aspects of more-complex mathematical structures. In other words, our successful theories aren't mathematics approximating physics, but mathematics approximating mathematics." Sounds like a great success for abstraction. The only problem is the failure to predict or reflect external reality.

What about the shape of the Universe? How long has the math approach been a failure in this regard? Let me quote from *The Preface of Nicolaus Copernicus to the Books of the Revolutions*: "Nor have they [mathematicians/philosophers] been able thereby to discern or deduce the principal thing namely the shape of the Universe and the unchangeable symmetry of its parts." Copernicus wrote those words some years before 1543 and yet they stand as a fitting testament of the 20th century.

Another example. According to Einstein's mathematical theory of gravity, a substance whose density is undilutable can inflate itself, doubling its volume at regular intervals, growing from a subatomic scale to a size vastly larger than our observable Universe in a split second and effectively putting the bang into Big Bang cosmology.[3] Welcome to the mathematical trick of *inflation*. Max Tegmark explains it for us: "Inflation is like a magic show where seemingly impossible tricks happen through creative use of the laws of physics. Indeed, inflation ... *can create an infinite volume inside a finite volume!* Specifically, it can start with something smaller than an atom and create an infinite space inside of it, containing infinitely many galaxies, without affecting the exterior space."[4] Impressive indeed! Pure magic. But in the search for the ultimate nature of reality —or at least the structure of the Universe— it is just fanciful conjuring seriously disconnected from reality.

Nevertheless, the bestowers of honors and awards were suitably impressed by the illusions of the experts. The failure of orthodox cosmology was overlooked; while praises were expressed and the tributes handed out.

For an insight into the motivation of the Nobel Awards Committee, let me remind you of the connection between *inflation* and *universe expansion*. Inflation is the math explanation behind the illusion of the accelerating expansion of the universe. The Awards Committee members were completely duped. Now for the award: The Nobel Prize in Physics 2011 went to Saul Perlmutter, Brian P. Schmidt, and Adam G. Riess, "for the discovery of the

---

[3] Max Tegmark, *Our Mathematical Universe: My Quest for the Ultimate Nature of Reality* (Alfred A. Knopf, New York, 2014); p101.

[4] Ibid., p114.

Seventh Interlude. Experts, Obstructers, and Fraudsters

accelerating expansion of the Universe through observations of distant supernovae." And let me point out how you know the selection committee was totally taken in by the illusion. Normally the word *universe* refers to a hypothetical model of the real world. But they capitalized the word and made it "the Universe"; in other words they actually believed that the real Universe is undergoing accelerating expansion!

In 2019 the Nobel Physics Prize was awarded "for contributions to our understanding of the evolution of the universe and Earth's place in the cosmos." One half of the Physics Prize went to Canadian-born James E. Peebles (b. 1935), "for theoretical discoveries in physical cosmology." Notice that this time the Committee was more careful; "universe" and "cosmos" were not capitalized.

James Peebles is the physicist who has contributed more than anyone else to the 20th-century standard cosmological model. The extended citation states,

> *Peebles advanced the concept of a dark matter component to the universe and its implications for the evolution of structure. Through this, and other work, he helped establish the theoretical framework for our picture of how galaxies have formed and evolved.*

But a closer look at his work reveals that he was unable to explain the basic cause of galaxy rotation! He had no explanation of how it is possible to have fully formed mature galaxies just half a billion years after the beginning of his big bang universe! The disconnect from reality is staggering: "a dark matter component" is predicted but none exists, none has ever been found; and there is an embarrassing lack of predictions on cosmic structure.

In 2020 a purely mathematical object was the focus of praise. Half of the Nobel Physics Award for that year went to Roger Penrose, "for the discovery that black hole formation is a robust prediction of the general theory of relativity." See the math-magic connection here? Einstein's general relativity is a mathematical theory that predicts, surprise, a purely mathematical object —a singularity black hole. If we learned anything from the previous chapters, it is this: The black hole idea is arguably the greatest conceptual disaster of 20th-century cosmology.

Historian Burton Feldman recognized the potential pitfall of mathematics. Interpretations may dazzle the audience, "But, as the history of science shows, elegantly mathematicised theories are not necessarily good science, or even science at all."[5]

---

[5] Burton Feldman, *The Nobel Prize, A History of Genius, Controversy, and Prestige* (Arcade Publishing, New York, 2000); p353.

Given this state of affairs in which a constructed worldview fails to reflect reality and fails to make valid predictions and, nevertheless, is well received, highly praised, and bestowed with honors; given such favorable treatment for such paucity of merit, then what reaction might we expect with a radically different theory in wonderful agreement with the real world? How would the scientific community react to a cosmology that makes predictions that actually turn out to be true?

Journals not interested in evidence-validated cosmology

When the research article on cosmic structure (the article on which much of the previous chapter is based) was sent for publication, it was repeatedly rejected. I was unable to find an astrophysics or cosmology journal willing to put it into print. Such was the case even when the reviews were favorable! ... They were presented with an article whose main theme is to demonstrate the remarkable and unprecedented agreement between existing theory (DSSU as found in the scientific literature) and existing observations (as reported first hand by astronomers). Yet it encountered one rejection after another —seven altogether. It experienced more refusals than any previous DSSU article, ever. Had I continued and sought other astrophysics and cosmology journals, the reaction would probably have been the same, just more rejections.

Was I surprised? Not in the least. Committed, as I am by principle and by practice, to perceiving reality in a purely objective manner, my attitude is reflected in a personal motto: "Never surprised, but *always* amazed!" It is a matter of recognizing a changing world —one in which subjective reality has taken hold and wormed its way into Western culture and corrupted rational thinking. All fellow truth-seekers should be concerned. Perhaps my experience will serve as an eye opener.

Anyway, time became a pressing factor. The longer publication was delayed, the greater becomes the risk of plagiarism and priority challenges. As they say in business and science, *time is of the essence.* After seven rejections, it was time to look elsewhere.

Finally, after eight months, the article, entitled *Steady State Cosmic Structure*, was published in a general science journal.

The following are the highlights of what transpired during several months of dealing with the gatekeepers of 20th-century cosmology. Throughout, as has been the case with all DSSU papers, the reviewers/examiners found no problems with the assumptions, no flaws in the logic, no disputes over the evidence. When objections were raised they were either general (never specific) or invalid (referring to something not in the paper).

Seventh Interlude. Experts, Obstructers, and Fraudsters 251

The *International Journal of Astrophysics and Space Science*, which prides itself on "covering the entire range of astronomy, astrophysics, astrophysical cosmology, and space science," responded with the editorial board's decision: "We think the subject of your manuscript is not fit for our journal." Why they think so was not reported.

The journal *Frontiers in Astronomy and Space Sciences* invites research across a range of specialties including cosmology and extragalactic astronomy. At least that's their claim. Interestingly, it seems my paper received special treatment by a "Specialty" editor, no less. I was informed that the manuscript was assessed by the Specialty Chief Editor (who was not named!), who determined the manuscript would not proceed into review. Why not? ... *"The conclusions are not convincingly supported by the evidence or theories presented."* –Unsigned!

The journal *Space Science International* is dedicated to the publication of research and review articles concerned with the study of the Universe.

The review report I received was most favorable:

> **Interesting logical arguments** resurrecting the aether in a system with dynamic gravity mechanics of a nonmaterial-and-dynamic aether, **assembled in a very well drafted and readable manuscript** describing a dynamic steady-state universe. Using Walker's "Physics of Consciousness" as the special mechanics to cite Einstein's General Theory was an interesting choice. ... It's [an] **excellent manuscript**.

But wait. A review like that, of course, requires a second opinion. It was quickly found. Here, reproduced in its entirety:

> *This manuscript should be rejected and it should not be considered for publication. It cannot be considered an article."* –anonymous reviewer

The editors ignored the first review and adopted the second review to arrive at their decision. The detailed assessment was trumped by an unsubstantiated opinion! These editor-obstructers unanimously decided that my manuscript should be rejected, it should not be published, in fact, it should not even be considered an article. ... Wow! It must be really disgustingly bad, a most miserable effort! Reading their assessment was a wonderful eye-opening moment.

Just below that decision statement came this: The editor, "Dr. Mehmet Tanriver and I (editor, Dr. Yu Liu) hope that this feedback will be helpful to you in developing your research."

Such words of wisdom from professionals. Priceless.

The *International Journal of Astronomy and Astrophysics* gave it a thumbs down on the basis of one reviewer's brief report. The report essentially said that the article is "based on untenable presumptions" and contains "nothing new, except perhaps to the author."

How can I get the message across? With bold flashing neon lights? Attention: WHAT IS NEW IS THAT THERE IS NO NET EXPANSION OF THE UNIVERSAL SPACE MEDIUM!

The report continued, "We all know that Hubble's law of universe expansion does exist and works ... I therefore see that this paper is itself preposterous."

Since the Editors did not invite me to respond, I'll present my rebuttal comments here instead (and delve more deeply into the issues raised):

The anonymous Reviewer for IJAA is mistaken and has simply not understood what was written. Quoting the Reviewer, "The concept that the expansion is not due to the velocity of recession [of galaxies] but is due to the expansion of space is nothing new except perhaps to the author."

NOWHERE in the paper is there a claim that there is any velocity of recession (of galaxies), nor is there a claim that the velocity of recession (of galaxies) is due to the expansion of space. (In fact, there is no cosmic recession at all.)

The Reviewer believes that the cosmological redshift is caused by the expansion of space between us and the distant galaxies. That is what the Reviewer teaches students. This educator is unaware that the cosmic redshift is a *velocity differential* redshift (a proven consequence of the velocity differential propagation of light) and that it occurs even in regions of contracting space!

Most disturbing is that the Reviewer seems unaware of the profound difference between *the expansion of the space medium*, on the one hand, and *the expansion of the Universe*, on the other.

My Article is written in clear easy-to-understand English, no jargon, no obscure terminology, no irrelevant details ... and yet there are reviewers who do not seem to comprehend what is actually written. For some perverse reason, they direct critical comments to things that are not at all stated in the Article!

Is it incompetence, ignorance, or censorial bias? Who knows. One message, however, comes through clearly. The dismissive attitude —suffused in remarks such as "This article does not fit into the body of knowledge that we can all share and use reliably together" and so "this paper is itself preposterous"— makes it quite clear that they (editors and reviewers) really are not interested in what does not fit into their *old body of knowledge of the previous century*.

And so grows my collection of fake reviews.

The journal *Advances in Astronomy* claims that it "publishes in all areas of astronomy, astrophysics, and cosmology, and accepts observational and theoretical investigations into ... the wider universe." Unfortunately, the editors did not honor their crystal-clear mission statement.

My manuscript, narrowly confined to *comparing theory predictions with observational evidence*, was deemed to lack a clear focus. "... the manuscript has not a clear focus."

The revolutionary cosmology presented was deemed to be "more an overview of current theories with a reinterpretation of some concepts."

My significant scientific findings were deemed to lack novelty and merit. I quote from the report, "Please note that we only consider novel manuscripts with significant scientific merit."

That's all they reported. No other comments or questions.

Although the editor claimed to have read the manuscript, I honestly don't think anyone actually read beyond the title of the paper.

Thus, delayed by another fake review.

The Astrophysical Journal is devoted to "developments, discoveries, and theories in astronomy and astrophysics" and welcomes "significant new research that is directly relevant to astrophysical applications, whether based on observational results or on theoretical insights or modeling." Unfortunately, it is still under the editorship of Old-Physics-follower Ethan T. Vishniac.

A few years ago (in 2015, May) I received an email from Dr Vishniac in which he revealed his incognizance of the fact that General Relativity utterly fails as a cosmology theory and made an interesting admission:

> "The standard model of cosmology is based on General Relativity, a theory which is internally self-consistent and has passed all available experimental tests. Naturally, **it could still be wrong. However, we are not interested in exploring alternative theories** unless they do equally well, ... consistent with the evidence. Only at that point would it be worth examining the cosmological implications of the alternative theory."

General Relativity is plain-as-day inconsistent with cosmic structural evidence.

Back then, he rejected a vastly superior cosmology theory actually consistent with the evidence.

He did the same with the more recent article with its even more powerful substantiation. I suspect Dr Vishniac recognized that the evidence presented is incontrovertible —and, worse, is absolutely devastating for the orthodox paradigm! Imagine, the internal confusion of someone who feels it is his duty to withholding a powerful new interpretation of the findings of his fellow astronomers/astrophysicists from his journal's readership —and above all, shield them from the reality that our steady state Universe is not expanding.

Galaxies MDPI promotes itself as an "Open Access Cosmology, Astronomy, and Astrophysics Journal." The reports that emanated from this publication were priceless.

The First Reviewer flippantly stated "The DSSU model proposed by the author ... cannot explain all cosmological observations." An example, please. Please? No. Not one observation is cited that the DSSU model fails to explain! I'm always hoping that someone will present a solid challenge, an accusation with meaningful specifics. Still waiting —waiting since the year 2001.

The accusation that the DSSU model cannot explain *all* cosmological observations is utterly irrelevant and pathetically amateurish. Think about it. The Darwinian theory of evolution does not explain all observations. According to the criterion this Reviewer applied to the cosmic structure Article, *that* very deficiency would have deemed Darwin's work unworthy to be published in a scientific journal! What rubbish!

The big bang hypothesis cannot explain lots of things: It can't explain cosmic cell structure, periodicity of galaxy clusters (like the Abel-85 system), the cause of galaxy ellipticity (of nonrotating galaxies), the root cause of galaxy rotation (of spirals), initial conditions, and on and on. Yet big bang research papers ... speculations on a preposterous scenario (according to Sean Carroll) are most welcomed and promoted in science magazines and broadcast media.

Another objection from this Reviewer: "The cosmic microwave background radiation ... cannot be satisfactorily explained in a steady state model," referring, of course, to the historical Steady State Expanding model. The Reviewer seems unaware that we are dealing with the steady state *Cellular* model. In this steady state cellular nonexpanding model the background radiation can be, and has been, admirably explained.

Even the foundation is rejected: "The model proposed by the author is not based on sound postulates." Which postulates? ... I have used but one, an empirical one, and I've never heard of anyone objecting to its reasonableness. I would love to know why the postulates are declared to be unsound! If the

Seventh Interlude. Experts, Obstructers, and Fraudsters 255

postulates really are flawed, why not specify the flaws? Why the secrecy?

And just to make sure that nothing escapes criticism, the report pointed out that the English (grammar and style) is inadequate.

The Second Reviewer simply parroted the standard nuggets of indoctrination relating to cosmology and space sciences and revealed an inability or unwillingness to see things beyond the narrow confines of the Big Bang myth! I'll try to keep this as brief as possible, since this person displayed an obstinate knack for obfuscating otherwise simple issues.

- Objected to the space medium aether —believes it doesn't exist. Believes the Michelson and Morley experiments failed to detect the relative motion of the so-called aether. (The Reviewer seems unaware of the historical scientific fact that Michelson and Morley reported a relative motion of about 5 to 7 kilometers per second!)
- For this Reviewer, nothing is too trivial to be the object of disdain: "The author begins his article with a controversial statement that cosmologists accept the expansion of space because one particular accredited cosmologist has said so. That's incorrect. Cosmologists accept this premise because it is evidenced by enormous amount of observable data after tremendous scrutiny." (Who cares why the premise is accepted! The important point is that the premise IS accepted. My statement in the article IS correct. In fact, I call it an *empirical premise* precisely because it is evidence supported.)
- Objected to the scientifically proven velocity differential mechanism by wrongly calling it an "incorrect premise." (See next item.)
- Ridiculed the basically simple *velocity differential redshifting* mechanism —something this "expert" utterly failed to understand. (Good grief, the proof involves only add-and-subtract algebra. Redshifting, and the Blueshifting discussed in the cosmic-structure article, occurs during the photon's propagation and accrues for as long as propagation continues. A photon can, therefore, stretch many times its original wavelength, even ~1100 times as for cosmic microwave photons.)
- Objected to the failure to use relativistic methodology. (But that is something quite irrelevant to the article's subject matter or the particular situation.)
- Questioned the creation of cellular structure: "Creating such regularly spaced structures would also require the initial [big bang] fluctuations to be not random, but exhibit the same symmetry." ... (Revealed here is a mind stuck in a myth. As I have stated over and over and will repeat yet again, cosmic cell structures are not created and were never created — they are perpetually *sustained*. They comprise a true steady state systematic structure. Forget about initial fluctuations and initial conditions. There weren't any.)
- Protested that the evidence does not support regular cellular structure: "This is not supported by observations of the large-scale structure in the

universe. In particular, this breaks the isotropy of the universe ... Therefore, this is nothing more than a falsified conjecture." (This is the strategy of a true believer —reject the evidence because it spoils the isotropic feature of Big Bang dogma.)
- The Reviewer disliked my rhetorical question relating to an infinite-and-expanding universe: "In physics, we do not bother with the question of the 'intentions' of the universe and do not attempt to answer questions such as *why the universe would bother to become more infinite*." (In that case, let's just invoke Ockham's razor and cut out the nonsense — excise the myth of universe-wide expansion. There is absolutely no need for the universe to expand itself. Observations can be explained without such a radical concept.)
- Complained there is not enough evidence to support the claims! (It's doubtful that more evidence would help. No. More evidence will not make any difference. The Reviewer is suffering from big-bang tunnel-vision. Just remove the blinders and look around.)

The Third Reviewer, without giving any examples, mentioned insufficient background, non-clarity of presentation, and unsupported conclusions; and claimed there is nothing new here, "there is no novelty", and therefore should be rejected.

Incidentally, there was no opportunity to respond to the issues raised in the three reviews, in fact, the Journal's website blocked that option (normally available to authors).

Obstructers

So there you have it. That was how the DSSU article on connecting predictions (theory-derived) with evidence (real measurements) became the most rejected ever. Rejected by *seven* astrophysics journals!

The good news is that the experts, the examiners, the obstructers, were still unable to find flaws on legitimate scientific grounds!

Scientists that they are, they know what they are doing, many are outstandingly brilliant. They know all about the crucial aspect of science, the connection between predictions and observational confirmation.

> *The goal of science is the extension of knowledge, that is, empirically confirmed and logically consistent statements of regularities (which are, in effect, predictions).* –From *The Great Betrayal, Fraud in Science*[6]

They did not forget. That's what should amaze, and why I am always amazed.

---

[6] H.F. Judson, *The Great Betrayal, Fraud in Science* (Harcourt Inc., Orlando, 2004); p32.

Seventh Interlude. Experts, Obstructers, and Fraudsters

Yes they know, a *theory* is a connection maker. And yet many cling to bankrupt theories and hypotheses that can so easily be demonstrated to be wrong, or simply unnecessary; meanwhile, and worse, they obstruct others who have uncovered new and better-fitting connections.

I am reminded of a saying.

> *When a man who is honestly mistaken hears the truth,*
> *he will either quit being mistaken or cease to be honest.*

What are the obstructers afraid of?

The material of the previous chapter explaining cosmic structure patterns, in its original manuscript form, was the most rejected work I have ever encountered in advancing DSSU theory. Yet all it does is present the extraordinary match between what the Theory predicts and what has actually been observed. Journals claiming to be subject-focused on astrophysics, cosmology, and gravitation declared the Article unacceptable! —even when, under peer-review, it was praised and judged to be "an excellent manuscript."

It was not that the editors/reviewers recognized any failure in the physics, or any flaws in the logic, or any other deficiency. Those, of course, could be corrected. No problem. But from their perspective, it was something far worse. They recognized the awesome power of the Theory. How could they not? Here was a cosmology that could predict and, therefore, *explain* all the major phenomena that their own masterpiece-of-misconception model could not! (And, as a bonus, it could do this in terms that non-experts can easily understand.) Yes, they fully understood what it was about and its devastating implications.

The previous chapter is primarily a presentation of irrefutable evidence. It is the exposure and promotion of such cited evidence —evidence they can't explain— they fear the most! Best not to draw attention to those. Theoretical points that challenge the official narrative are different. Theoretical points they can argue away, cleverly or crudely, like shyster lawyers with their twisted logic, non sequiturs, obfuscations, irrelevancies, and *ad hominems*.

What has these obstructers paranoid? It is the ascendency of the most powerful cosmology ever developed (a theory that threatens their sacred Big-Bang dogma). Moreover, it is the evidence of orderly structure (evidence which not only doesn't fit but actually contradicts their chaos-rooted big-bang hypothesis)?

> *If the past is a guide to the future, our modern beliefs might be greatly mistaken, and one day a new universe might arise, grander than our present [20th-century] model. Those living in the future*

will look back in history and see our universe as out of date as all the rest. ... they might wonder what we were doing, or not doing, with our large brains. –Edward Harrison[7]

✠ ✠ ✠

---

[7] E.R. Harrison, *Masks of the Universe, Changing Ideas on the Nature of the Cosmos*, 2nd ed. (Cambridge University Press, Cambridge, UK, 2003); p6.

Law of Gravity Unification

8. Aether Theory of Unified Gravity

How aether is responsible for mass and the five manifestations of gravity

> *Newton and Einstein were supreme as unifiers. The great triumphs of physics have been triumphs of unification. We almost take it for granted that the road of progress in physics will be a wider and wider unification bringing more and more phenomena within the scope of a few fundamental principles.*
> –Freeman J. Dyson, Infinite in All Directions

> *We thus rather follow Nature, who producing nothing vain or superfluous often prefers to endow one cause with many effects.*
> –Nicolaus Copernicus (1473-1543)

The closest we can come to the Copernican ideal of having one cause able to produce many effects is to, again, turn to our familiar universal space medium —aether.

In the endeavor towards "a wider and wider unification bringing more and more phenomena within the scope of a few fundamental principles" everything depends on the nature of aether. With a proper explication of aether's properties and processes, we are able to explain all the manifestations of gravity. This means gaining a deep understanding of (1) **The primary cause** of ordinary convergent gravity; (2) **The secondary convergent effect**, also known as the gravity produced by the gravity field itself; (3) **The divergent gravity effect**, known informally as antigravity and formally as Lambda (the cosmological constant); (4) **The vorticular effect**, described as an aether-dragging effect and a dual stress effect; manifests as an important gravitational amplification effect; (5) **Gravity waves**, a unique acoustic-like effect that spans across two domains of existence. But there is more. Even the very cause of the property of *mass* —aether's crucial role in the bestowal of mass— is a part of this marvelous unification.

Background

Historically, gravity theories based on aether have been a failure. There are two main reasons: First, there was the insistence of attributing energy to the constituent aether entities —or whatever it was that defined the specific aether. In some cases, the aether "particles" were said to possess mass. Second, in an effort to make the aether dynamic, it was deemed to vary in density and manifest density gradients. Essentially, theorists failed to recognize the need for a subquantum level of existence; **failed to appreciate its importance**; and failed to adopt the aether as *that* subquantum level —a substrate below the conventional mass-and-energy domain. Moreover, they failed to appreciate the pitfall associated with the variable-density hypothesis.

However, by avoiding the energy-mass allure and the variable-density trap, a powerful unified gravity theory has been constructed. This unprecedented approach has made possible a conceptually simple way to bring all manifestations of gravity together under the aether umbrella —under the distinct aether that serves as the foundation of DSSU theory. For the first time in the scientific quest to understand gravity and to go beyond merely describing gravity, a process has been identified that underlies, severally, the primary cause of gravity; the normal-type convergent gravity; the divergent-type gravity (known variously as *Lambda expansion*, *dark energy*, and *antigravity*); the spiral-flow amplification of gravity (manifesting in rotating structures); and gravity waves. In each and every case, the energy can be linked to aether; but not in the way that others have been attempting for centuries.

There are two kinds of theories —descriptive and explanatory. Descriptive theories detail the apparent aspects, specify the measurable features, formulate their relatedness and interactions. Explanatory theories go deeper; they spell out the underlying causes of the interactions. Explanatory theories are powerfully predictive and capable of providing profound illumination. I raise this point in order to contrast the difference between our aether theory of gravity and other theories of gravity, namely, Newton's force theory and Einstein's geometric theory.

The difference of greatest importance, for the workings of any gravity theory, has to do with the meaning of *space*. In the case of Newtonian gravity, space has no properties; however, it is said to be absolute. What does that mean? It means that you should be able to use space to reference locations; you should be able to label absolute positions within absolute space. But if Newtonian space has no identifiable properties, *there is no way to make use of its absoluteness.* Consequently, absolute space serves no functional purpose. For Einstein, space was a mathematical canvas, some sort of smooth continuum. With respect to Einstein's space, nothing was absolute; which, of course, makes it difficult to explain the absoluteness of rotational motion. Everything else, all other forms of motion, was irrespective of space. That is, motion was relative —relative to whatever frame of reference one chooses or one happens to occupy.

Space plays an extensive role in Einstein's gravity theory with its space dynamics, and spacetime, and space curvature (all mathematical concepts). In contrast to Newton's do-nothing static space, Einstein made his into "something" dynamic. *Space dynamics* means that space has the ability to expand or contract. But what is space? No real answer. What about *spacetime*? It is a mathematical concept-of-convenience and has no connection to reality. The last term, *space curvature*, is the mathematical (geometric) identifier of space expansion or contraction. We have no problem with expansion and contraction; they make sense. But, again, what is space? No one seems to know. But whatever it is, it has the ability to do amazing things —supposedly even expand the entire universe, even cause the acceleration of *that* expansion! Needless to say, in order to perform such fantastic feats, the space of general relativity theory must possess energy. This in turn means the problem of understanding "space" grows even deeper, because now one also needs a fundamental definition of energy.

The popular approach is to say, as Brian Greene does in his book *The Fabric of the Cosmos*, that space (or spacetime) is a fabric. A fabric?! ... So then, it is the "fabric" that curves, stretches, and shrinks; and, like space, it is said to be saturated with energy. Obviously, space is not literally a fabric. The popular approach is to focus attention on metaphors, which then help to camouflage the inability, in the context of general relativity, to meaningfully deal with the question of *what is space*?

While no one has a clear explanation of the space of Einstein's gravity theory, almost everyone agrees on the presence-of-energy aspect. In physical cosmology and astronomy, it is called *dark energy*, an unknown form of energy hypothesized to permeate all of space and to accelerate the expansion of the universe. But what does it mean for space to have energy? Notice the depth of the problem of understanding gravity —as something conveyed by the fabric-like space. The figurative fabric is a math canvas and space is an abstraction filled with phantom vacuum energy. The failure of the explanatory side of the theory is exemplified by the mismatch between the predicted enormous vacuum-energy density and the negligible amount of it actually in evidence. Experts in the field have long considered this to be an embarrassing failure.

The extremely important question of *what is space?* is addressed in a comparative way in **Table 8-1**. A deep understanding of space and what it contains leads to an explanatory theory of gravity as well as a problem-free cosmology.

Table 8-1. What is space? It depends, as revealed by the comparison of three dramatically different views. Newton's space is said to be absolute (like an empty container for the stuff of the universe), does nothing, and has no causal connection to gravity. Einstein's space is a mathematical abstraction; its geometrical dynamics define gravity. For the DSSU*, space is a nothingness, a metaphorical, empty vessel filled with uniquely-defined aether. Three-dimensional space has no properties; all the action, everything that happens, is rooted in aether, the universal medium.

Worldview	What is Space?	Defining feature	What is Gravity?
NEWTON:	Absolute	Space is STATIC	Force
EINSTEIN:	Abstraction (coded in 4-dimensional geometry)	Space is DYNAMIC and saturated with energy	Spacetime curvature
DSSU*:	Nothingness vessel permeated by defined aether	Space has NO PROPERTIES	Aether (acceleration of flow)

* *The Dynamic Steady State Universe* —the cosmology theory that holds that a nonmaterial aether is the ultimate bedrock of Nature, and further, that aether expands and contracts *regionally and equally* resulting in a cosmic-scale cellularly-structured universe. It is a model based on the premise that all things are processes.

Under the DSSU's aether paradigm, space is an empty nothingness; other than having three dimensions, it has no properties; it merely serves as a container. However, because nothingness is inconceivable —or let me say, Nature does not permit a state of nothingness— space must be "filled" with a substrate of some sort. This ethereal medium, it turns out, is what supports (sustains) the existence of mass, matter, and energy. It makes possible the existence of the material world. *That* substrate is what is being called aether.

It works as a wonderful simplifying factor. This approach greatly simplifies our endeavor to understand the Universe, since aether acts as the cause *and* mechanism of gravity. Space curvature is totally unnecessary. All the action is with the aether (aether as a mechanical substrate) and not with space (space as an esoteric abstraction). This, in turn, means that space (space as a container) is Euclidean. In other words we can use this background space as a working canvas —an uncomplicated 3-dimensional volume in which aether flow patterns can be modelled and analyzed.

The three key features of our subquantum mechanical aether are as follows: The entities that comprise the aether do not possess mass; moreover, these entities, in their vibratory activity, do not represent energy. (The energy aspect, at the fundamental level, will be explained in due course.) The second key property is that the spacing density of aether, the spacing density of the discrete aether units, is constant within a narrow tolerance range. The third property is that aether is dynamic —in fact, it is *uniquely* dynamic. With this new understanding of space and aether we are led directly to a powerful aether-based gravity. And gravity based on aether is a broader theory —a theory able to accomplish considerably more than Newtonian gravity and more than Einstein's gravity. Newtonian gravity does not work for cosmic Voids and spiral galaxies; general relativity fails for spiral galaxies and black holes. And they both fail when it comes to explaining cosmic-scale cellularity. Aether gravity works for all cases.

Aether, as the central player of the theory, is subject to various types of stress. The stresses, of which there are five, and their association to gravity are as follows:

- Excitation stress. It is active in the Primary cause of gravity.
- Convergent-compression stress. It plays the key role in normal (everyday) gravity.
- Tension stress. It is important for antigravity manifesting in the cosmic Voids.
- Shear stress. It plays the crucial role in modifying the gravity associated with rotating structures.
- Radial-compression stress. A minor player responsible for the acoustic aspect of gravity (gravity waves).

The following sections will explain how these stresses are connected to the five gravity effects mentioned above.

We begin with a discussion of the underlying driver of gravity. The aim is to gain understand of gravity's root cause.

Cause of mass property and its connection to Primary gravity

In addition to its uncomplicated space-as-a-container aspect, the universe has another wonderfully simplifying feature. As we learned in Chapter 6, all matter consists of electromagnetic waves, all matter is light. This has been understood for a very long time. It was understood by Nikola Tesla (1856-1943), the great *master of light* himself, when he famously said, "Everything is the light, everything is electricity. I am electricity in the human form."

British physicist and astronomer Sir James Hopwood Jeans (1877-1946), back in 1931, stated:

"[T]he tendency of modern physics is to resolve the whole material universe into waves, and nothing but waves. These waves are of two kinds: bottled-up waves, which we call matter [i.e., mass], and unbottled waves, which we call radiation or light. The process of annihilation of matter is merely that of unbottling imprisoned wave-energy and setting it free to travel through space. These concepts reduce the whole universe to a world of radiation, potential or existent, ..."[1]

Jeans reiterated the point when he wrote, "... we may think of [material] matter and radiation as two kinds of waves —a kind which goes round and round in circles, and a kind which travels in straight lines."[2]

In agreement with what Tesla had long believed, James Jeans concluded, "... all physical phenomena are ultimately electrical." That is, everything is some manifestation of electromagnetics.[3]

He clearly meant that all matter, all transformations, everything perceptible, is the result of photons in motion —moving in a confined state or a free state. Even the phenomenon called *mass* is ultimately "electrical." In other words, if we focus just on the mass aspect, anything that is baryonic, anything that possesses the quality of mass, is simply an organized package of confined photons. Undoubtedly, Jeans' photon theory of particles and phenomena provides a powerfully unifying framework. In its 21[St]-century incarnation, the theory found support among physicists such as John Graeme Williamson and his colleagues.[4,5,6]

Now for the link with aether. Understand what is about to happen here. By taking the universe's sole fundamental particle that we have identified as the electromagnetic photon and linking it to the universe's essence substrate that we have recognized as the aether, we are connecting everything —literally *everything* — to this aether.

Photons, as quantum particles, exist as excitations of the subquantum

---

[1] James H. Jeans, *Chap. 3: Matter and Radiation*, The Mysterious Universe (Cambridge University Press, 1931); p69.

[2] Ibid., p51.

[3] James H. Jeans, *Chap. 4: Relativity and the Ether*, The Mysterious Universe (Cambridge University Press, 1931); p65.

[4] J. G. Williamson and J.M.B. van der Mark, *Is the electron a photon with toroidal topology?* Annales de la Fondation Louis de Broglie, 1997, V.**22**(2), 133–146.
(Posted at: https://www.researchgate.net/publication/273418514)

[5] J. G. Williamson, *On the nature of the electron and other particles*, The Cybernetics Society 40th Anniversary Annual Conference (2008) in London.
(Posted at: https://www.researchgate.net/publication/267370968)

[6] J. G. Williamson, *On the nature of the photon and the electron*, Conference Proceedings of the International Society for Optics and Photonics (SPIE 9570, The Nature of Light: What are Photons? VI. September 2015).
(Posted at: https://www.researchgate.net/publication/281749668) (Doi:10.1117/12.2188259)

entities of aether. Photons exist purely as a process of aether —an excitation. This represents a profound connection because it means that the photon, in and of itself, is not something beyond a manifestation of an aether activity. In other words, without the excitation of aether, no photon can exist.

And finally, there is the link between aether excitation and gravity —the root cause of convergent gravity.

When aether undergoes electromagnetic excitation, it is invariably accompanied by its own annihilation. When photons excite the subquantum aether "particles," the affected particles vanish. They literally vanish from the universe. No! This is not a violation of thermodynamic law. The aether particles, being *subquantum*, cannot and do not possess energy. Although, they *are* mechanical. They are mechanical (in the sense of being discrete), but they are not physical (in the sense of having mass or energy). So, when photons propagate through aether (that is, are conducted by aether), whether in self-looping patterns or free ranging, they will continuously annihilate a certain small quantity of aether. The significance of this process cannot be overstated. Extensive research has shown that *the aether destroying propagation mode of photons is the secret of the Universe.*[7]

It follows that mass particles and objects will absorb and annihilate aether —tiny amounts on the particle scale, astronomical amounts on the astronomical scale. Thus, in order to sustain the existence of the mass, a converging flow of aether is required.

Convergent flow, by virtue of being compelled to merge, means that the aether flow must accelerate. And there lies the direct link to gravity, the link we need to connect aether and gravitational attraction. Gravity is defined as the effect produced by the dynamics of the space medium —*the accelerated motion of aether*. This is sometimes stated as the inhomogeneous flow of aether towards, and into, matter.

The *direct* cause of gravitation, as it relates to planets, moons, and stars, is simply the combined absorption effect of all the mass and mass equivalences that such bodies represent. The cause of gravitation is the activity of the multitude of confined photons as they excite-annihilate aether. Surrounding any large mass accumulation, there is always a bulk flow of aether —an inflow absolutely necessary in order to feed a truly insatiable demand. This bulk-flow aspect of gravitation is simply the side-effect of the relentless demand by mass (and its equivalences) for the essence medium.

Summarizing the crucial links connecting mass/matter, the photon, aether, and gravity. All matter and all phenomena are linked to the fundamental energy particle, the photon, via a unique process linked to aether. And aether,

---

[7] C. Ranzan, *DSSU Validated by Redshift Theory and Structural Evidence*, Physics Essays, Vol.**28**, No.4, pp455-473 (2015). (Doi: http://dx.doi.org/10.4006/0836-1398-28.4.455)

via its motion, is linked to gravity.

Primary convergent effect

The simplest way to gain insight into Primary gravity is by assuming, for the moment, that the aether is an unalterable fluid. We pretend that it cannot contract or self-dissipate. The strategy is to analyze the aether flow without taking into consideration the stress effect —without letting the stress of convergence alter the aether in any way.

Imagine a motionless planet-size mass surrounded by a concentric sphere —an imaginary outer surface with a fixed radius r_o, as shown in **Figure 8-1**. In order for the aether to reach the mass body, it must pass through this outer "surface." Let the radially inward flow-speed, at the instant of entry, be v_o. In order to see how the speed of our idealized fluid changes, an inner concentric sphere with arbitrary radius r is added (**Figure 8-1**). Our attention is focused on the shell defined by the two spheres. The idea is to compare the flow entering the shell (through the outer shell "surface") and the flow exiting the shell (through the inner shell "surface"). For a precise comparison, all we need is the standard fluid-flow continuity equation as may be found in any physics textbook:

$$\begin{bmatrix} \text{area of concentric} \\ \text{outer sphere} \end{bmatrix} \times \begin{bmatrix} \text{flow velocity at} \\ \text{outer sphere} \end{bmatrix} \times \begin{bmatrix} \text{fluid density at} \\ \text{outer sphere} \end{bmatrix}$$
$$= \begin{bmatrix} \text{area of concentric} \\ \text{inner sphere} \end{bmatrix} \times \begin{bmatrix} \text{flow velocity at} \\ \text{inner sphere} \end{bmatrix} \times \begin{bmatrix} \text{fluid density at} \\ \text{inner sphere} \end{bmatrix}. \quad (1)$$

Don't worry; most of the terms cancel out. The final expression turns out to be quite simple.

Since aether density by definition is constant, the two density terms cancel. Then using the symbols from the figure, the areas and velocities are related as follows:

$$\left(4\pi r_o^2\right) v_o = \left(4\pi r^2\right) v ; \quad (2)$$

$$v = v_o r_o^2 \frac{1}{r^2} . \quad (3)$$

Since v_o and r_o are treated as "fixed" values, they are lumped together as a single constant. We can then express the *primary flow* as a function of the radial position as,

$$v_\text{p}(r) = -|\text{constant}| \frac{1}{r^2} . \quad (4)$$

This says that the aether flow is proportional to the inverse square of the

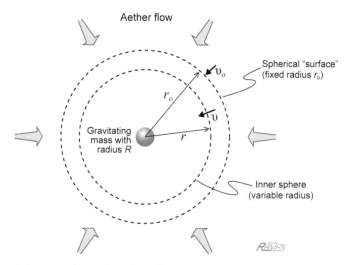

Figure 8-1. In order to formulate the Primary convergent flow of aether, we imagine two spherical boundaries as shown. The outer sphere is held constant, which means the aether speed (v_0) there does not change. The idea, then, is to find the relationship between the radial distance r and the aether speed v (at r). We just want to know how flow speed changes with distance, under the hypothetical condition of treating aether as a non-compressible, non-contractile fluid.

radial position. The radius r, of course, has to be greater than (or equal to) the central body's own radius R. As was expected, the magnitude of the inflow *increases* with proximity to the gravitating body.

Now, in order to make the connection to gravity, all we need is the acceleration of that flow. It just involves some basic calculus steps (see footnote [8]). By taking the *time* derivative of the above expression, the acceleration, and hence the *Primary-gravity* intensity, is found to be proportional to $1/r^5$:

$$a_{\text{primary}}(r) \propto \frac{1}{r^5}. \tag{5}$$

---

[8] **Primary acceleration:** $a_\text{p}(r) = \dfrac{dv_\text{p}(r)}{dt} = \dfrac{d\left(-|\text{constant}|r^{-2}\right)}{dt}$;

$a_\text{p}(r) = \text{con}\left(\dfrac{dr^{-2}}{dr}\dfrac{dr}{dt}\right) = \text{con}\left(-2r^{-3}v_\text{p}(r)\right) = \text{con}\left(-2r^{-3}\text{con } r^{-2}\right) = -|C|r^{-5}$.

Thus, (Primary acceleration) $\propto (1/r^5)$, where $r \geq R$.

This expresses the acceleration proportionality an object would "experience" under the described exploratory situation.

Remarkably, the primary acceleration varies inversely with the *fifth power*! It means the Primary-gravity effect is extraordinarily weak. Putting this into proper perspective, the intensity of gravitation, according to Isaac Newton and real-world experience, varies in agreement with the inverse-square law —and not as $1/r^5$. The Newtonian gravitational attraction between two bodies diminishes with increasing distance between them as the inverse of the square of that distance; if the distance is doubled the force declines by a factor of four. However, if only the *Primary gravity* effect were the active mechanism, then a doubling of distance between two masses would *decrease* their mutual gravitational attraction by an astonishing factor of thirty-two.

The importance of the Primary gravitation effect is that it serves as the indirect cause of *Secondary gravitation* —a far more powerful effect.

Secondary convergent effect

Let us see how aether responds to the stress of convergence and how it undergoes compression without changing its number density.

The primary flow speed, as determined above, is proportional to $1/r^2$. But the actual flow rate, at some radial distance, is proportional to $\sqrt{1/r}$. See Appendix A for the derivation. The actual flow rate turns out to be much greater than what was determined for non-contractile aether. This can be seen graphically in **Figure 8-2**. Notice, in comparing the two velocity-magnitude curves, not only is the actual inflow speed greater (at whatever radius it is examined) but also its rate of increase is greater.[9]

Ponder the situation for a moment. Is the intensification of the flow the consequence of additional aether entering the system (say by some sort of expansion)? Or is it a consequence of a loss of aether (by some sort of contraction)? There is an intuitive answer —in keeping with the aether's defining properties— but it is not immediately obvious. Before providing the intuitive rejoinder, a proper proof is in order.

For the investigation of Primary gravity, a stable non-contractile aether fluid was assumed. Then, by constructing an imaginary shell around a gravitating body and applying the fluid flow equation (1), it was found that

(Area$_{entry}$ × velocity$_{entry}$) = (Area$_{exit}$ × velocity$_{exit}$); or

---

[9] Strictly speaking, however, the relationship breaks down for distances fairly close to the surface (closer than 2.52 R). But this is not important for the demonstration because there is really no way of determining the Primary flow at the body's surface. There is just no way to remove, from the formulation, the secondary effect occurring in the interior of the mass body; and so, the calculated surface inflow is not a pure Primary flow.

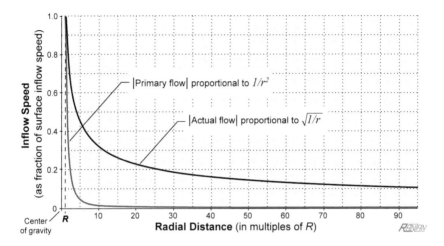

Figure 8-2. Primary inflow versus actual inflow. The Primary flow is that of aether if it were not altered in any way —no shrinkage, no loss, no density change. It is simply treated as a stable fluid. The actual flow is a combination of the primary and secondary flows; the acceleration of this combined flow accurately models Newtonian gravity. (Note, the gravitating mass is assumed to be at rest within the aether medium.)

$$((A_e v_e) - (A_x v_x)) = 0. \tag{6}$$

Essentially, the volume entering the shell was equal to the volume exiting the shell (per unit of time) —*which meant that there was no fluid loss.*

So let us see (with the help of **Figure 8-3**) what actually happens if the aether behaves in a manner that accords with ordinary Newtonian gravity. For the sake of simplicity, we continue to assume that the gravitating body is at rest with respect to the aether medium.

The argument goes as follows:

From **Figure 8-3**, we see that r_{entry} is *greater* than r_{exit}.

We can therefore write, $(r_e^2)(\sqrt{1/r_e}) > (r_x^2)(\sqrt{1/r_x})$. (7)

The relationship remains valid if we add a few constants and end up with

$$(4\pi r_e^2)(C\sqrt{1/r_e}) > (4\pi r_x^2)(C\sqrt{1/r_x}). \tag{8}$$

The first terms on each side represent the spherical areas of the respective entry and exit "surfaces". The second terms are the respective flow speeds. As detailed in Appendix A (equation 6), the flow speed at *r* is given by the

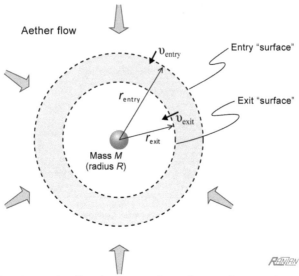

Figure 8-3. Converging flow leads to aether volume reduction. The volume of aether flowing out from the shell is LESS than the volume flowing in. The proof is given in the text. We interpret this not as a change in density of aether but, rather, as *an absolute loss of aether*. Symmetrical flow is assumed.

expression $C\sqrt{1/r}$. (It is the velocity necessary for agreement with Newtonian gravity. It is the velocity expression that produces a Newtonian *acceleration* proportionality of $1/r^2$.) The constant "C" depends on the quantity of mass M and includes the Newtonian gravity constant.

The equation simply states,

(Area$_{entry}$ × Flowspeed$_{entry}$) > (Area$_{exit}$ × Flowspeed$_{exit}$).   (9)

Equivalently,

(Vol. flowrate @entry) > (Vol. flowrate @exit).   (10)

The volume flowing into the shell region is greater than that flowing out (per unit of time). There is the answer. This proves that volume-of-aether reduction does occur.

There are basically two ways to bring this about.

One way is to adopt the constant density property as a foundational principle. We would then say, the argument just presented proves that aether has somehow been lost during the converging flow. It has to be an out and out loss; it cannot merely be some shrinkage by compaction; it has to be the one, and cannot be the other, because of the constant-count-density condition. The cause of the loss is naturally attributed to the compression stress attendant with the convergent flow.

The other way (to account for the volume reduction) is to allow the aether density to vary. Under this interpretation, the compressive stress of convergent flow is allowed to alter the density of whatever it is that comprises (or defines) aether. It is a pitfall that has repeatedly trapped theoreticians. Volume reduction by increasing density, not surprisingly, can be made to work mathematically; but has not led to the vital connection with reality. The method is straightforward. Recall, when *we* used the fluid continuity equation, the density terms were kept equal and dropped out. But with the traditional approach, the density terms are adjusted to effect the appropriate change in volume and thereby directly achieve the flow acceleration that matches Newtonian gravity. The distinction between Primary and Secondary gravity is then not needed and not recognized. However, serious problems arise when attempts are made to expand the concept into a broader theory.

The two methods for effecting volume reduction are outlined in **Table 8-2**.

Table 8-2. Comparing the two ways of reducing the volume of the aether fluid. In an effort to achieve acceleration proportionality to the inverse square of the distance r, some theorists advance the notion of a variable-density aether (right-hand column). Whether the density change refers to inherent particulate nature or the energy content, the method is a pitfall. With DSSU aether (middle column), the volume reduction is achieved through the unique *process of self-dissipation* —a response to the stress of convergent-flow and the axiomatic striving to maintain a constant density. It is this vanishment that makes DSSU aether dynamic —and therefore gravitational.

	How to effect volume reduction	
	DSSU aether	Popular approach (pitfall)
Key property:	Spacing density remains CONSTANT	Density VARIES
Method:	Volume is reduced by aether loss (**SELF-DISSIPATION**)	Volume is reduced by compression (DENSITY INCREASE)
Test:	Conforms to reality	Works math-wise but fails reality

Primary gravity was explainable with a suppositional non-dynamic (stable) aether. Secondary gravity, on the other hand, could only be explained by dynamic (contractile) aether. Aether needed to have the ability to contract. But there had to be a constraint on this ability —we had to include the foundational requirement that DSSU aether is not compressible. More precisely, it is not compressible in the sense that the discrete aether entities cannot be packed closer together. Our aether always maintains a constant number density within some narrow range. It turns out, there is only one way for aether to be contractile (to reduce or shrink in volume) and meet those conditions.

For it to be both dynamic and count-density stable, *aether must be self-dissipative*.

Aether, when subjected to the compressive stress of being involved in a converging flow, suffers a quantitative proportional vanishment. Aether units are literally pressed out of existence. As pointed out earlier with the vanishment associated with excitation, this does not violate the First law of thermodynamics. *Self-dissipation* is one of the main processes that gives DSSU aether its dynamic quality. (Another one is discussed shortly.)

The key point is this: The actual flow rate includes the Primary flow portion as well as a LOSS portion (with significant consequences). Convergent gravity is the result of two flow components, or two effects.

It is interesting to note that under the general-relativity view of gravity there is, somewhat analogously, a two component effect and also a spacetime self-interaction; although it is by no means an accurate analogy. The way it is often described is to say that gravitating mass produces a gravity field (the primary effect), which then produces a secondary gravity effect, and even an additional effect. According to astrophysicist, Edward Harrison, "This self-interaction of spacetime is what is so important about general relativity. This self-interaction exists because the curvature of spacetime is itself a form of energy, which produces its own gravitational field, and is hence the source of further curvature. ... Thus curvature generates curvature."[10]

What about quantifying the aether dissipation (in terms of its volume loss)? It is not too difficult to do this. The easiest way to obtain a useful function is to start by making the shell, used in **Figure 8-3**, only one meter thick. Or we can work with an even thinner *elemental* shell, using the methodology of calculus (see Appendix C for the details). The resulting *aether volume-loss function*, which gives us the fractional loss of aether within a test volume located at radius r, may be stated as

$$\text{vol}_{\text{unit loss rate}} = \tfrac{3}{2}(2GM)^{1/2} \frac{1}{\sqrt{r^3}}. \qquad (11)$$

This is our equation for **aether self-dissipation** —expressed here as a function of radius r and M (the total mass within the limits of *that* radius). There are two ways to interpret the results of its application: One is to treat the volumes in terms of cubic meters so that units are m^3/s per m^3, which is then understood as cubic meters of aether loss per second, per cubic meter of Euclidean space (background container space). The other interpretation is

---

[10] E. R. Harrison, *Cosmology, the Science of the Universe* (Cambridge University Press, Cambridge, UK, 1981); p170.

simply as the fractional volume loss per second of time (within any arbitrary small volume). The smaller the radial distance (that is, the closer to the gravitating mass) of the examination point, the greater will be the loss. The peak loss occurs at the surface. For the Earth (radius 6.37×10^6 meters and mass 5.98×10^{24} kilograms), it is about 0.00264 m^3/s per background cubic meter, or 0.264% each second. For an extreme example, when the above equation is applied to a Terminal neutron star (mass 3.4 Suns) near the surface (radius 10,000 meters), the loss to self-dissipation is a significant 45,000 cubic meters per second within each cubic meter of background space.

Here, incidentally, is an intuitive explanation of why aether loss results in self-acceleration. Think of a leaky bucket in which the number of holes increases with proximity to the bottom. There is a gradation of few holes at the top and many holes at the bottom. Imagine the water in this porous bucket being continuously resupplied. A comoving particle carried downward with the water (toward the bottom of the bucket) will accelerate; whereas, with only one hole (or several holes) at the bottom there would be no acceleration.

Summing up, Secondary gravity is the result of

The connection between aether flow and empirical gravity

Aether gravity theory says the converging flow is proportional to $\sqrt{1/r}$ (Figure 8-2). Basic calculus says the rate of change of that flow is proportional to $1/r^2$ —which is exactly in agreement with Newtonian gravity. The acceleration of aether is directly connected to ordinary empirical gravity. Consider a completely isolated Earth (not under the influence of any external gravitating body). The acceleration of the aether flow towards the Earth turns out to be exactly the same as the acceleration of a freefalling object (falling from some great distance). Upon nearing the Earth's surface, it would be accelerating at 9.8 meters per second squared (ignoring air resistance); and exactly the same as the acceleration of aether there. This means the object would be comoving *with* the aether. In fact, it is the acceleration of the aether driving the object towards the Earth. In general, motion tends in the direction of the maximum gradient of acceleration —of aether or of gravity, it makes no difference, they are, functionally, one and the same. □

aether's response to stress. This amplifying aspect of convergent gravity is related, in a causative way, to *the sensitivity of aether to compressive stress* — the stress that accompanies convergent flow toward the central mass. Simply stated, aether is not compressible; but neither does it resist compression. It cannot resist compression, for to do so would require an ability to sustain stress, which ability it does not possess. Secondary gravity is the direct consequence of a quantitative loss of aether.

Divergent gravity, the antigravity effect

The major processes that operate in the universe do so as opposite pairs or as reverse sets. The Positive energy process is countered by various negative energy processes, positive electric charge is in some way symmetrically countered by negative charge, energy-loss processes are countered by an energy-gain process. And so it is with gravity, convergent gravity has its opposite in divergent gravity (sometimes called the antigravity effect).

Cosmic tension, a cause of divergent gravity

Convergent gravity involves the quantitative reduction of aether. Divergent gravity is an opposite effect, as it involves the quantitative *increase* of aether. On the subquantum scale, it means the coming-into-being of additional aether units —new discrete fluctuators. On the scale of the cosmic cells, it means the bulk exponential *expansion of the space medium*. The nominal cause of this emergence of new aether is the cosmic tension associated with the large-scale structure of the universe.

Here is where we enter the realm of cosmology and draw attention to the distribution of galaxy clusters. It is the distribution of clusters that determines where the tension —and, hence, the expansion of the medium— occurs. Long confirmed by astronomical observations, clusters are arranged at the boundaries of cosmic-scale voids. Our universe is a cellular construction, with vast empty regions surrounded by significant clusters of galaxies along with dust-and-gas clouds and other debris —material inevitably attracted to the galaxies. These galaxy clusters, as major centers of gravitation, are the source of cosmic tension.

Consider how the clusters, positioned as they are on opposite sides of a Void, respond to each other. Each is gravitationally "pulling" on the other across a vast barren region. Each pair of clusters produces a *negative cosmic stress* in the region that separates them (a region approximately 350 million lightyears across). Moreover, all the galaxy clusters comprising a typical cosmic structural cell can be paired in this way. And there are seven such pairs active in every 3-dimensional cosmic cell (shown in simplified cross-section in **Figure 8-4**). The result is a vast zone in which aether is under t e n s i o n. Realize that opposing clusters cannot come together to relieve the tension. Every cluster is simultaneously being pulled from the opposite directions. In fact, it is being pulled, more or less symmetrically, from several cells (either 4 or 6), in each of which it is an intimate member.

Interestingly, a cosmic region that is under tension behaves much like Einstein's Lambda force or *cosmological constant* —they share the ability to produce expansion. Under our reality-based view, aether expands; under the general-relativity view, the fabric of space expands. But then, Conventional Astrophysics makes the mistake of extrapolating the effect —extrapolating it

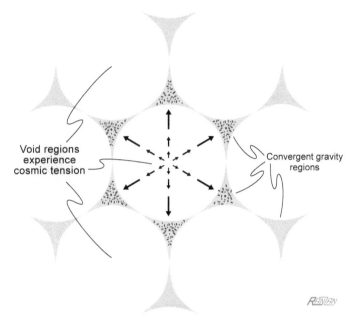

Figure 8-4. Divergent gravity is rooted in the interrelationship among *major mass distribution*, *cosmic tension*, and *aether emergence*. The cellular network consists of rich galaxy clusters and aether-filled voids (and presents a pattern shown here as an idealized cross-section view). As major centers of gravitation, the clusters "pull" on each other in, more or less, symmetrical pairs, as indicated by the gravitation/acceleration vectors (not to scale). Consequently, the central regions of the cosmic cells are regions of negative pressure —equivalent to *cosmic tension*. The tension promotes the emergence of new aether. This expansion of aether, in turn, produces void-centric divergent gravity —a radially outward comoving acceleration of aether. (Cells are typically about 350 million lightyears in diameter.)

beyond the great voids. The effect is correctly interpreted as the expansion of the space fluid, but this expansion is then wrongly interpreted, according to the conventional wisdom, as something leading to the pushing-apart of galaxy clusters (and isolated galaxies) and the growth of the voids. There is a struggle for dominance between the opposites of gravity and Lambda. The expansion "force" is treated as something independent with no causal connection to the galaxy clusters —a big mistake. Astrophysicists, having adopted a picture of Lambda as a wholly-independent force/energy, have no reason to constrain its extrapolation and have naively hypothesized its ability to expand the whole universe!

Under the DSSU worldview there is no such extrapolation. Nature reacts to the cosmic tension by bringing forth more aether. *Major clusters do not move apart*; they remain stationary. They remain stationary because *primary* and *secondary* gravitation processes continuously consume the new aether. There exists a wonderfully natural harmony of opposites.

It is the cross-Void tension that is an important factor in the cause of divergent gravitation. But it is not the only factor.

The axiomatic aspect

The emergence of aether is nominally caused by the tension across the cosmic Voids —by the tension produced by galaxy clusters "pulling" from opposite sides of a Void. But at a deeper ontological level there is more to it. More accurately, the emergence of aether is a causeless process. It occurs without any prior cause. It is axiomatic. Nevertheless, the emergence/expansion cannot take place just anywhere in the universe. The cosmic cellular structure determines where the expansion of the medium occurs (even while the structure itself is determined, in large part, by the axiomatic emergence process). Here is what the stress of cosmic tension does: It allows newly emerged aether units to remain extant. It reduces the probability of fundamental fluctuators from vanishing.

The significance of the causeless process, the process of perpetual emergence of the essence medium, may be understood in a philosophical or a metaphysical context. Essentially, it means that if it were possible (which it is not) to isolate a volume of our defined aether, it would gradually expand and grow.

Quantifying the divergent effect

It is assumed that the aether emerges uniformly within Voids. Since there is no reason to believe otherwise, the expansion is considered to be homologous.

There is a very simple way to work out the rate of expansion. All we need to do is construct a graph of the aether flow that may be found across a cosmic cell.

The first step is to establish a horizontal axis —to serve as our axis of location or radial distance. Since opposite nodal galaxy clusters are of key importance, we run the axis through their centers and through the Void that separates them. See **Figure 8-5**. We arrive at a suitable scale be estimating a reasonable cluster-to-cluster distance —this can be the same as the diameter of a cosmic cell, a distance commonly taken to be 350 million lighyears (Mly). This means that the distance from the core of a galaxy cluster to the Void center is 175 Mly.

What about the mass and diameter of each cluster? Based on available astronomical observations, a reasonable mass value is 3×10^{15} solar masses[11]

---

[11] One solar mass is equal to about 2×10^{30} kilograms.

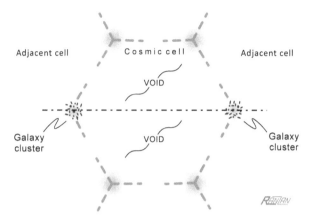

Figure 8-5. Schematic cross-section of a cosmic cell. An axis has been drawn so that it passes through the centers of opposite galaxy clusters and the intervening Void.

and a credible radius is 10 million lighyears.

Next, we graph the aether velocity profiles for the two galaxy clusters. This is accomplished with the aether flow expression $v = -\sqrt{2GM/r}$ (from Appendix A) with the cluster mass M set to 3×10^{15} solar masses (and converted to kilograms). Since the expression does not work for the interior of a cluster, it's a good idea to specify a suitable restriction. The domain of the graph, that is, the useful portion of the radius r, should be set to 10 Mly (as the minimum) and 175 Mly (as the maximum). The minimum refers to the approximate distance from the cluster's center of gravity to the cluster's nominal "surface"; the maximum refers to the distance from the cluster's center of gravity all the way to the center of the Void. See **Figure 8-6a**.

Next, we make use of a simple but important fact as well as the reasonable assumption mentioned earlier. The fact is that *the aether flow at the center of a void is zero*. In other words, the curve we seek must touch the horizontal axis, there, at the Void center. The reasonable assumption is that aether expansion is a uniform (homologous) growth process. Those two factors provide the justification for the following final step. We draw a straight line, as shown in **Figure 8-6b**, through the cosmic cell center-point and tangent to the basic aether-flow curve. The point where they touch indicates where the Newtonian *inflow* EQUALS the expansion *outflow* —the outflow from the Void.

The point of tangency occurs at the radial distance of 58 Mly from the cluster's center. At this tangent point, the aether flow has a magnitude of 1200 kilometers per second. We can now reasonably argue that the tangent point (at 58 Mly) represents the limit of the galaxy cluster's *convergent gravity* region. This means that the extrapolated portion of the Newtonian-gravity graph (between radii 58 and 175 Mly) can be removed, leaving in its place only a simple straight line graph (**Figure 8-6b**).

The linear graph makes it easy to calculate the rate of expansion. All we have to do is interpret the slope of our tangent line. It says, the speed of the aether increases from zero (at the Void center) to 1200 kilometers per second over a distance of 117 Mly. This means, for every 1 million lightyears that the aether moves on its way to the cluster along the distance-axis, the speed increases by 10.3 kilometers per second. (Simply divide 1200 by 117.) This, in turn, means that within the zone of divergence, every 1 million lightyears of distance adds about 10 kilometers of new aether every second.

Stated another way, two test objects placed one kilometer apart within the expanding zone will move away from each other (while comoving with the aether) at a rate of only 3.44 centimeters per million years.

In terms of volume, this means that the universal space medium contained within a one-kilometer-sided cube will expand by 103,000 cubic meters during every million years.

Under the aether paradigm, when we speak of gravity in a fundamental way, we mean the actual acceleration of aether. As I have stated many times, this is the immediate cause of gravity (whether attraction or repulsion). So, if we want to properly define the divergent gravity effect, we need to examine the acceleration aspect.

We start with a velocity graphic. Working with a Void-centered coordinate system, we plot the aether velocity function (the linear portion from **Figure 8-6b**). This time the velocity of the flow is positive (being in the outward direction from the origin placed at the Void center). See **Figure 8-7**. The velocity is a linear function of the radius —referenced to the center of expansion.

$$\upsilon(r)_{\text{outflow}} = (10.3 \ km/s/Mly) \ r, \qquad (12)$$

where r is expressed in million-lighyear units.

The rate of change of this velocity, the acceleration of the flow, is obtained by extracting the derivative with respect to time.[12] The result is another linear function of the radius.

---

[12] Divergent acceleration equals $\dfrac{d\upsilon(r)_{\text{outflow}}}{dt} = \dfrac{d\upsilon(r)}{dr} \dfrac{dr}{dt}$,

$$= \dfrac{d(10.3 km/s/Mly)r}{dr}(\upsilon(r)_{\text{outflow}}),$$

$= (10.3 \ km/s/Mly) \ ((10.3 \ km/s/Mly) \ r)$.

Thus, $\qquad a(r)_{\text{outward}} = +(10.3 \ km/s/Mly)^2 \ r$.

Note, the radius here is the distance from the Void center (NOT from the cluster center).

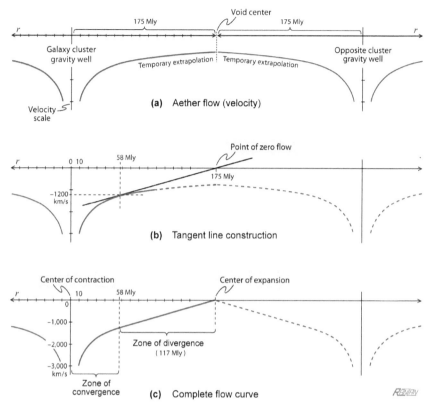

Figure 8-6. Graphic method for finding the aether-flow curve and the expansion rate associated with divergent gravitation. (a) The basic convergent-flow curves for two representative galaxy clusters located on opposite sides of a cosmic cell have been extended to meet at the Void center. The graph is based on the velocity flow equation from Appendix A applied to a cluster mass equivalent to 3×10^{15} Suns. (b) A tangent line is drawn so that it passes through the Void center point, where aether flow is known to be zero. The magnitude of the slope of this tangent gives the rate of expansion. (c) The complete aether-flow curve consists of the contractile portion and the expansion portion. The expansion portion is associated with the *divergent gravity* effect. Velocities have been algebraically referenced to the center of mass. (For a Void-centered coordinate system, the curves would simply be mirror reflected above the horizontal axis. The comoving velocities would then all be positive.)

$$a(r)_{\text{outward}} = (10.3 \ km/s/Mly)^2 \ r. \qquad (13)$$

Here again, r is with respect to the Void center and is expressed in millionlighyear units.

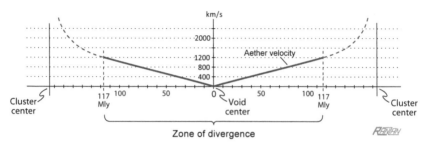

Figure 8-7. Graph of positive aether flow within the zone of divergent gravity. The origin marks the Void center, which of course is the center of expansion. The linear rate of aether expansion determines the slope of the graph —previously found to be 10.3 km/s/Mly. Expansion ends at approximately 117 million lightyears, after which the aether undergoes contraction.

There is an easy way to check the graphic method used to determine the acceleration expression (and the divergent velocity from which it was derived). At the point of tangency (**Figure 8-6b**), the acceleration of aether due to the *convergent* gravity of the cluster must be equal to the acceleration of aether due to the *divergent* gravity of the Void (calculated along the axis joining the two centers). The convergent gravitational acceleration, at the 58 Mly r-location, works out to be 1.3 picometers per second per second (1.3×10^{-12} m/s^2).[13] The divergent gravity expression, at 117 Mly from the Void center, gives the same acceleration 1.3 picometers. (The arduous part is making the proper conversion of the various units.)

A general definition. Divergent gravity is the consequence of aether —by virtue of its axiomatic nature and of being subjected to cosmic tension— expanding/growing in volume. This expansion causes the aether to accelerate in accordance with the general expression,

Acceleration$_{outward}$ = (constant expansion rate)^2 × (distance from void center).

A few related things worth noting:
- The expansion rate (within Voids) that agrees with a wealth of astronomical observations is surprisingly small. It just takes aether expanding in volume at a fractional rate of 0.0001 per million years. This is a volume growth of just 0.01 percent during 1,000,000 years. In

---

[13] Based on the equation $a(r) = -GM/r^2$, which is the derivative of the inflow expression $v(r) = -\sqrt{2GM/r}$. (G is the Newtonian gravitational constant and M is the mass of the galaxy cluster.)

- somewhat more easily understood terms, all that is required is for each kilometer of aether to "grow" by 3.44 centimeters every one million years.[14]
- What greatly simplified our graphic analysis was the use of a speed-of-expansion function that is actually a linear expression. Things would have been quite different if we had used time-dependent functions. The same motion, velocity and acceleration, *when expressed with respect to time*, are in fact exponential; but the motion equations with respect to radial distance are **linear**. Fortuitously, this feature allowed for an intuitive graphical analysis of the relationship between convergent and divergent gravity —an otherwise challenging association (made more so by the fact that the actual gravity domains are not spherical).[15]
- An interesting question. How long would it take to "drift" from the Void center all the way to the edge of the galaxy cluster's gravity well? — drifting with the expanding aether? ... The time-dependent function, is $r(t) = r_o e^{Xt}$, where r_o is the initial distance (say from somewhere near the Void center) and X is the expansion rate.[16] Solving it for time t and applying appropriate unit conversions, one finds that it takes **138 gigayears** (138,000,000,000 years) to drift from a starting position 1.0 Mly from the Void center out to the 117 Mly position, the "tangency point" 58 Mly from the cluster center.
- An overlooked steady state quantitative balance. We learned from Chapter 7 that nodal galaxy clusters are surrounded by either four or six Voids. It's always one or the other. The Voids supply all the aether "fluid" that the clusters consume. This harmony of opposites was something not recognized by the community of theory unifiers of the 20th century —and therefore, entirely missed was its profound cosmological implication. The quantitative balance between emergence (expansion) of aether in the divergence regions and the primary- and self- extinction in the convergence regions, is the reason why *The Universe* does not expand. There is no net expansion.

Vorticular effects

There are two separate effects to be discussed here. Both relate to structure rotation. One is the aether dragging effect, the other is the vorticular stress

---

[14] C. Ranzan, *DSSU Validated by Redshift Theory and Structural Evidence*, Physics Essays Vol.**28**, No.4, pp455-473 (2015). (Doi: http://dx.doi.org/10.4006/0836-1398-28.4.455)

[15] C. Ranzan, *The Nature of Gravity –How one factor unifies gravity's convergent, divergent, vortex, and wave effects*, International Journal of Astrophysics and Space Science Vol.**6**, No.5, 2018, pp.73-92. (Doi: http://dx.doi.org/10.11648/j.ijass.20180605.11)

[16] Ibid.

effect. In the context of the aether theory of gravity, both are gravity boosters, that is, they act as gravitational amplification effects.

Aether dragging effect

When a gravitating body rotates, it causes the inflowing aether to be dragged along with the rotation. This, in turn, can cause a reduction in the centrifugal effect. Why aether drag can lead to a reduction in centrifugation was explained in Chapter 5. Following a brief review, we will examine how centrifugal effect reduction contributes to gravitational amplification.

An expedient way to discern the relationship between the centrifugal "force" and the aether drag is with a thought experiment. The idea is to compare the various velocity vectors of a test object and of the aether —and identify the key vector that determines the intensity of the centrifugal phenomenon.

We'll need a gravitating body. Other than possessing rotation, the body is assumed (as usual) to be at-rest within the universal medium.

Relevant facts to keep in mind:
- The speed of aether flow is something that is not subject to *special relativity*; however, the inflow component perpendicular to the gravitating body's surface most definitely *is*. It follows that the perpendicular inflow speed cannot exceed lightspeed.
- Under the rules of conventional 20th-century physics, the centrifugal effect is directly proportional to the speed of motion (about the axis of rotation) at any selected radial distance from the axis.
- However, in real-world physics where gravity is an aether phenomenon, the centrifugal effect depends strictly on the velocity *through* aether.
- Background container-space will again serve as a useful conceptual frame of reference.

Thought experiment procedure:
- Our approach involves sequentially and significantly increasing the gravitating body's mass content —BUT without changing its physical size and without changing its rate of rotation. In other words, only the density increases while the radius and the equatorial speed of rotation are held constant. (Within a non-aether theory, this would imply a constant centrifugal force. Using the stated conditions, the Old Physics predicts that there will be no change in the centrifugal effect. Such prediction, however, does not match reality.)
- What about the angular momentum? As the material is deposited, there will, of course, be an increase in rotational energy. However, if the rate of rotation is maintained (imagine the mass being added in such a way so that there is to be no change in the rotation rate), the change in angular momentum is not important. Thus, the angular momentum is of no concern to the experiment.

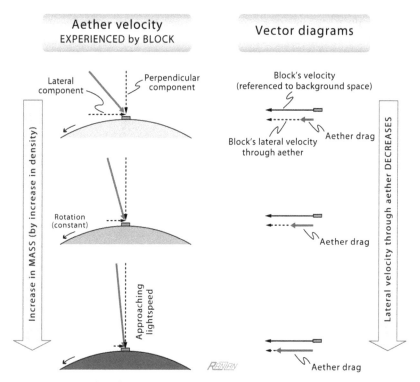

Figure 8-8. Thought experiment examination of the relationship between rotational motion *through* aether and aether drag —and the implications for the centrifugal effect. The procedural sequence involves significantly increasing the gravitating body's mass content —without changing its physical size and without changing its rate of rotation. Consequently: (i) the quantity of aether flowing in must increase. (ii) The test block experiences an increase in perpendicular aether flow and a decrease in lateral flow. (iii) The test block's velocity with respect to background space is equal to the sum of its lateral velocity through aether and the aether drag (the lateral velocity of the aether at the same location). (iv) Since the centrifugal effect depends on the lateral motion/velocity through aether (and not on the motion through background space, which motion remains constant throughout the experiment), the block experiences a diminishment of the centrifugal effect.

The changes we expect to occur are as follow (**Figure 8-8**):
- In order to sustain the existence of the additional mass, the quantity of aether flow must increase. Since the total surface area remains unchanged, there has to be an increase in the magnitude of the aether velocity. There is no alternative.
- The surface, including the test block, experiences an increase in perpendicular aether flow; and most significantly, a decrease in lateral flow.

- Although the block's motion through background space remains constant throughout the experiment, there is a substantial change in its lateral speed through aether.
- The test block maintains its constant background circular motion by partly cutting through aether and partly dragging along with aether. As the drag component increases, the other component decreases (that is, the lateral through-aether component decreases).

The key observation is that the aether drag causes a reduction in the magnitude of the rotational velocity *through* aether. From the snapshots as the experiment progresses, we see that the block's tangential vector diminishes (its lateral speed through aether decreases). And, by definition, this means a decrease in the centrifugal effect.

Why is all this considered an amplification effect? Aether drag makes the rotational motion through aether less than what it would otherwise be. Motion through aether is what is restricted by special relativity; and rotational motion through aether is what determines the intensity of the centrifugal effect. And so, an increase in aether drag causes a reduction in the centrifugal effect. Finally, because a reduction in the centrifugal effect simply means that the mass body can tolerate a higher spin rate before overcoming gravity and flying apart, it follows that aether drag increases the intensity of gravity. Simply put, aether drag causes mass to become more tightly bound together —it acts as an amplifier of gravity.

The rotational drag effect is summarized in **Table 8-3**, which includes a comparison with the general view for non-aether theories.

Table 8-3. Comparison of the rotational drag effect as interpreted by aether and non-aether theories. In non-aether gravity theories, it is called the frame-dragging effect, but is not associated with an increase in the intensity of gravity.

Aspects of structure rotation	Aether drag (Aether vortex)	Frame drag (non-aether theories)
Centrifuge mechanism:	• Reduction in centrifugal effect • In the extreme case, a total cancellation	No reduction
Rotation (and drag) limit:	No limits for Superneutron Stars	Equatorial speed limited to the lightspeed constant
Implication for gravity:	Tends to increase the intensity of gravity	No increase in intensity

The next experiment examines the situation of a planet in circular orbit around a white-dwarf star (very massive, very dense). Two configurations are considered and subjected to analysis: In one, the star is *not* rotating and the

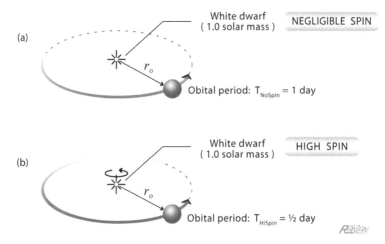

Figure 8-9. For the negligible-spin configuration (a), Newtonian and Keplerian equations work admirably. From the details of the orbit, the mass of the star can be calculated. Not so for configuration (b). When the star spins, thereby generating an aether vortex, conventional methods allow only the calculation of the *apparent* mass. High spin rates can significantly change the orbital period and radically increase the *apparent* mass (compared to the actual mass value). As described in the text, the calculated apparent mass in part (b) turns out to be four times the actual mass.

planet's orbital period is one whole day. In the other configuration, an identical dwarf star is rapidly rotating and the planet's orbital period happens to be only one-half day. Essentially, the central mass and the orbital radius are the same for both situations —only the star's spin and the orbital period are different, both properties being hypothetically observable. See **Figure 8-9**.

In order to make a meaningful comparison of the two configurations, we need a formula for the central mass M expressed in terms of orbital radius r and orbital period T of a small planetary companion. The required mass formula is, $M = \dfrac{4\pi^2 r^3}{G \cdot T^2}$. For the simple derivation, see Appendix D. For the situation shown in **Figure 8-9a**, this can be expressed as

$$M_{NoSpin} = \frac{4\pi^2}{G} \frac{r_o^3}{T_{NoSpin}^2}. \tag{14}$$

This allows us to accurately determine the star's mass, provided the axial rotation is negligible.

However, when a combination of large mass and high spin generates an aether vortex, the above formula can give significantly inaccurate results. It will not work for the "HiSpin" star shown in **Figure 8-9b**. Of course, we already know that M_{NoSpin} equals M_{HiSpin}. We did, after all, specify that the two stars have identical mass in both configurations. But if we did not know this, there would be no way to calculate the mass with the conventional equations of gravity. Only the star's *apparent* mass value can be explicated in this way.

So, for situation (b) in **Figure 8-9**, we write

$$M_{\text{apparent}} = \frac{4\pi^2}{G} \frac{r_o^3}{T_{\text{HiSpin}}^2}. \tag{15}$$

By combining the last two equations (dividing one into the other) and canceling the shared terms, we end up with a simple ratio,

$$\frac{M_{\text{apparent}}}{M_{\text{NoSpin}}} = \left(\frac{T_{\text{NoSpin}}^2}{T_{\text{HiSpin}}^2} \right). \tag{16}$$

We solve this for the two situations shown in **Figure 8-9**. The orbital period for the *NoSpin* state is 1.0 day; and for the *HiSpin* state it is 0.5 day.

$$\frac{M_{\text{apparent}}}{M_{\text{NoSpin}}} = \left(\frac{1 \text{ day}}{\tfrac{1}{2} \text{ day}} \right)^2 = 4.0. \tag{17}$$

Based on the hypothetical orbital data, the apparent mass is 4 times as great as it would be in the absence of rotation. If for example we assume a mass for the white dwarf equivalent to 1.0 solar mass, then the Newtonian law of gravity will predict an *apparent* mass-value of 4.0 Suns.

In other words, Keplerian or Newtonian calculations tell us that the spinning solar-mass white dwarf will have the same gravitational footprint as a stationary four-solar-mass body. Vorticular drag can have a powerful gravitational amplifying effect indeed.

The thing to note is that the orbital periods astronomers may observe says nothing about how much of the motion is *through* aether and how much is *with* aether. If the central body's rotation rate and associated drag effect are unknown, then knowing the orbital period or the apparent orbital velocity cannot be trusted to give the actual mass value. (Nor can the centrifugal effect be accurately determined.) Moreover, all this applies not only to contiguous bodies but also to large-scale gravitating structures.

It is this unknown aspect of vorticular motion that manifests as the biggest problem in the analysis of spiral galaxies. The difficulty is with interpreting the observed rotation curves, the rate at which stars orbit their galactic core. The orbital-speed data suggest there should be much more mass present than is actually observable. In the words of science writer Tim Folger, "*In every single [rotating] galaxy ever studied, the stars and gas move faster than*

Newton's laws say they should, as if gravity from a hidden mass in ... the galaxy were yanking them along, boosting their speed."[17] Instead of recognizing the gravity amplifying aspect of the aether-vorticular effect, astronomers have reified what used to be called "missing mass" by concocting *Dark Matter*. In an effort to match the quantity of gravitating mass that should be there according to Newtonian gravity, the star-stuff accounting ledgers are made to balance by cooking up huge amounts of mysterious matter.

The vorticular effect was never properly understood by the gravity experts of the 20th century. They missed the important underlying aspects of frame dragging. When an astronomical object, or structure, is surrounded by aether in a state of vortex motion, the potency of its gravity increases in proportion to the speed of the vortex. The spiral inflow of aether into the structure acts as a gravitational amplifier making it seem as if more mass is present than is actually the case.

Centrifugal effect cancellation

Turning now to what is probably the most fascinating feature of the aether theory of gravity. The intensity of gravity, and speed of aether inflow, can attain a maximal point where there is a total cancellation of the centrifugal effect.

The most fundamental condition for the manifestation of the centrifugal effect was famously demonstrated by Isaac Newton (1643-1727) with his simple but ingenious bucket experiments. As we learned in Chapter 5, his rotating and swinging water-filled buckets established the fact that the effect only manifests when there is rotation relative to some undefined background. Clearly, there was something special about the surrounding world. Motion relative to this "something" imparted an absoluteness quality to the motion and bestowed centrifugal forces.

Ernst Mach (1838-1916) reaffirmed the surrounding-world idea, but a true causal mechanism failed him. The centrifugal force is produced only if the rotation is relative to the surrounding universe; in his words, "relative to the fixed stars." He stated, "For me only relative motion exists ... When a body rotates relative to the fixed stars centrifugal forces are produced; when it rotates relative to some different body and not relative to the fixed stars, no centrifugal forces are produced."[18] Clearly, for Mach, absolute motion only had meaning in the sense of being relative to the universe as a whole.

Consider the obvious implication of Ernst Mach's claim. Assume a body

---

[17] T. Folger, *Nailing Down Gravity*, Discover, (2003 October) p36.
[18] E. Mach, *The Science of Mechanics* (Open Court, LaSalle, Illinois, 1942), as in E. R. Harrison, *Cosmology, The Science of the Universe*, 1981.

rotates; large or small, it does not matter. If the universe were to rotate around the body, *at the same rate* about the same axis, then there would be no centrifugal effect. There would be no rotation "relative to the fixed stars"; and, it follows, there would be no outward tendency from the rotation axis. The rotation would be undetectable, even meaningless.

Ernst Mach's assumption that the universe-as-a-whole is the special determining factor was wrong; and yet, the argument of centrifugal cancellation was correct. Mach and his contemporaries had overlooked the essential intermediary element.

In the modern view, there is a new determining factor, one that changes the way matter relates to the rest of the universe. It is not the distant stars that are important but, rather, the evanescent medium between the stars and between all bodies (and all particles). A body's entire "sensory" connection with the surrounding universe is by way of the universal space medium —the aether that empowers gravity. The entire universe need not rotate in lockstep for centrifugal cancellation to take place. Only the body's local universe, the surrounding aether, needs to rotate in sync, in order to produce the same negating effect. If aether and body were to rotate tightly together, the body would "believe" itself to be stationary. It would manifest no centrifugal effects.

Something equivalent to this arises with a type of collapsed star we studied earlier. Total centrifugal cancellation occurs with end-stage collapsed stars — stars of degenerate matter, stars that cannot undergo further collapse, not by the mechanism of gravity and not by any other mechanism. We called this unique type of object a Terminal star. Its total mass and radius are fixed; only the rate of rotation and polar emission may vary. Its most relevant aspect as far as the centrifugal effect is concerned is that it has a critical-state surface, meaning that the aether inflow at this surface equals the speed of light. Recall, a Terminal star has a no-escape boundary (except at the poles). And remember this important distinction: it has no empty region within —it is NOT a black hole, although it may look like one.

An enclosing lightspeed boundary —an unique situation for contiguous mass— means there is almost no way for mass or radiation to escape from a Terminal star. Wherever the boundary is active (which is everywhere except at the magnetic poles), there can be no centrifugal effect. Regardless of the structure's spin rate, the effect is negated totally. See **Figure 8-10**.

With the Terminal star, the intensity of gravity attains its ultimate manifestation. Nothing in nature equals or exceeds its surface gravitational acceleration.

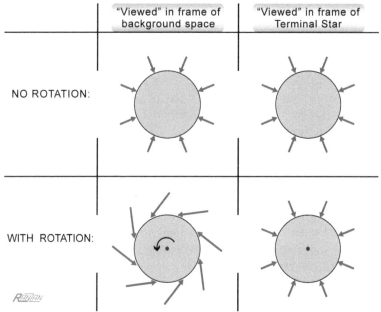

Figure 8-10. Unique in its ability to cancel centrifugal effects, the Terminal star is shown from two perspectives. The first row shows a nonrotating Terminal star along with its aether flow vectors, from the perspective of the background-space frame and also in the star's own reference frame. The second row shows a rotating Terminal star with its aether flow, again from the perspective of the background frame and also in the star's own frame. From the perspective of the Terminal star, the direction of the aether streaming onto its surface is *always perpendicular to the surface* —regardless of rotation. The space medium inflow is perpendicular to the surface exactly as it would be if the body were *not* rotating. Hence, it is a situation of *rotation with no centrifugal effect whatsoever*.

Vorticular stress effect

The universal medium interacts with the observable world via stresses placed upon it. In the context of the aether theory of gravity, this means that the medium becomes dynamic[19] in response to the stresses of the material world; then, in its dynamic state, the aether interacts with the material world. Simply put, the physical world causes aether to expand, to flow, to contract; and the aether, in turn, influences the arrangement of the physical astronomical realm. As detailed earlier, the stress of mass (the stress that mass itself produces) causes primary gravitational acceleration; the stress of convergent flow causes additional gravitational acceleration; and the stress of cosmic

---

[19] Aether reveals its dynamic aspect when it expands or contracts.

tension causes the gravitational acceleration to be extended beyond the Newtonian domain (**Figure 8-6**). In addition to the compression and tension stresses, there is also a shear stress. When aether is dragged into vorticular motion it suffers shear stress, or what is sometimes called a torsion stress.

Aether is a fluid, albeit a nonmaterial one. It is the nature of fluids that when undergoing vorticular or torsional motion there will be laminar flow — layer upon layer each moving with slightly different speed. For an analogy, just look at the shear disturbances generated in the atmosphere of the planet Jupiter (witness the shear turbulence produced by the rotational differences in latitude). When one layer of the aether fluid is made to slide over another, each will exert an action against the other in opposition to the relative movement — that is, a tangential force-like disturbance will act against the sliding motion. It is this tangential disturbance that is identified with the shear stress suffered by the aether. Another attribute of the laminar flow is that the stress is proportional to the speed with which the successive layers slide over each other. The faster a gravitating body rotates, the greater is the shear stress. And, as is the nature of aether, it responds to the stress by intensifying its self-dissipation.

Thus, there is an augmentation of the basic self-dissipation —the contraction caused by convergent-flow stress (compression stress). This is supplemented with the dissipation caused by torsional-flow stress (the shear stress). Although it is being called a shear stress, it is, on the microscopic scale, really not much different than a compression stress —a condition whereby aether units are being pushed together.

Before explaining how shear stress affects orbital motion, it is worth pointing out that the previously described vortex-drag effect does not involve an energy change, that is, it does not entail any additional change in the self-dissipation of aether; however, the vorticular stress effect does. Aether dragging, in and of itself, does not involve a change in energy at the fundamental level; but the shear stress that accompanies it does.

The shear-stress-induced additional self-dissipation of aether causes an intensification of gravity and this, in turn, causes otherwise circular and elliptical orbits to become inward-spiral orbits. It is this added boost to gravity that causes the inward spiral of stars in rotating systems and the merger spirals of close binaries.

The classic example of this is the exceptional neutron binary first observed in 1974. The pair has a mutual orbit that is about the same as the diameter of our Sun and has an orbital period of only 7.75 hours. Precise orbital timing measurements were made possible because one of the neutron stars, named PSR 1913+16, is a pulsar. Over many years since the discovery in 1974 (for which Joseph H. Taylor, Jr. and Russell A. Hulse were awarded the 1993 Nobel Prize in Physics), the stars in this binary have been observed spiraling inward at an increasing rate. The mutual orbital period has been shortening as the separation distance has been diminishing.[20] "Taylor and his colleagues have been able to follow the evershortening [sic] separation of the two stars and the ever higher speed they attain as they slowly spiral in toward an ultimate catastrophe some 400 million years from now."[21] Assuming that the significant spin is being generated solely by the PSR 1913+16 pulsar, then it may be said that the pulsar is inducing the shear stress and is therefore responsible for the spiral orbit. If its partner also has a high spin rate (which for some reason is not detectable), then it will share the responsibility.

Binary systems, it should be mentioned, are responsible for producing another gravity effect —gravity waves. We'll discuss these in a moment.

Let me add these little-known historical notes on torsion stress. The discovery of torsion is generally credited to research done by Russian professor N. P. Myshkin (with the physical-chemical society at the time) in the late 1800s. He called it the "fifth force."[22] Years later in 1913, the historical record notes that the first theoretical works devoted to a theory of gravitation containing a force relating to a twisting movement through the fabric of spacetime was performed by Dr Eli Cartan. He termed this force *torsion*. Cartan's gravitation theory, however, never obtained support.[23] Although Eli Cartan was a colleague of Albert Einstein, the torsion idea did not catch on; and so, Einstein's gravitational theory turned out to be free of torsion.[24] Even though General Relativity describes gravity as a curvature of space resulting from stress induced by various forms of energy, torsional stress is missing. It stands as another example of the incompleteness of the theory.

Recapping: Both the vorticular-drag effect and the vorticular-stress effect tend to amplify the intensity of the gravity of the rotating body or structure; they just do it in different ways. One does it by weakening the centrifugal tendency; the other does it by intensifying the aether self-dissipation. Both are clearly predicted by the aether theory of gravity.

Non-aether gravity theories make no such predictions. There are three main reasons? (i) Rotation speed. While the aether theory sets the rotation speed limit of mass, with respect to the aether medium, which itself partakes in

---

[20] Malcolm S. Longair, Chap.6 "An Introduction to Relativistic Gravity," in *Galaxy Formation 2nd Ed* (Springer, Berlin, 2008); p192.

[21] J. A. Wheeler, *A Journey into Gravity and Spacetime*, Scientific American Library (W. H. Freeman & Co., New York, 1990); p205.

[22] Yu. V. Nachalov, *Theoretical Basics of Experimental Phenomena*, Web-article: http://www.rexresearch.com/torsion/torsion2.htm (2018-8-14).

[23] Yu. V. Nachalov and A. N. Sokolov, *Experimental investigation of new long-range actions*, Web-article: http://amasci.com/freenrg/tors/doc17.html (2018-8-14).

[24] H. Kleinert, *Gauge Fields in Condensed Matter* (World Scientific, 1990).

the rotation to varying degrees; the non-aether theories set the rotation speed limit with respect to undefined space. In other words, the aether theory has virtually no rotational speed limit with respect to background space; non-aether theories, on the other hand, are strictly limited to the speed of light. (ii) Stress induced contraction. Without aether, there obviously can be no space-medium self-dissipation. And without stress-induced vanishment, there can be no torsional stress effect. General Relativity, notably, has no torsional property pertaining to the "distortion" of its space. (iii) The most fundamental reason —aether is absolutely essential. No realistic models for gravity (or the universe itself) are possible without incorporating a subquantum process of energy. This requires a subquantum medium. It requires the aether we have been working with. But as things stand, the conventional wisdom has no such medium and no conception of a subquantum process of energy —particularly, no process for aether self-dissipation.

Gravity waves

An axiomatic property of DSSU aether is that it strives to maintain a constant spacing density of its constituent subquantum entities. But it is not an absolutely inflexible constant density. Some very small deviations do occur. For instance, when aether is stressed towards a greater-than-normal spacing density, the self-dissipation process becomes active and maintains the spacing in conformity within a narrow tolerance range. The greater the stress, the greater is the quantitative vanishment of aether units. On the other hand, when aether is stressed towards a lower-than-normal spacing density, there will be an emergence of new aether units in an ongoing effort to maintain the normal spacing value. This is what happens in the cosmic Voids where new aether emerges in response to the tension stress induced by the surrounding galaxy clusters.

As long as there is this tolerance range in the spacing density, it follows that the universal medium has the ability to convey waves of compression and rarefaction. Although these are waves in a nonmaterial medium, they are analogous to those of a material medium. And in this sense, they are the acoustic waves of our universe; and are, no doubt, extraordinarily weak.

The two mechanisms for the generation of these acoustic-like waves: One involves the implosion and explosion of a star. This is, more or less, a one-time event that generates spherically propagating waves of compression and rarefaction. The other mechanism requires orbital motion, most commonly a close-binary system. The orbital motion produces outwardly spiraling acoustic waves in aether that are most intense in the orbital plane of the source and diminish to irrelevance along the axis of revolution.

Gravity waves are longitudinal waves of compression and rarefaction in the aether medium itself. Continuous wave patterns are generated by

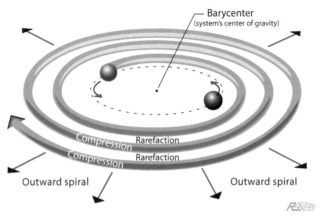

Figure 8-11. Orbital motion generates acoustic-like waves in the universal space medium. These gravity waves propagate outward as a spiral pattern of compression and rarefaction (of the aether). The waves, which presumably travel at the speed of light, are most intense in the plane of the orbital system and diminish to zero at the axis of revolution. (For the sake of clarity, only the waves generated by one of the stars are shown.)

gravitating bodies in orbital motion (**Figure 8-11**). Being the weakest of all the gravity effects, gravity waves require significantly massive binaries with extreme rates of barycentric spin, or mutual orbital motion. For the waves to carry energy and have any effect on the immediate surroundings, the masses need to be neutron stars or supermassive structures.

Aether gravity theory predicts that these waves have a limited ability to propagate energy. Gravity waves carry energy only to the extent that there is aether self-dissipation within the compression phase of the wave. This is an inviolate requirement of the DSSU fundamental definition of energy. If the waves become so weak that there is no loss of aether, then the gravity wave can no longer be said to carry energy. If the density of the compression phase lies within the aether's spacing tolerance, then there will be no aether vanishment —meaning, no energy manifestation. This, however, does not in itself mean that the waves have become undetectable. Nor does it mean that the waves stop propagating. It simply means that if such waves are detected, then it would represent an observation at the subquantum level of existence. It would be the detection of the acoustics of our subquantum medium.

In 2015, on September 14th, two highly sensitive Earth-based interferometers, known as the Advanced LIGO gravity wave detectors, recorded the signature signals of the merger of two large-mass objects, naïvely interpreted as black holes. One had a calculated mass equivalence of 36 Suns and the other 29 Suns, as determined from a computer simulation of the merger Event. Researchers necessarily turned to the simulation, which provided an idealized reconstruction of the many high-speed cycles of the spiral orbits, because only a small part of the merger was actually detected. There simply was not enough useable data.

It was reported in the journal *Nature* that "LIGO saw only just over one cycle of the Event's ringdown waves before the signal became buried once more in the background noise ..." The hope is for "LIGO [to] detect black-hole mergers that are larger than this one, or that occur closer to Earth than the Event's estimated distance of 1.3 billion lightyears, and thus give 'louder' waves that stay above the noise for longer."[25]

Following the initial 2016-February publication of the Event, there have been a number of additional reported gravity wave observances, including one supplemented with optical data. As the evidence mounts and its quality improves, it seems reasonably safe to assume the phenomenon is real.

Given the vast distance of the source at 1,300,000,000 lightyears; and the diminishment of the effect by the inverse square rule over that great distance; and the extraordinary weakness of the waves in the first place; and the relative quickness with which the compression wave would weaken and settle within the stress-tolerance range; it seems highly probable that the detected waves were not gravitational *energy* waves but rather *subquantum-level* gravity waves.

The 2015 Advanced LIGO event may very well have been the first direct detection of a subquantum non-energy phenomenon —the reception of a signal outside the normal domain of energy. The source event, the spiral merger, was an energy generating/transmitting phenomenon; however, during the transmission of the signal (the gravity waves), its intensity fell below the range of definable energy. From that point onward, the waves' compression phase no longer induced any self-dissipation and, therefore, by definition, the gravity waves carried no energy. Nevertheless, the remnant gravity waves continue to propagate and, if the instrumentation is sufficiently sensitive, may be detected at the subquantum level.

Rotating systems may generate gravitational energy waves, but what our instruments detect are merely the ghostly non-energy remnants.

Key points regarding gravity waves as they pertain to the aether theory of gravity

First and foremost, they are spatial-density waves of the nonmaterial aether generated by orbital motion and by cataclysmic nova-type events. Gravity waves are longitudinal waves; they propagate as phases of compression and rarefaction in the aether medium itself.

Second, they "carry" energy only to the extent to which the process of

---

[25] D. Castelvecchi, *The Next Wave*, Nature Vol.**531** (2016 March); p431.

aether self-dissipation accompanies the compression. They are energy waves only while there is a loss of aether units.

Third, energy conveying gravity waves represent a THIRD form of stress-induced self-dissipation of aether. (The other two are convergent stress, and vorticular/torsion stress.)

Fourth, as wave intensity decreases with distance from the source they eventually become gravity waves at the sub-energy level.

Summary of stresses, the unifying factor, and theory subsummation

Stress and strain in gravity theory

According to Einstein's gravity theory based on his spacetime fluid, matter produces a stress in the spacetime region over which it has influence. As any structural engineer knows, stress is accompanied by strain —meaning that there is some form of deformation. The deformation that the spacetime fluid of general relativity theory undergoes comes in the guise of *curvature*. The presence of mass somehow determines, in a systematic way, the numerical values of the spacetime coordinates of a region. Those values represent curvature —a purely geometric concept— and are interpreted as being equivalent to gravity.

Quoting textbook author, Edward R. Harrison, "The Einstein equation of general relativity states that the curvature of spacetime is influenced by matter; or the strain of spacetime is stress produced by matter. ... We can interpret the Einstein equation to mean that curvature is equivalent to gravity."[26]

In the age-old tradition of Pythagoras and Plato, the modern-day curvature numbers, the abstractions of spacetime geometry, are elevated to a status more real than the mass objects and the space medium and its motions.

Mass produces stress. Under the general relativity paradigm (**Figure 8-12**, left-hand column), the resulting strain manifests as spacetime curvature. But in the Dynamic Steady State Universe the strain manifests as space-medium contraction. Moreover, if we consider an entire cosmic gravity cell, then we are justified in saying the strain manifests as either medium contraction *or as medium expansion* —depending on location within a total gravity cell. See **Figure 8-12**, right-hand column.

When the vacuum of outer space (*spacetime* in one theory, *aether* in the other) is undergoing strain, it means there is a self-interaction taking place. Edward Harrison again: "*Self-interaction of space is the essence of general*

---

[26] E. R. Harrison, *Cosmology, the Science of the Universe* (Cambridge University Press, Cambridge, UK, 1981); p169.

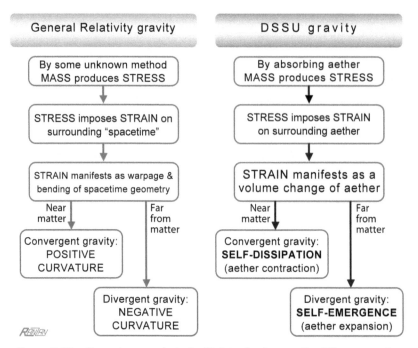

Figure 8-12. How stress and strain fit into fundamentally different gravity theories. General Relativity treats matter as a form of stress that produces geometric strain —the strain of curvature of abstract space. Under this geometric view the strain manifests as either positive or negative curvature. The DSSU aether-gravity theory treats matter as the primary stressor of aether. The resulting strain on the surrounding aether manifests as medium **self-dissipation** (aether contraction) or as medium **self-emergence** (aether expansion).

relativity."[27] It is something the vacuum is doing on its own. While mass imposes stress, the vacuum responds accordingly —according to the respective mechanisms of self-interaction. In the theory Harrison is describing, it is an interaction of geometric curvatures —abstract strain of a mathematical spacetime continuum. In the theory that underlies the DSSU, it is space-medium contraction, the *self-dissipation* of aether; it is also medium expansion, the *self-emergence* of aether.

Summary of Stresses

The following table summarizes the major stresses to which the universal medium is subjected.

Radiation particles (neutrinos and electromagnetic particles) represent the

---

[27] E. R. Harrison, *Masks of the Universe*, 2nd ed. (Cambridge University Press, Cambridge, UK, 2003); p183.

greatest stress "experienced" by aether. When they are confined they become mass —that is, they manifest as mass. Mass represents the most concentrated substantiation of stress in aether. Because of its tendency to aggregate, mass is the primary stressor and therefore the primary cause of gravitation. Aggregated mass causes a secondary stress on the surrounding aether; and on the largest scale, it causes a tertiary stress across cosmic Voids.

Table 8-4. Summary of the major stresses to which the universal medium (aether) is subjected.

Primary stress (Concentrated stress)	Secondary stress (on surrounding aether)	Tertiary stress (across cosmic Voids)	Torsion stress (Shear stress)
↓↓	↓↓	↓↓	↓↓
Manifests as MASS (& energy particles): PRIMARY ABSORBER of aether	Manifests as negative aether strain: SELF-DISSIPATION	Manifests as positive aether strain: SELF-EMERGENCE	Manifests as negative aether strain: More SELF-DISSIPATION
↓↓	↓↓	↓↓	↓↓
PRIMARY CAUSE of GRAVITY	CONVERGENT GRAVITY	DIVERGENT GRAVITY	GRAVITY INTENSIFIER

Rotating mass augments the secondary stress with the additional stress of torsion.

In each and every one of these stress situation there is a volumetric change of aether.

The unifying Factor

The DSSU aether theory of gravity is based on a mechanical (meaning discretized) aether consisting of non-mass, non-energy, fundamental units — entities described as subquantum pulsators. The theory has a unique property not found in any previous model. This property is rooted in the belief that any theory of gravity founded on aether or any other type of "fluid" medium, whether material or evanescent, must incorporate some mechanism for a volume reduction when modeling convergent gravity. DSSU gravity does this by way of a stress-induced process of self-dissipation in combination with a postulate of *constant spatial density*. By the process of self-dissipation, aether maintains an innate spatial density (among its discrete units). Every cubic centimeter of space (space as a 3-dimensional contained) contains about the same number of aether units as any other —regardless of location in the universe.

But the maintenance of aether density goes beyond familiar convergent gravity. It is the determining Factor for explaining how the different stresses manage to produce the various gravity effects listed in **Table 8-5**.

Table 8-5. Aether theory of gravity. The various gravity effects are identified with stresses in the aether and the aether's response to those stresses. One factor acts as a unifier. It cannot be overstated, but the unifying Factor of *spatial density constancy* is the key to the success of our aether-based gravity.

Gravity component	Type of stress involved	Aether response
Convergent effect:	Compression (convergent flow stress)	Vanishment (self-extinction)
Divergent effect:	Tension (cosmic tension)	Emergence (expansion)
Vorticular drag effect:	n/a	Spiral inflow
Vorticular stress effect:	Shear (laminar flow stress) (compression at the microscale)	Vanishment
Wave effect (energy carrying):	Compression in compaction phase	Weak vanishment
Wave effect (no energy):	Subquantum	No vanishment

While for convergent gravity, the Factor causes aether contraction; for divergent gravity, the Factor causes aether expansion. In the case of the vorticular stress, the Factor causes an amplification of gravity (by additional aether contraction). Lastly, the Factor makes possible gravitational energy waves and sub-energy gravity waves (depending on the intensity of the cyclic stress-disturbance of aether).

Such is the nature of gravity where we have this one Factor underlying gravity's convergent, divergent, vortex, and wave effects. It is the key to the unification and the success of our aether-based gravity.

Theory subsummation

Aether-based gravity is a broader theory —a theory able to accomplish considerably more than Newtonian gravity and more than Einstein's gravity. Newtonian gravity does not work for cosmic Voids and spiral galaxies; general relativity fails for spiral galaxies and black holes. And they both fail when it comes to explaining cosmic-scale cellularity. Aether gravity works for all cases.

The DSSU aether theory of gravity is not only a beautiful construction in its own right, but it also reveals how wonderfully simple the workings of Nature can be. All the known gravity effects are brought about by natural simple processes of aether.

It is the antidote for the bafflement-burdening cognoscenti implementing the incomplete, poorly constructed, overly extrapolated, and even plainly wrong, theories of the previous century. Physicist Janna Levin expressed the predicament this way:

> "Is there a theory beyond Einstein's that will avoid the ugliness of infinite singularities; a theory that can handle ... black holes without becoming singular? ... We're inspired by the predictions of relativity to look for an even greater theory, a theory that looks like Newtonian gravity when gravity is weak and looks like general relativity when gravity is strong, but may look entirely different when gravity is strongest." –J. Levin, *How the Universe Got Its Spots*

When gravity is based on aether, it leads to a deeper understanding of the Universe itself. It provides the explanatory details underlying the observable systematic patterns of our Cosmos —the grand structural features of our *steady-state cellular universe*. Its applicability extends to infinity, yet avoids the ugliness of the infinity paradox.

DSSU gravity encompasses a full range of manifestations, from autonomous cosmic gravity domains, down to the maximum compaction and the ultimate gravitational intensity of Terminal stars —all predicted and realized *without becoming singular*.

✠ ✠ ✠

Epilogue

Extraordinary Connectedness of the Grand Design

Aether is the essence of existence. It is the foundation underlying everything that exists. Everything *in* and *of* the Universe is, one way or another, tied to the *essence medium* (DSSU aether).

Aether is the key to the fundamental definition of energy, yet, remarkably, is itself not a form of energy. The aether of our Dynamic Steady State Universe is the first ever dynamic aether consisting of non-energy, non-mass, discrete entities. In other words, the aether has the ability to manifest energy —yet its discrete units (when in the unexcited state) do not! All manifestations of energy *always* involve a quantitative change of the essence medium.

Turning to matter. All matter exists as the excitation of the essence medium. But what makes all the difference in the world is that it is an excitation involving the *consumptive annihilation* of the raw essence. Ontologically, matter may be in the form of "free" lightspeed particles — photons, electromagnetic pulses, and ghostly neutrinos. Or it may be in the form of "bottled" lightspeed particles (self-looping photons) which we recognize as elementary mass particles —electrons, positrons, neutrons, protons, and so on, particles that constitute our physical realm. All radiation and mass particles are sustained by the destructive consumption of the essence medium

This ongoing consumption of aether is then connected to gravity. The consumption is the root cause of gravitation. Aether is the driver of gravity by the very process of sustaining the existence of matter.

Aether axiomatically continually perpetually replaces itself thereby compensating for all the consumptive processes (sustaining the existence of mass being only one of them). No theory of the universe is complete without at least one Causeless process. Without a Primary cause, a theory inevitably falls into an old metaphysical trap of an endless regression of causes —the search for ever more abstract and esoteric initial drivers, in the sense of a first cause. The axiomatic emergence of aether is our necessary Causeless process.

On the grand scale, then, aether emerges/expands and undergoes consumption/contraction. These processes occur concurrently and regionally. They are not components of a cycle. Rather, the pair form a harmony of opposites manifesting on the cosmic scale as vast Voids and immense galaxy clusters. Thus, aether sustains the steady state cosmic cellular structure —as astronomers have long observed.

Aether is the ultimate limiter of the size (in terms of mass content) of any contiguous gravitating structure. It works this seeming magic by means of the *Law of mass extinction/vanishment by aether deprivation* (Chapter 4). The most important mechanism in the grand design, our Universe, is of course gravitation with its five manifestations. And again, all depend on the existence of aether and its unique properties.

How important is the explanation of the gravity effects? It is critically important. As the saying goes, *If your theory of gravity is wrong (or incomplete), then so will be your entire cosmology*. The models of the 20th century totally failed to predict the observable cosmic structure.

Our aether-based cosmology works. We have successfully connected everything of importance. Aether sustains the existence of all mass and radiation. This mode of existence underlies gravitation. Gravitation is responsible for astronomical objects and systems, and for sustaining cosmic structure (Chapter 7).

Understand this:

The connectedness brought about by the existence and workings of aether means that our world is a cellularly structured Dynamic Steady State Universe.

But what is incomparably profound is this:

Remember the two processes that define those most wondrous objects, the Terminal stars? ... Yes, the *velocity differential Blueshifting* of trapped radiation (and subsequent escape) and the *mass extinction by aether deprivation*. Because of those two processes of aether, one depending on its inflow the other on its absence, the entire grand design is sustained in perpetuity! And so, the Universe in its cellular steady state has always existed as such and will always exist as such. *The Universe is perpetual.*

It is in this way —by endlessly, eternally, sustaining the Universe in perpetual existence— the essence medium is the connection to infinity. This recognition I find intensely profound.

Finally, there is the deepest connection of all:

There are two levels of existence: One is the sub-physical domain of aether —this is the domain of aether itself, of its *emergence/expansion*, and its *vanishment/contraction*. The other is the material realm of mass and energy particles. The material realm exists by virtue of that remarkable process by which the Universe's one-and-only fundamental energy particle, the photon, excites/absorbs/consumes the Universe's fundamental sub-physical particle. In other words, the material realm exists because the photonic energy entity constantly "feeds" on the aether fluid. Aether is the ultimate unifier (Chapter 6) *connecting the physical and sub-physical realms.*

In 1998, three years before the DSSU discovery, Edward Osborne Wilson, the brilliant Harvard University professor, wrote a remarkable book about connectedness. It was titled "Consilience, the Unity of Knowledge" and revolved around Wilson's great confidence in the rational scientific approach for solving the deepest of mysteries. Within its pages he proposed that all of existence can be organized and understood in accordance with a few fundamental natural laws. Included was this optimistic prediction:

"Science offers the boldest metaphysics of our age. ... if we dream, press to discover, explain, and dream again, thereby plunging repeatedly into new terrain, the world will somehow [be]come clearer and we will grasp the true strangeness of the universe. And **the strangeness will all prove to be connected and make sense.**"

The dream began in 2001, the discovery was announced in 2002, the terrain was explored repeatedly, the awesome complexity of the landscape was examined piece by isolated piece, year after year the world become clearer, one after another mysteries were resolved, the former strangeness was revealed to be interconnected and found to make perfect sense. A revolutionary new understanding of the Universe had emerged.

The strangeness all proved to be connected and to make sense.

✠ ✠ ✠

Appendices

Appendix A: Basic aether flow velocity equation

Derivation of the equation that relates the velocity of the inflowing aether to the size and mass of an isolated nonrotating gravitating body:

Consider the test mass shown in **Figure A1** resting on the surface of a planetary body. Although seemingly motionless, the object is "experiencing" acceleration —it's called gravitational acceleration. (The object "feels" the familiar gravity effect.) Two accelerations are involved: the platform on which the test mass rests is accelerating it upward; while the inflowing aether is accelerating it downward. The two are perfectly balanced, as evident by the lack of motion (with respect to the surface).

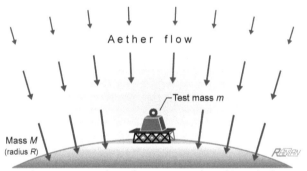

Figure A1. Aether streams and accelerates towards and into the large mass. The stationary test mass "experiences" the inflow acceleration as a gravity effect. And it "experiences" the inflow speed as aether-referenced motion according to the formula $(2GM/r)^{1/2}$. (The large body is assumed to be at rest within the universal medium.)

It is easy enough to prove that the platform on which the mass rests causes the test mass's upward accelerating: just remove all supports and watch the object as it freefalls. To convince yourself that the aether is responsible for the downward acceleration, just place the test mass anywhere into the aether's downward stream and observe its freefall acceleration.

(Recall, matter cannot exist without the presence and absorption of aether, the matter-sustaining essence medium.)

The goal is an expression for the aether velocity; we start, however, with the aether acceleration. We take advantage of the fact that the two accelerations, that of the object in freefall and that of the aether flow, are equal; and also make use of the fact that the acceleration is directly proportional to the mass M of the planetary body and inversely proportional to the square of the distance from the center of the planetary body (this would be R^2 at the body's surface and r^2 farther out). Expressed symbolically

$$a \propto -\frac{M}{r^2}.$$

Add a constant of proportionality to obtain the acceleration equation:

$$a = -(\text{constant}) \times \frac{M}{r^2}. \qquad (1)$$

The negative sign is there to indicate the motion's downward direction. The constant —whose value, 6.67×10^{-11} N m^2/kg, was originally determined experimentally by the Englishman Henry Cavendish back in 1798— is symbolized by G. For any location at, or above, the planet's surface, this acceleration expression describes a body in freefall as well as the aether inflow.

$$a = -G\frac{M}{r^2}, \qquad \text{where } r \geq R. \qquad (2)$$

Replace a with its definition the derivative of velocity, dv/dt, and apply the chain rule:

$$\frac{dv}{dt} = \frac{dv}{dr}\frac{dr}{dt} = -\frac{GM}{r^2}. \qquad (3)$$

Then replace dr/dt with its identity v, rearrange terms, integrate, and solved for the velocity:

$$\int v\,dv = -\int \frac{GM}{r^2}dr, \qquad (4)$$

$$\frac{v^2}{2} = -\frac{GM}{-r} + C. \qquad (5)$$

Now, since the test mass in the figure is stationary, its distance from the center of the planetary body is fixed, the velocity in the equation must therefore be related to the aether. It must be related to the radial inflow of aether. Notice, there are two perspectives here: The aether is streaming *downward* past the test mass; but one could also say, the small mass is traveling *upward* through the aether. Both interpretations are embedded in the equation (and are made explicit in the next set of equations). In order to simplify the equation further, note that when the radial distance is extreme then obviously the aether inflow due specifically to mass M must be virtually zero. (Keep in mind, the assumption is that the large body is comoving with the cosmic background flow; meaning that there is zero relative aether flow.)

Mathematically, this means C in the above equation equals zero. Thus, the basic aether flow equation is

$$v^2 = \frac{2GM}{r} \quad \text{or} \quad v = \pm\sqrt{2GM/r}, \tag{6}$$

where G is the gravitational constant and r is the radial distance (from the center of the mass M) to any position of interest, be it at the surface of M or external to M. The positive solution expresses the upward motion of the test mass *through* the aether (in the positive radial direction). The negative solution represents the *aether flow velocity* (in the negative radial direction) streaming past the test mass.

The negative solution represents a spherically symmetrical inflow field — giving the speed of *inflowing aether* at any radial location specified by r (and M represents the total mass within the radius r).

In vector form: $\vec{v}_{flow} = -\sqrt{2GM/r} \times (\vec{r}_{unit})$. (7)

When a background aether flow is also present, as happens with objects within galaxies, the expression is

$$\vec{v}_{net\,flow} = -\sqrt{2GM/r} \times (\vec{r}_{unit}) + (\vec{v}_{background}). \tag{8}$$

A more detailed analysis of aether flow, in which a second gravitational constant "α" is included, is available in the works of physicist Reginald T. Cahill.[1]

Appendix B: Finding the radius of the Terminal structure

Any contiguous mass structure that has reached the critical state must continue collapsing until halted by the ultimate density barrier. In terms of the graphical representation (see **Figure 4-4**), this means the slope of the linear portion (the part for the interior of the structure) must increase.

Why this is so, is easy to demonstrate.

By inspection of the **Figure 4-4b** graph, the Slope $= \dfrac{v_{surface}}{R_{surface}}$. (1)

Customize the expression from Appendix A so that it applies to the aether inflow at the surface: $v_{surface} = \sqrt{2GM/R_{surface}}$, (2)

Combine equations (1) and (2).

---

[1] R.T. Cahill. *Dynamical 3-Space: Alternative Explanation of the 'Dark Matter Ring'*, Progress in Physics Vol.3, Issue 4, pp13-17 (Oct 2007). Posted at: http://www.ptep-online.com

$$\text{Slope} = \frac{\sqrt{2GM/R_{surface}}}{R_{surface}}. \qquad (3)$$

Next, mass M can be expressed in terms of volume and density to give,

$$\text{Slope} = \frac{\sqrt{2G \cdot \tfrac{4}{3}\pi R_{sur}^3 \rho / R_{sur}}}{R_{sur}} \text{ ; which reduces to}$$

$$\text{Slope} = \frac{v_{surface}}{R_{surface}} = \sqrt{\tfrac{8}{3}\pi G \rho}. \qquad (4)$$

Thus, the slope is proportional solely to density.

For the *Terminal state* situation, $v_{surface}$ is lightspeed c and density is ρ_{max}:

$$\text{Slope}_{end\text{-}state} = \frac{c}{R_{surface}} = \sqrt{\tfrac{8}{3}\pi G \rho_{max}}. \qquad (5)$$

The radius of the final collapsed structure is found by setting c equal to 3×10^8 m/s; and ρ_{max} equal to 1.60×10^{18} kg/m³; and $G = 6.673 \times 10^{-11}$ N·m²/kg²; then solving for $R_{surface}$.

The radius, then, of the Terminal star is: $R_{surface} = 10$ kilometers.

And the linear slope shown in **Figure 4-4c** is $\dfrac{1c/c}{10km}$.

Appendix C: Quantifying the self-dissipation of aether (the cause of the secondary gravitation effect)

The following shows the derivation of expressions of the volume loss of aether during its convergent flow. The premise is that DSSU aether cannot be compressed —specifically in the sense that its density cannot be altered. It can however contract, that is, it can self-dissipate.

We start by constructing a *thin-shell* sphere to enclose a central gravitating body, having mass M, as shown in **Figure A2**. The radius of the shell is held constant. Aether flows into this *elemental* shell with speed v_2, undergoes a certain amount of contraction-dissipation, and then passes through the inner shell wall with speed v_1.

(Change in flowrate) = (flow rate out) – (flowrate in);

$\Delta \text{Vol}_{FLOW} = (v_1 \times \text{Area}_{INNER}) - (v_2 \times \text{Area}_{OUTER}),$ (1)

where the expected negative value will represent a loss of aether (while an unexpected positive would indicate a gain).

We know that the speed of the flow v_1 (at radius r) is equal to $(2GM/r)^{1/2}$, per Appendix A equation (6).

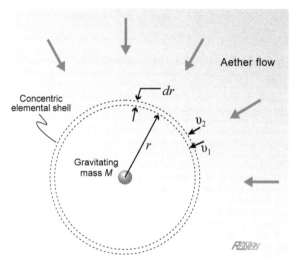

Figure A2. An imaginary thin-shell (shown in cross-section) is constructed concentric with a mass body M. The difference between the speed v_2 of aether entering the shell and the speed v_1 of aether leaving the shell is used to derive an expression for the volume contraction (self-dissipation) of the space medium.

Similarly the speed of the flow v_2 (at radius $r+dr$) is equal to $(2GM/(r+dr))^{1/2}$.

And so, we replace the speeds in equation (1) with $v_1 = C/r^{1/2}$ and $v_2 = C/(r + dr)^{1/2}$, where $C = (2GM)^{1/2}$.

Then, the instantaneous change in the volume flow rate is

$$dV_{FLOW} = \left((C/r^{1/2})\, 4\pi r^2\right) - \left((C/(r + dr)^{1/2})\, 4\pi(r + dr)^2\right), \quad (2)$$
$$= \left(4\pi C\, r^{3/2}\right) - \left(4\pi C\, (r + dr)^{3/2}\right),$$
$$= 4\pi C \left(r^{3/2} - (r + dr)^{3/2}\right),$$
$$= 4\pi C \left(r^{3/2} - r^{3/2}(1 + dr/r)^{3/2}\right),$$
$$= 4\pi C\, r^{3/2} \left(1 - (1 + dr/r)^{3/2}\right), \text{ (since } dr/r \ll 1, \text{ the binomial}$$

theorem approximation applies)

$$\approx 4\pi C\, r^{3/2} \left(1 - (1 + 3dr/2r)\right).$$
$$= -4\pi C (3/2)\, r^{1/2}\, dr,$$
$$dV_{FLOW} = -4\pi (2GM)^{1/2} (3/2)\, r^{1/2}\, dr, \quad (3)$$

where the negative sign confirms a loss of aether volume.

Thus the volume contraction rate within the thin shell of radius r is

$$dV_{LOSS\,RATE} = 6\pi\, (2GMr)^{1/2}\, dr. \quad (4)$$

And the unit contraction rate at distance r, obtained by dividing (4) by *the shell's Euclidean volume $(4\pi r^2 \, dr)$*, is

$$\text{vol}_{\text{unit loss rate}} = 6\pi \, (2GMr)^{1/2} \, dr \, / \, (4\pi \, r^2 \, dr),$$

$$\text{vol}_{\text{unit loss rate}} = \tfrac{3}{2}(2GM)^{1/2} \, r^{-3/2}. \tag{5}$$

This is our equation for **aether self-dissipation** —expressed here as a function of radius r and M (the total mass within the limits of *that* radius). The units are m^3/s per m^3 and are interpreted as cubic meters of aether per second per cubic meter of Euclidean space; or simply as the fractional volume loss per second.

What if the mass body is moving uniformly with respect to aether? In that case, the derivation should still remain valid. A constant bulk flow of aether does not stress the aether; therefore, there is no additional self-dissipation. Such background aether flow (as a constant vector field) can simply be subtracted, leaving behind just the body's own convergent vector field, which actually manifests the aether dynamics —namely, contractile gravity. It does not matter in the least if mass M is traveling uniformly *through* the aether sea.

Appendix D: Central mass expressed in terms of its companion's orbital radius and period

A star having mass M is orbited by a planet whose circular path has a radius r. The orbiting planet is subject to a centripetal force (directed towards the axis of the orbit). For the moment, the cause of this force is of no concern. Force is defined as the quantity of mass multiplied by its acceleration, $F = ma$.

For uniform circular motion, as is the case here, the acceleration magnitude is the square of the orbital velocity divided by the distance to the star's center. Expressed with symbols, $a = v^2/r$.

The centripetal force experienced by the planet, therefore, can be expressed as $F_{\text{centripetal}} = ma = mv^2/r$.

The centripetal force is produced by gravity acting on the planet according to the Newtonian formula, $F_{\text{gravity}} = GMm/r^2$. If gravity is treated as a force (and no other forces are acting on the planet), then logically the centripetal force must be equal to the gravitational force, so that

$$\frac{GMm}{r^2} = \frac{mv^2}{r}.$$

The orbital velocity v can be expressed in terms of the orbital period T:

$$v = \frac{(\text{orbit circumference})}{(\text{orbit period})} = \frac{2\pi r}{T}.$$

Combining the above two equations, gives us

$$\frac{GMm}{r^2} = \frac{m}{r}\frac{4\pi^2 r^2}{T^2},\text{ which simplifies to } M = \frac{4\pi^2 r^3}{G \cdot T^2}.$$

Now, when gravity is based on dynamic aether and significant rotation is involved, this formula takes on a new meaning. It no longer represents the actual mass. It then represents only the *apparent* mass. See Chapter 8, **Figure 8-9**. When vorticular motion of aether is involved (as with high-mass, high-spin bodies), appropriate subscripts should be added, as done here:

$$M_{apparent} = \frac{4\pi^2}{G} \cdot \frac{r_o^3}{T_{HiSpin}^2}.$$

This concept of *apparent mass* serves to link the observable gravity-effect to the unobservable vorticular effect (associated with aether being dragged by rotating mass). As it turns out, the apparent mass can be much greater than the actual mass present.

✠ ✠ ✠

Glossary

absorption lines: The lines appearing in a spectrum, caused by atoms or ions absorbing particular colors (particular frequencies) from a continuous (usually thermal) spectrum. Absorption lines (or bands) of a star are produced when elements, compounds, or ions present in the outer layers of the star absorb radiation from a continuous distribution of wavelengths generated at a lower level in the star. When the light passes through external gas clouds, additional absorption lines may be introduced into the finally observed spectrum. *Also see* emission lines.

accretion disk (gaseous): A disk of gas rotating about a central object. As the gas slowly spirals inward, it releases gravitational energy. If the central object is compact (e.g., a neutron or *Superneutron star*), the accretion disk becomes very hot and luminous.

aether: (1) The original *ether*: In Aristotelian physics, the fifth element, the *quintessence*, of which the 'heavens' are made. In classical physics, the invisible medium that diffuses all space.

(2) The historic: The *material* medium that fills the apparent emptiness of the universe. Invented by René Descartes and by Isaac Newton; reinvented by many others, including James Clerk Maxwell who used it for his electromagnetic theory; but was discredited and discarded by Einstein.

(3) The DSSU aether: The subquantum medium that permeates all space. It is the *nonmaterial essence* of the Universe and consists of discrete units —fundamental fluctuators, or essence oscillators. As a basic space medium, it serves as the propagator of electromagnetic waves. As a space-permeating dynamic medium, it manifests gravitation; its nature is responsible for the several guises of gravity. Aether was detected and verified in at least six separate experiments during the 20th century.

aether deprivation: is an absence of aether. It is simple a chocking-off of aether flow. It is the essential condition whereby matter is extinguished. Totally. Since matter cannot exist without aether, it vanishes. The condition occurs only in the interior of critical-state contiguous mass.

aether deprivation annihilation: a process of total destruction of matter that takes place deep inside extreme mass concentrations. It occurs when mass aggregation reaches a state at which an insufficient quantity of aether reaches the core; and since matter cannot exist in the absence of aether, the aether deficiency results in the *terminal annihilation* of the affected matter. (When a neutron star, for instance, gains too much additional mass, its core will become a region of *terminal annihilation*.)

aether vortex: The inward spiraling motion of bulk aether in response to the rotation of gravitating mass. (Sometimes referred to as frame dragging.) Its importance lies in the fact that it imposes an additional stress on the aether; which in turn acts to amplify the potency of gravity and, therefore, is key to understanding the nature of gravity within spiral galaxies.

acceleration: In reference to the motion of an object, or a medium (such as aether), it means a change in its velocity (a change in either speed or direction).

angular momentum: A measure of an object's tendency to keep rotating and to maintain its orientation. Mathematically it depends on the object's mass M, radius r, and rotational velocity v, and is proportional to Mvr.

annihilation: When applied to particle-antiparticle collisions (or mutual destruction), the term involves the conversion of the particles to pure energy, usually high-energy photons. For example, when an electron and positron annihilate, they produce gamma photons.

antigravity: The cosmic "repulsion" effect produced by the emergence/expansion of the *aether* medium.

antiparticles: The by-products of collisions of particles in high energy interactions (occurring, for example, in terrestrial particle accelerators, and near compact neutron and *Superneutron stars*) and often detected in cosmic rays.

astrophysical jet: A long thin linear feature of bright emission extending from compact structures such neutron and *Superneutron stars*, and also from the rotating nuclei of galaxies.

background space: It is the Universe's 3-dimensional emptiness-space; it has no properties whatsoever and serves merely as a "container" —a repository permeated by a universal essence (a nonmaterial medium) commonly called aether.

barycenter: The center of mass of a system of gravitationally linked astronomical bodies.

baryon: Any massive atomic particle made up of three quarks. Neutrons and protons are baryons.

Big Bang: The popular name for any expansionary model in which an explosion-like event initiated the entire universe. Hot Big Bang: A mythical creation model of the universe which begins at infinitely high density and temperature, expands explosively, and cools to become like the universe we observe now.

big bang: The hypothetical event that, according to some astronomical theories, created the Universe and propelled its expansion. It supposedly occurred about 15 billion years ago.

binary star: Two stars in orbit around each other, held together by their mutual gravity.

black hole (basic): According to the 20th-century view, it is any gravitating object, or region, possessing an *event horizon* (a "surface" from which the *escape velocity* exceeds the speed of light). In terms of *general relativity*, the space around a black hole reaches infinite curvature, and the interior tends to infinite density.

black hole (mathematical): A black hole is a mathematical construction associated with a point mass of some specified magnitude —a point mass called a singularity. Differs from the usual treatment of mass in the following way: In conventional gravitation calculations, the mass body is assumed to merely act as if it were concentrated at a point (its *center of mass*); but for a black hole, the mass supposedly exists, in its entirety, at the point!

black hole (singularity): A black hole for which all of its mass is concentrated at a single central point. It does not exist except as a mathematical object.

boson: A particle with integer spin. Hypothetical carriers/intermediaries of the four forces of nature within the Old Physics model of particles. (Fermions, in contrast, are fundamental particles with half-integer spin.)

cD galaxy: A supergiant elliptical/spherical galaxy found at the center of a galaxy cluster.

celestial sphere: Apparent sphere of the sky; a sphere centered on the Earth observer and having celestial poles aligned with Earth's polar axis. Directions of objects are denoted by *right ascension* (the angle measured eastward along the 24-hour celestial equator from the vernal equinox) and by angular *declination* (above or below the equatorial plane).

cellular universe (DSSU): The Dynamic Steady State Universe is intrinsically partitioned into observable dodecahedral structural cells (delineated by the distribution of galaxies) and, simultaneously, into gravity cells with invisible-but-definable boundaries and ruled by a galaxy cluster at the center. Every node of the dodecahedron cell is the site of a rich galaxy cluster; and every nodal galaxy cluster is the center of an autonomous gravity domain (cosmic gravity cell).

centrifugal effect: *See* centripetal effect.

centripetal effect: A force or effect, such as gravitation, that causes a body to deviate from motion in a straight line to motion along a curved path, the effect being directed toward the center of curvature of the body's motion. The pseudo-force acting against this constraint, that is the effect equal in magnitude but opposite in direction, is the *centrifugal effect*. The centrifugal "force" results from the inertia of all material bodies, their resistance to acceleration, and unlike gravitational or electrical forces, cannot be considered a real force.

compact stars: Very small stars consisting of extremely dense *degenerate* matter. These stars include white dwarfs, brown dwarfs, neutron and superneutron stars.

conservation of angular momentum: The principle in physics stating that the angular momentum of a rotating body remains constant unless forces act to speed it up or slow it down. Stated mathematically, mvr remains constant, where m is a mass element (of the rotating body) moving with a speed v in a circle of radius r. An extremely important consequence of this principle is that when a rotating body shrinks, its rotational velocity must increase.

conservation principle (or conserved quantity): Take any of the basic measurable physical quantities (like mass, charge, spin, position, and velocity) and combine them mathematically according to some fixed formula. If the resulting combination does not change when the measured particles interact, then the lack of change represents a conservation law/principle. Examples are the total energy and total momentum of a system.

contractile gravity: *See* convergent gravity.

convergent gravity: *Convergent or contractile gravity* is the acceleration of aether flow towards, and into, mass bodies (where the local flow converges). It manifests as an *apparent* force of attraction. It is the ordinary gravity of our everyday experience. On the larger scale, convergent/contractile gravity maintains cohesion within each nodal galaxy cluster. *See also* divergent gravity and cosmic gravitational cell.

cosmic cell (cosmic structural cell): The dodecahedral-shaped structural unit of the DSSU. It represents the universe's largest structure.

cosmic evolution: Systematic change of the universe over time (*not* a Darwinian process). The term applies to the various hypothetical Big Bangs. It does *not* apply to DSSU cosmology.

cosmic gravitational cell: The autonomous gravitational domain centered around each nodal galaxy cluster and extending to the *Void* centers of the four (or six) surrounding structural *cosmic cells*. All trajectories within the gravitational cell ultimately end at the central *Supergalaxy*. Essentially, every node of a dodecahedron structural cell is the center of an autonomous gravity domain.

cosmic microwave background (CMB or CMBR)**:** The "light" from ultra-distant stars. This background radiation is starlight that has undergone extreme redshifting because it originated from extraordinary cosmic distances. It is the radiation that gives the Universe its background temperature of 2.7 kelvin degrees.

cosmic redshift: The most important measure for determining cosmic distance of far off galaxies. Twentieth-century cosmology (unlike DSSU cosmology) interpreted the *cosmic redshift* as the evidence of actual receding velocities (of the source galaxies) and then extrapolated this evidence to support the speculation of the expansion of the whole universe.

cosmic redshift cause: In expanding-universe models, the cause is primarily the *expansion of space* (i.e., expansion of the vacuum or *quantum foam*). In the DSSU, the cause is essentially the *velocity-differential spectral shift* that radiation acquires as it transits across gravity wells and particularly as it traverses the great *cosmic gravitation cells*.

cosmic theory: The attempt to explain the existence and nature of the Universe, as well as our own existence and experiences, in terms of observed and unobserved entities and processes.

cosmic Void: The mostly-empty region of the interior of the dodecahedral-shaped cosmic structural cells. It is here that aether emergence/formation occurs.

cosmological constant: (1) Traditionally, it is the multiplicative constant (denoted by Lambda Λ) for a term proportional to the metric in Einstein's *general relativity* equation relating the curvature of space to energy-momentum. When positive it represents space expansion and potentially leads to an acceleration of the expansion of the universe. (2) In modern mytho-science usage it is identified with vacuum energy and dark energy. (3) In *DSSU cosmology* it represents the emergence/expansion of aether, but has nothing to do with Universe expansion. In other words, the *essence medium* expands, but the Universe *does not*.

cosmological principle: A principle of spatial homogeneity that states there are no preferred places in the Universe. Another version states that the universe, on the large scale, is homogeneous and isotropic; that is, uniform in all places and in all directions. Since our Universe is distinctly cellularly structured, it arguably does not conform to this principle.

cosmologist: One who studies *cosmology*, the science of the universe.

cosmology: (1) The general science of the cosmos or material universe, its structures, its composition, and its laws. Combines astronomy, astrophysics, particle physics, and mathematics to assemble the knowledge into a world picture. (2) A particular cosmological theory. The DSSU is a steady-state cellular cosmology; in the chronology of worldviews, it is the 5th *cosmology*.

Cosmos: The Universe as an embodiment of a system of order and harmony. (The term is from the Greek word *kosmos* for order and beauty)

critical-state neutron star: *See* Terminal star and Superneutron star.

critical-state boundary: In the context of gravitational coalescence and collapse, it is the location where the aether inflow attains lightspeed *with respect to the center of gravity*. Theoretically, such a boundary could surround multiple collapsed objects.

critical-state surface: The surface of a gravitationally collapsed star where the aether inflow attains lightspeed *with respect to the body's surface*. The boundary that encloses mass existing in its ultimate density state. *See* Terminal star.

curvature of space: In conventional astrophysics, it refers to the mathematical representation of the distortion of abstract space (spacetime); the 3 types of curvature are spherical (denoting space contraction), flat (Euclidean), and hyperbolic (denoting space expansion). In Natural astrophysics, because space is considered to be merely a 3-dimensional volume of nothingness, the term is meaningless.

dark energy: *See* cosmological constant and vacuum energy.

dark matter: The exotic ingredient required by Big Bang models to explain the rotation curves of galaxies and the structural cohesion of galaxy clusters. Supposedly, a form of matter that does not emit, absorb, or scatter any light. Its only interactions are said to be gravitational. Has never been detected and remains elusive. It is the Achilles heel of the expanding-universe hypothesis. (*Also see* rotation curve/graph.)

degenerate gas/matter: Generated during the final stages of gravitational collapse, degenerate gas is an extremely dense state of matter in which the electrons and nuclei are tightly packed. The pressure of a degenerate gas does not depend on its temperature. All compact stars, from white dwarfs to superneutron stars, are composed of degenerate matter.

density limit: In the context of *total* gravitational collapse, Nature has an ultimate maximum state —taken to be 1.6×10^{18} kilograms per cubic meter. *See also* ultimate density state.

divergent gravity: The acceleration of the aether flow in the cosmic Voids where the dynamic flow has a radially diverging pattern. *Divergent gravity* acts in that portion of the *cosmic gravitational cell* where comoving trajectories (of objects) are diverging away from the centers of the surrounding Voids. It is often called the Lambda force/effect. *See also* convergent gravity.

dodecahedron: A twelve-sided polyhedron. A *regular* dodecahedron has identical pentagonal faces. The *rhombic dodecahedron* is irregular, but symmetrical, and has identical rhombus faces.

Doppler effect (Doppler shift): The change in the measured/observed frequency (and wavelength) of an acoustic or electromagnetic wave due to relative motion of source and observer. With sound the change is in the pitch; with light the change is in color. Named after the 19th-century physicist credited with its discovery.

dwarf star: *See* white dwarf.

Dynamic Steady State Universe (DSSU): The cosmology theory founded on the view that the space medium (a nonmaterial aether) is dynamic and that the medium expands and contracts *regionally and equally* resulting in a cosmic-scale cellularly-structured universe. It is defined by four key processes which provide a rationally coherent account of the major phenomena of our Universe.

Einstein shift: *See* gravitational redshift.

electromagnetic field: A region, surrounding a positive or negative charge, in which a radial pattern of aether excitation is accompanied by a process of aether absorption and vanishment.

electromagnetic force: The force arising between electrically charged particles or between charges and magnetic fields (and between magnetic poles). This force, intermediated by the photon, holds electrons to the nucleus of atoms, holds atoms together in molecules, makes moving charges spiral around magnetic field lines, and causes the alignment of magnetic dipoles.

electromagnetic wave: A disturbance or excitation consisting of oscillating electric and magnetic fields (lines of force) with directions at right angles to each other and to the direction of propagation. It is a form of radiation that transfers energy and momentum.

electron: A negatively charged subatomic particle with rest mass 9.1×10^{-31} kilogram (rest energy 0.511 MeV). Structurally, it is a double loop of a single-wavelength self-orbiting photon —a self-looping quantum of electromagnetic radiation.

elliptical galaxy: One of the primary kinds of galaxies. It is a smooth distribution of stars, spheroidal in shape (usually prolate along the axis of motion), and has practically no rotation.

emission lines: An emission line in a spectrum indicates a narrow range of wavelengths that is brighter than neighbouring wavelengths of the broader light-source spectrum. Emission lines are made by atoms or ions in a hot gas; often seen in the light from certain astronomical objects such as quasars. *Also see* absorption lines.

energy: (1) The capacity to do work. (2) Manifestation of a particular kind of force.

energy process: Any localized quantitative change in aether units. Energy, both mass-energy and radiation-energy, at the most fundamental level is manifest in the absorption-annihilation of units of the *space medium* (defined as *a nonmaterial aether*). Without this active process, neither mass nor radiation can exist.

entropy: is an increase in disorder, a trend towards thermal equilibrium. It represents a decrease in the useable forms of energy. For a closed or an isolated system, entropy is *not* conserved; it is *increasing* all the time.

(1) In Big Bang cosmology the source of low entropy is the expansion of the entire universe. The entropy is said to be forever increasing for the universe as a whole, however, this increase is incompatible with accelerated expansion and actually leads to a paradox.

(2) In DSSU cosmology the Universe is *not* a closed system: the source of low entropy is the perpetual emergence/expansion of the space medium. Entropy *increases* in the usual manner, while entropy simultaneously *decreases* via the process of aether deprivation mass extinction and the process of Blueshifting, both occurring within *Terminal stars*. Each cosmic cell of the DSSU behaves as an autonomous thermodynamic system.

escape velocity: The minimum speed, with respect to the gravitating body, that will allow an object to escape its gravitational "pull" (or gravitational field). For a primary body of mass M and radius R, and ignoring any frictional resisting force, the *escape velocity* magnitude is $\sqrt{(2GM/R)}$.

essence oscillators: The discrete units of the essence medium —a *nonphysical aether*. Crucially, these units represent neither mass nor energy. (A vitally important concept in DSSU theory.)

essence medium: A synonymous term for nonmaterial aether. It is the *subquantum* medium that permeates all space; that consists of non-mass, non-energy, oscillators.

ether: See *aether*.

Euclidean space: (1) Background nothingness space —space as an empty container. (2) Traditionally, flat space —space that is not curved (not distorted). (3) Analogously, a region in which *aether* is neither expanding nor contracting.

event horizon: The boundary at which the speed of aether flow, with respect to the center of the gravitating region and with respect to the background Euclidean space, is equal to the speed of light. Outside the boundary (that is, on the external side of the *horizon* surrounding the gravitating masses) the inflow of aether is less than lightspeed; while immediately inside the horizon, the inflow is *greater* than lightspeed. Event horizons are associated with *supermassive black regions*, often misleadingly called *supermassive black holes*.

expansion redshift: Among the various explanations hypothesized for the observed *cosmic redshift*, the expansion redshift was the one favored by most 20th-century cosmologists. The way it works is straightforward, lightwaves are stretched (redshifted) due to the expansion of the vacuum (or space medium) through which they are propagating. Incomplete as it was, this explanation was used to support the disastrous expanding-universe paradigm.

fermion: A particle with half-integer spin. The fundamental fermions are the matter particles such as electrons and quarks.

fundamental energy: See *energy process*.

fundamental fluctuators / oscillators: *See* essence oscillators.

fundamental force: The basic physical force of the Universe (the DSSU) is the *electromagnetic force*. And the fundamental physical particle, the carrier of the electromagnetic property, is the photon.

fusion: The combination of atomic nuclei to make heavier nuclei. If the collisions result from high-temperature interactions/collisions, it is called thermonuclear fusion.

galaxy: A basic cosmic structural unit containing roughly between 10^7 and 10^{11} stars, with gas and dust, all bound by gravity. There are five types: dwarf, elliptical, spiral, irregular, and Supergiant (found at the center of every *cosmic gravity cell*).

gravitational energy: The energy in the gravitational fields acting between masses. For two masses M and m a distance r apart, the gravitational energy is $-GMm/r$. Gravitational energy is considered to be negative; the conventional reason is that one must put energy into these objects to separate them to infinity where their energy would be zero. Under the DSSU paradigm, there is an additional reason (a fundamental one) for treating it as negative —there is an ongoing loss (a self-dissipation) of aether within contractile gravity regions.

gravitational lensing: The effect produced when the gravitational attraction of a massive object in deep space bends the light coming from a more distant object and focuses the light to create a magnified image (often distorted) of the more distant source. In the DSSU, this bending effect is caused by the *aether contracting "field"* surrounding the intervening galaxy or object.

gravitational redshift: Photons lose energy (increase their wavelength) when they move "uphill" away from a mass concentration. The effect is an example of the *velocity differential spectral-shift*, in which the front and back ends of the wave are propagating at slightly different speeds.

gravitation processes: (1) The *direct absorption / assimilation* of aether by all mass and all radiation; this process is the **primary cause of contractile gravitation**. (2) A process of the self-extinction of the space medium; the stress-induced vanishment of aether within contractile gravitation "fields"; this process is the secondary cause of gravity. (These processes also produce the property of mass or inertial mass.)

gravity: The effect that causes the acceleration of all entities towards the center of mass, and is nominally proportional to $1/r^2$ (i.e., the inverse square of the distance to the mass center).

(1) In general relativity it is the effect of the curvature distortion of spacetime produced by the presence of mass and energy.

(2) In a quantum description, it is a force mediated by gravitons —hypothetical massless bosons— acting as the force carriers.

(3) In DSSU cosmology it is the effect of the dynamics of the essence medium — specifically the accelerated flow, or inhomogeneous inflow, of aether towards, and into, matter. In the context of processes, gravity is the side effect of, primarily, the mass-manifesting process.

gravity cell (cosmic gravity cell): The cosmic-scale domain of a single unified gravitation region —a *gravity superdomain*. It is the autonomous region defined by a patterned flow of aether toward and into the nodal galaxy cluster that sits at the center (the cell's center of gravity). A cosmic gravity cell *defines the limits of the range of gravity* (or dynamic influence) of a nodal galaxy cluster. It is a region within which all objects stream toward the core of the galaxy cluster, and all matter (except escaping radiation and matter that has undergone Terminal annihilation) ultimately falls into the central supergiant galaxy. Gravity cells exist in two shapes, the tetrahedron and the octahedron.

hadron: Particles that are made of quarks; particles such as protons, neutrons, and mesons, and their antiparticles.

hertz (Hz): Unit of frequency; measured in number (of cycles) per second.

Higgs boson, Higgs field: In Old Physics, a hypothetical subatomic force-particle is believed to bestow the property of mass to all particles exhibiting this property; in aggregate, these Higgs "particles" constitute a related Higgs force field. As a mass acquisition concept, it is fatally flawed. It fails to explain how the Higgs particle acquires its own self-mass which is supposedly quite substantial.

homogeneity: The view, under Old Physics, is that the components of the universe are evenly and randomly distributed within the universe on the large scale. The New Physics places the emphasis on the Universe being systematically cellular.

Hubble constant: (1) In expanding-universe cosmology it is symbolized by H_o and represents the rate at which the recession velocity of galaxies increases with distance. The present value is roughly 22 km/s per million lightyears of distance; but since the expansion rate varies with the age of the Big Bang universe the Hubble expression is often written as "H" (without the naught) and is then called the Hubble parameter. (2) In the DSSU, H is the parameter that measures the rate of emergence/expansion of aether (as defined) within cosmic Voids and is expressed as the speed with which two comoving points, 1 million lightyears apart, undergo separation by the emergence of new aether. The value is about 10 km/s per Mly, but varies with the location within the Void environment.

Hubble's law: *See* redshift-distance law; and velocity-distance law.

hypothesis: A supposition or tentative explanation made on the basis of limited evidence (some set of observations) as a starting point for further investigation.

inflation: A mytho-science concept appended to the Big Bang creation hypothesis. It denotes a brief initial period of extremely fast (exponentially accelerating) expansion, involving a super-dense vacuum, and was supposedly the event or process that launched the expanding universe and its contents.

interface: The common region between the cosmic structural cells; the region of aether-streaming confluence; the region where comoving material collides. Interfaces surround each cosmic Void.

ionized: A condition in which one or more electrons have been removed from atoms, usually by energetic photons or collisions with other particles.

isotropy: The property of being the same in all directions. The applicability of this property depends on the cosmology model. *Also see* homogeneity.

Lambda force, +Λ: It is considered in conventional astrophysics as the 5th force of nature. It is often described as being a kind of *antigravity*, but its cause is unknown. This 5th force is the property of the *vacuum* or *aether* which, in an amazing coincidence, the Greek philosopher Aristotle called the 5th element. In DSSU cosmology: It is the force/effect that manifests within the cosmic Voids; and is caused (in part) by the fact that the *essence medium* across each Void is under tension. The positive Λ force is responsible for all large scale motion and subsequent angular momentum. Equivalently, it is the *negative pressure* present in the interior of each Void.

lepton: A lightweight subatomic particle that does not feel the strong force of the Standard model of particle physics. The known leptons are the electron, muon, tau, and their associated neutrinos; and their antiparticles.

lightyear: A measure of astronomical distance —the distance light travels in one year. 1 lightyear equals 9.47×10^{12} kilometers. (When converting to parsecs: 1 parsec = 3.26 lightyears.)

linear galaxy cluster: The filamentous aggregation of galaxies that exists at any *triple boundary region* where three cosmic cells meet each other. Along such 'meeting line,' galaxy structures from three neighbouring cells aggregate to form a concentration that extends from one node to another. Most often a *linear cluster* is observed as a branching arm of a *nodal galaxy cluster*.

Mach's principle: states that inertial mass and all inertial forces are due to the existence and distribution of all the matter in the universe. However, *cosmic gravitation cells*, because they limit the range of gravitation, make this an obsolete concept.

mass: An epiphenomenon of a more basic entity, an energy particle (or particles) in a state of spatial confinement (confinement in the sense of self-looping). Magnitude of the mass particle depends on the degree of the spatial confinement.

mass-energy: The energy $E_o = mc^2$ associated with a particle or object's mass. It is also called *rest energy*, because it is the energy of an object when it is stationary, with no kinetic energy.

matter density limit: *See* ultimate density state.

matter extinction: The extinction of matter, the total vanishment of matter, by the process of *aether deprivation*. According to the *matter extinction law*, when matter (mass and energy) is subjected to *aether deprivation*, it ceases to exist.

Mly: Mega lightyears or million lightyears; a measure of cosmic distance.

neutrino: The neutrino is a neutralized packet of radiant energy —a subatomic particle consisting of intertwined photons with a phase offset of π radians (180°). The offset is such that it results in the effective cancellation of their normal electric and magnetic effects.

meson: A class of subatomic particles which, according to the Standard model, contain two quarks (a quark and an antiquark).

MeV: Million electron Volts; a unit of energy.

microwave background: *See* cosmic microwave background.

nodal galaxy cluster: the multi-branched galaxy aggregation that occupies each vertex of the cosmic-scale structural cells (shaped as rhombic dodecahedra). Each nodal galaxy cluster is the dense central region of a *cosmic gravitation cell*.

nodal structure: the immense matter concentrations at the cosmic cell vertices. There are basically two types, characterized by the number of "arms" (*filament-like clusters*) that meet at a vertex. Minor nodes have 4-armed structures, Major nodes have 8 arms.

nova (event): A process in which a layer of hydrogen builds up on a white dwarf star and then fuses explosively into helium. Nova explosions do not destroy the star and may be recurrent.

nucleon: A proton or a neutron.

Old Physics: A term used to characterize the physics (the branch of science concerned with the nature and properties of matter and energy) that is specifically burdened by three major omissions. (i) It fails to recognize the existence of the mechanical space medium —a non-energy, non-mass, particulate aether. (ii) It fails to incorporate the velocity-differential propagation of light; and fails to employ the velocity-differential redshift for cosmological distances. (iii) It fails to recognize the unity underlying the several manifestations of gravity.

photon: A wavelike particle of radiation energy; it is the carrier of the electromagnetic force. (It is an energy particle that may be thought of as a laterally oscillating excitation of the aether while traveling in the longitudinal direction.)

photon conduction process: A wavelike conduction-excitation-disturbance of aether.

This "conduction" is unlike any other. The photon is conducted *by* aether in a manner that is destructive *of* aether.

Platonic solids: The five regular polyhedra —the tetrahedron, hexahedron, octahedron, dodecahedron, and icosahedron —esteemed by Plato and the Pythagoreans as embodying aesthetic and rational ideals.

polar portals: Openings in the energy surface of a Terminal star; openings through which photons and neutrinos can escape and manifest the Terminal star's polar emission beams.

positron: Another name for an antielectron.

Primary gravitation: The weak accelerating effect produced by the direct absorption of aether by mass —the conduction-absorption process that sustains mass particles and radiation particles and electromagnetic fields. The weakly accelerating inflow of aether so generated, in turn, is the cause of the considerably more potent *secondary gravity* effect.

proton: A subatomic particle (mass 1.67×10^{-27} kg; rest energy 938 MeV) with positive charge, consisting of three quarks (according to the Standard model). The number of protons present in an atomic nucleus defines the chemical element.

pulsar: First detected in 1967, pulsars are rotating neutron stars (as first proposed by Thomas Gold in 1968). The "pulses" are detectable when the emission beams repeatedly sweep across the line of sight to Earth.

radiation: Usually refers to any form of electromagnetic energy (photons) of any wavelength; as well as neutrinos.

radiation pressure: The force exerted mainly by photons when they strike a surface. In the terrestrial context, it is negligible. In the extra-terrestrial context, however, it clears the gas and dust from the stellar environment allowing stars to be observable in the optical spectrum.

reality: All reality is the interplay between two particles, photons (confined, self-looping, or freely propagating) and subquantum units (i.e., discrete aether oscillators).

redshift: A shift in the wavelength of electromagnetic radiation to a longer wavelength. For visible light, this implies a shift toward the red-colored end of the spectrum. Technically, it is the index z defined as the ratio $\Delta\lambda/(\lambda_{source})$, where λ is the wavelength. The redshift is used as a prized measure for determining cosmic distance. The five most often cited causes are Doppler (recessional motion), gravitational, space-medium expansion, velocity differential, and tired light. *See also* cosmic redshift.

redshift-distance law: (redshift index) = $(1/c)$ × (empirical constant) × Distance

$$z = (HD)/c.$$

redshift-distance relation: The correlation between redshift in the spectra of galaxies and their distances. The equation used by astronomers and theorists depends on the particular cosmological model. The Big Bang, a single-cell construct, and the DSSU, a multi-cellular model, use distinctly different formulae.

relativistic: Refers to particles or objects moving close to the speed of light, resulting in their total energy being much greater than their rest mass. In a relativistic gas, the particles have kinetic energy greater than their mass-energy.

relativity: Theories concerning the transformation of fundamental properties (space, time, length, mass) between different observers.

relativity, general theory of: Einstein's mathematical theory, incorporating the gravitational effect, in which space and time are geometrized. Applicable to accelerating frames of reference.

relativity, special theory of: Einstein's theory of the electrodynamics of uniformly moving frames of reference.

rhombus: A parallelogram with all sides of equal length.

Right Ascension (R.A.): A coordinate, measured in hours and minutes of time, for locating objects on the celestial sphere, analogous to longitude on the Earth's surface. The longitude-like position lines project onto the celestial sphere, dividing it into 24 slices, each 15 degrees wide.

rotation curve/graph: The graph of rotation speed (y-axis) versus distance from the center of the system (x-axis). When interpreted strictly within Newtonian physics, rotation curves yield information on the mass distribution within the system. The failure of the rotation speeds to drop at large radial distances in spiral galaxies leads to two radically different interpretations: (1) In Old Physics, it is evidence of the presence of additional mass, believed to be in the form of a *dark matter* halo. (2) In DSSU/cellular cosmology, it is evidence of the gravity-amplifying effect caused by *aether vanishment* associated with the shear stress of rotation; based on the validated aether theory of gravity.

Schwarzschild radius: A theoretically predicted dimension for the lightspeed boundary of a nonrotating mass body shrunk down (with a corresponding increase in density) to meet the "lightspeed surface" condition. It is a mathematical construct devoid of any connection to reality. When the concept is applied to a pre-collapsed mass/body, it fails as a prediction of the actual radius of the lightspeed boundary; it fails because it ignores Nature's density limit and overlooks *mass loss via aether deprivation*.

self-dissipation process: When stressed, aether, as defined within DSSU theory, undergoes a process of self-dissipation —a proportional vanishment— as it "strives" to maintain a constant *spacing density* (density in terms of the number of discrete aether "particles" per unit of background space).

singularity: A concept of 20th-century cosmology used to describe a point entity of infinite density —a point location where standard theories break down. Singularities do not exist in the real World.

space: A general term, in astrophysics, for *the vacuum, the quantum foam, the cosmic fabric*, etc. It is an ambiguous term for the ubiquitous medium of the universe.

space (DSSU): In the astrophysical sense, space is an empty container, the background nothingness, so to speak. It is the 3-dimensional background and is completely permeated by the universe's essence medium —a non-mass, non-energy, discretized aether. Space is a nothingness volume; it has no properties; none whatsoever. Its only function is to serve as an empty vessel of three spatial dimensions.

space curvature: *See* curvature of space.

space-medium contraction: All matter (mass and radiation) contracts the universal medium (DSSU aether), (1) directly through a process of *excitation-assimilation-*

annihilation and, (2) indirectly through a process of self-dissipation within a *contraction field* that surrounds each and every object or particle, individually and collectively.

space-medium dynamics: Refers to the *emergence/expansion* and *vanishment/contraction* aspects of the aether medium, as well as its flow. These aspects constitute a complete conceptual description of what sustains the cellular structure of the infinite Universe.

space-medium expansion: The *emergence/expansion* of aether within the cosmic Voids. Research into the DSSU points to a rate of expansion of about 10 kilometers/second across the span of one million lightyears distance.

special relativity speed rule: the rule that nothing can travel faster than about 300,000 kilometers per second *through* vacuum, or *through* aether.

spectrometer: An instrument that spreads the light from stars, or from any other source, into its different wavelengths.

spectrum: A graph of the intensity of light (or other electromagnetic radiation) against wavelength (or frequency).

speed of light: $c \approx 3.00 \times 10^8$ meters/second through vacuum or DSSU aether.

speed of light constancy: The speed of light is *absolutely* constant with respect to the space medium (aether) AND is *relatively* constant with respect to inertial (stationary of uniformly moving) observers. The inertial observer's measurement of the speed of light is always the same because his clocks slow down and distances appear compensatingly smaller. The measure of speed is simply distance over clocktime; and this ratio always remains the same under uniform motion.

spiral galaxy: One of the main (and most majestic) kinds of galaxy, with bulge and disk and spiral arms. Essentially, it is a large-scale *rotating cluster* of stars (along with gas and dust).

Steady State Expanding universe: A speculative universe (originally advocated by Hoyle, Bondi, and Gold in the 1940s) that undergoes perpetual uniform expansion while maintaining constant density and physical properties. Matter must be continually created to offset the expansion-caused dilution and maintain the constant density.

Steady State Nonexpanding universe: An infinite universe with constant density on the largest scale and constant physical processes and properties. Matter is continually being formed *and* annihilated. The space medium itself is continually emerging in certain regions *and* vanishing in other regions. Categorically, our DSSU is a Steady State Nonexpanding universe.

strong nuclear force: One of the conventional four forces of nature. According to Standard particle physics, the strong nuclear force holds the particles in the nucleus of atoms together.

superdomains of gravity: *See* gravity cell.

superfluidity: A condition in which a fluid has no friction. Technically, it is the absence of viscosity.

supermassive black hole: (A term not used in DSSU theory.) *See* Supermassive region.

Supermassive region: An extremely massive, extremely compact, *cluster of gravitationally collapsed objects* such as white dwarfs, neutron stars, and, of course, Terminal stars. (It is equivalent to what is conventionally, but wrongly, believed to be a *supermassive black hole.*). It is a rotating *noncontiguous* mass structure (essentially a *cluster* of discrete masses) surrounded by prodigious aether inflow. It can exist and persist only if there is significant rotation, for otherwise it would collapse into a single Terminal star within a matter of seconds.

Superneutron star: *See* Terminal star.

supernova: A cataclysmic stellar explosion. The two most common types are the thermonuclear detonation of a white dwarf star (used to determine distances to remote galaxies) and the core collapse of a high-mass star.

Terminal annihilation: the process of non-interaction vanishment of matter —the total negation of the affected mass/energy. The necessary and sufficient condition that brings about *Terminal annihilation* is *aether-deprivation*.

Terminal star: The structure resulting from a *total* gravitational collapse. It is a contiguous mass with a *critical state* boundary —meaning it has a lightspeed surface-boundary (where aether flows in at the speed of light but, because the "surface" has a pure energy layer, there is no violation of special relativity). It stands as the Universe's most unusual type of star —an *end state* neutron star. Once such a star forms, it can neither grow larger nor smaller. Its volume and mass-content remain forever fixed; its perfectly spherical shape is unaffected by rotation; its density is Nature's absolute ultimate. Synonymous terms: *Terminal neutron star, Terminal-state star, Superneutron star, end-state neutron star.*

Terminal star (in actively terminating state): A mass-accreting *Terminal neutron star*. The *terminating* process happens when, because of excess mass accretion, an insufficient quantity of aether I able to reach the core —making it an active *aether deprivation* core. Essentially, the Terminal star is absorbing more matter than it can sustain with a strictly limited supply of aether inflow. With an insufficient quantity of aether reaching the core, the matter within is subjected to extinction/annihilation.

Terminal state: The state that exists when we have the greatest quantity of contiguous matter within the least volume. It is the state of bulk matter, with the greatest density Nature will permit, enclosed by the least surface area (in compliance with special relativity).

theory: A rational self-consistent account of a wider range of phenomena than is ordinarily accounted for by a hypothesis.

thermonuclear fusion: The type of reaction in which atomic nuclei collide and combine, their collision energy (kinetic energy) arising from high temperature (thermal motion).

ultimate density state: The mass density existing within a totally collapsed star (*end-state neutron star*). It manifests when matter is in its degenerate state, when mass particles have lost all kinetic energy, when there is a total absence of thermal energy, when neutrons are in direct contact with other neutrons. Nature's maximum density state is taken to be 1.6×10^{18} kilograms per cubic meter.

unified gravitation region (unified gravitation cell): A cosmic region systematically maintained by three gravity effects. The central mass —a nodal galaxy cluster— serves as the *primary cause of gravity*. The cluster and its extended vicinity is the region where *contractile gravity* rules (where aether contraction occurs). This, in turn, is surrounded by regions where *divergent gravity* rules (where aether expansion/emergence occurs). It is called "unified" because all three effects are tied to the acceleration of the aether flow within the cosmic cell. All the flow trajectories end at the core of the central cluster. *See also* cosmic gravitational cell.

unified theory: In general, a theory that gathers a wide range of fundamentally different phenomena under a single precept. The DSSU *fundamental process of energy* is such a theory. The DSSU *aether theory of gravity* is another.

vacuum: In modern astrophysics, it is a generic term for the medium that permeates all space. (1) Twentieth-century orthodoxy deemed it to be a *quantum foam* with quantum fields that continuously spawn virtual particle-antiparticle pairs. (2) In the DSSU, it is a sub-physical *aether*, a subquantum essence.

vacuum energy (vacuum energy density): Quantum theory requires empty space to be filled with particles and antiparticles being continually created and annihilated. This leads to a net mass density of the *vacuum*, hence an energy density. The prediction for such hypothetical energy density is breathtakingly enormous. The problem is no such energy exists; there simply is no evidence.

velocity differential redshift: This refers to a change in the wavelength of light caused by the small difference in the flow of the space medium (the aether) in which the light is propagating. It is a wavelength change attributed to a small velocity difference between the front and back ends of the light wave. This spectral-shift mechanism is not merely a new discovery of the way light propagates across, or in and out of, gravity wells; it involves radical changes in our understanding of the Cosmos.

velocity-distance law: Big Bang's mathematical rule of cosmic expansion:

(recession velocity) = (empirical constant) × Distance

$cz = H \times D$,

where z is the measured redshift index, c is the speed of light, H is the Hubble constant, D is the distance of the light source at the time of reception (the "now" time). It applies only in expanding-universe models.

virtual quantum foam (or virtual foam): The sub-microscopic description of space, according to Old Physics, consists of virtual-real quantum particles (i.e., mass particles) and energy oscillators. Since space is said to be saturated with these mass and energy entities, the concept leads to the embarrassing prediction of unimaginably enormous energy density for the vacuum. The prediction is demonstrably wrong. (In contrast, with the sub-microscopic description of the space medium of the –DSSU, the discrete units of aether possess neither mass nor energy.)

weak nuclear force: One of the conventional four forces of nature. The weak nuclear force is responsible for radioactive decay as well as the fusion reactions in the Sun that provide heat and light for the Earth.

white dwarf: A dense star with a radius approximately the same as the Earth's but whose mass is comparable with the Sun's. White dwarfs burn no nuclear fuel and shine by residual heat. They are the end stage of stellar evolution for Sun-like stars. If, however, a white dwarf is in a binary system and the other star dumps matter onto it, the white dwarf can undergo thermonuclear detonation as a type-1a supernova explosion.

✠ ✠ ✠

Index

Bold page numbers indicate Glossary entries.

A

Abel-85 system, 208
absolute motion, 166, 167, 168
absorption lines, **312**
academic apostasy, 129
acceleration, **312**
acceleration of mass, 99
  See also energy triangle.
accretion disk, **312**
Adair, Robert K., 212
aether, **312**
  absorption/annihilation, 108, 148
  absorption/consumption, 99, 150
  acceleration, 34, 265, 273
  as ethereal medium, 262
  as fundamental unifier, 200, 259
  as sub-physical medium, 204
  as *subquantum* medium, 147, 263, 292
  as ultimate unifier, 204
  axiomatic emergence, 147, 148, 200, 276
  converging flow, 108, 270
  definitions, 241, **312**
  detected, 15, 46
  discrete units, 34, 132
  dismissed, 46
  drag, 163, 168, 169, 171, 172, 282
  dynamic, 271
  Einstein's, 34, 45, 46, 155, 243
  emergence/expansion, 217, 292
  flow, 34, 35
  flow profile, 63, 109, 136
  gain-and-loss balance, 204
  gain-loss balance Table, 205
  historic, 242, **312**
  in DSSU, 34, 132, 147, 242, 260
  neglected, 46
  nonmaterial fluid, 132
  nonphysical, 242, 243
  profound connection, 242
  self-dissipation, 148, 175, 199, 205, 271, 272, 290, 308, **323**
  self-dissipation equation, 272, 310
  self-dissipation, proof, 308
  spacing density, 263, 292
  types of stress, 263, 297
  units/entities, 190
  velocity equation, 34, 305
  velocity gradient, 38
  velocity profile, 69
  volume contraction rate, 309
  volume loss equations, 308
  volume reduction Table, 271
  vortex, **312**
aether deprivation, 10, 141, 142, 146, 148, 150, 225, 226, **312**
aether deprivation annihilation, 140, 142, 144, 147, 149, **312**
aether deprivation zone, 141, 142
aether emergence as *positive energy*, 200
aether flow
  convergent, 265, 270
  differential, 36
  equation, 305, 307
  greater than lightspeed, 169
  lightspeed, 70, 95, 108, 110, 135, 138, 169
  perpendicular inflow, 169, 282
  primary, 266, 267
  primary (graph), 269
  shear stress, 290

spiral inflow, 287
stress of convergence, 268, 271
tangential component, 160, 161
vorticular/torsional, 290
aether theory of gravity, 12, 94, 273
  gravity effects (Table), 298
  stress and strain (Table), 297
  the unifying Factor, 297
Albrecht, Andreas, 124
Aldebaran. *See* Taurus A.
angular momentum, 149, **313**
annihilation, **313**
annihilation of mass, 10
antigravity, **313**
antiparticles, **313**
Arp, Halton C., 26, 29
Arp's warning, 27
Asimov, Isaac, 214
astronomical redshift, 24
astrophysical jets, 9, 10, 60, 77, 116, **313**

B

background space, **313**
Bacon, Francis, 18
barycenter, **313**
baryon, **313**
Bellarmino bias, 122
Bellarmino, Italian cardinal, 122
biases and prejudices, 19, 252
Big Bang, **313**
binary star, **313**
black hole, 116, 127, 131, 149, 151, **313**
  "the nightmare legacy", 153
  angular momentum paradox, 116
  as mathematical object, 145, **313**
  event horizon, 116, 127
  formation, 249
  gravity paradox, 116
  illusory, 154
  nonexistence, 151, 152
  per Wikipedia, 117

  See also singularity.
  Stephen Hawking's view, 155
  supermassive, 153
black hole ... disaster, 249
Blueshifting mechanism, 70, 76, 80, 88
Blueshifting process, 9, 63, 64, 67, 219, 221
  and energy conservation, 76
  profound implication, 81
  summary, 81
boson, **313**
Brault, James W., 28
Brecher, Kenneth, 153
Burbidge, Geoffrey, 29

C

Cahill, Reginald T., 307
cancer of collectivism, 21
Capra, Fritjof, 179
Carroll, Sean M., 218
Cartan, Eli, 291
causeless process, 276, 301
Cavendish, Henry, 306
celestial sphere, **314**
cellular universe (DSSU), **314**
centrifugal effect, 13, 146, 157, 164, 166, 282, **314**
  attenuation, 162, 173
  attenuation graph, 163
  critical rotation graph, 159
  curtailment, 162, 164
  governing principle, 174
  key element, 173
  negation, 11, 167, 168, 169, 174, 287
  reduction, 284
  special relativity restriction, 159, 162
  strange historical perspectives, 179
  strange journalistic reviews, 181
  thought experiment, 172, 282, 283
centripetal effect, **314**
centripetal force, 310

CERN, 93
circular uniform motion, 310
civilization's cornerstone, 21
cognitive biases, 18
compact stars, **314**
confirmation bias, 18
connectedness, 301
conservation of angular
    momentum, **314**
conservation of matter, 149, 153
conservation principle, **314**
contractile gravity, **314**
contradictory conclusions, 83
convergent gravity, **314**
Copernican principle, 25
Copernicus, Nicolaus, 210, 248,
    259
cosmic background radiation, 39,
    53, 75, 85, 254, **315**
cosmic background temperature,
    48
cosmic cell, **314**
cosmic cell, schematic, 237, 277
cosmic cellular structure, 11, 238,
    240, 243, 255
cosmic climate change, 47, 48
cosmic conservation of energy,
    215
cosmic distance vs redshift
    graphs, 43
cosmic egg hypothesis, 47
cosmic emission jets, 80
    See also astrophysical jets.
cosmic energy conservation,
    highlights, 209
cosmic evolution, **315**
cosmic gravitational cell, **315**
cosmic jets. *See* astrophysical jets.
cosmic microwave background,
    315
    disputed, 89
    See cosmic background radiation.
cosmic ray particles collisions, 79
cosmic redshift, 8, 23, 24, 26, 27,
30, 53, 61, **315**
cosmic redshift interpretation
    "new principle of nature", 26
    "unrecognized principle of nature",
        29, 47
    basic Doppler effect, 24, 25
    controversy, 27
    dilemma, 26
    Einstein Shift, 27
    expansion of vacuum, 25
    flawed, 48
    gravitational, 27
    gravitational drag, 28
    overview Table, 29
    recession with expanding space, 26
    tired light, 27, 28
    velocity differential, 38, 40, **326**
    with fatal flaw, 29
cosmic structure
    "filamentary sine wave", 233
    "ribbon-like", 235
    "cold spot anomaly", 237
    cellular, 231, 275
    Coma Wall, 235, 236
    dodecahedral cells, 239
    extraordinary sequences of
        clusters, 238
    filamentary, 235
    holes, 235, 237
    Major node, 228
    Minor node, 228, 239
    parallel walls, 234, 235
    periodic linear clusters, 239
    predicted by DSSU, 229, 234
    regular cellular pattern, 233
    rhombic dodecahedra, 228, 229
    right-angled walls, 232
    Sloan Great Wall, 235
    sustainment, 229
    Veronoi honeycomb, 231
    void-cluster network, 230, 238
    wall-like, 233, 234
    walls of galaxies, 235, 236
    zone of divergence, 278

cosmic tension, 274, 275
cosmic theory, **315**
cosmic Voids, 187, 216, 219, 228, 235, **315**
  cross section, 188, 217
  sequence, 238
cosmic-scale energy conservation, 11, 201, 209
cosmic-scale gravity domains, 43
cosmological constant, 274, **315**
cosmological principle, 240, 244, **315**
cosmology, **315**
  big bang, 245, 248, 249, **313**
  expanding universe, 47, 228, **324**
  gatekeepers, 250
  illusions, 248
  masterpiece of misconception, 47, 48, 257
  mathematical, 247
  multi-cellular nonexpanding, 228
  nonexpanding universe, **324**
  orthodox, 247, 254
  preposterous, 53
  problem-free, 245
  relativistic, 47
  steady state, **324**
Cosmos, **316**
creation process, 80
critical-state boundary, **316**
critical-state structure, 87, 146
critical-state surface, **316**
curvature of space, **316**

D

dark energy mystery, 261
dark matter, **316**
dark matter mystery, 176, 287
Davies, Paul, 59, 106, 107, 213
degenerate gas/matter, **316**
density limit, **316**
Descartes, René, 185, 242
Dirac energy-state equation
  negative energy interpretation, 194, 195
  positive energy interpretation, 194
Dirac equation
  definition, 193
  flowchart, 194, 195
Dirac, Paul, 191, 247
divergent gravity, **316**
dodecahedral tessellation, 239
Doppler effect, 39, 40, **316**
DSSU aether theory of gravity, 298
DSSU cosmology, 189, **316**
DSSU theory, 84, 88, 120, 121, 122, 124, 125, 147, 209
Dyson, Freeman J., 51, 259

E

Eddington, Arthur S., 98, 116, 122, 150, 213
Einasto, Jaan, 231
Einstein shift. *See* gravitational spectral shift.
Einstein, Albert, 15, 98
  influenced by Mach's principle, 179
  on aether, 45, 150
  on aether motion, 46
  on dynamic vacuum, 46
  on light propagation, 46
  on rotation, 180
  on Schwarzschild radius, 122
Einstein-deSitter universe, 46
Einstellung effect, 18
electromagnetic field, **317**
electromagnetic force, **317**
electromagnetic wave, **317**
electron, 69, 91, 93, 100, 104, 196, 197, **317**
  magnetic dipole, 196
electron degeneracy pressure, 133
electron degeneracy state, 111, 133
emergence of matter, 154

emission lines, **317**
end-state neutron star, 71, 72, 78, 86, 108, 122, 137
  schematic, 107
  See also Terminal star.
end-state structure, 70, 105, 132, 136, 137, 144, 155
  See also Terminal star.
energy, **317**
  classification Table, 200, 201
  conversion event, 112
  electromagnetic, 190
  found in nature, 190
  foundational definition, 189
  fundamental process, 84, 99, 132, **317**
  had experts baffled, 189
  kinetic, 93, 95, 96, 99, 101, 187
  limitless energy flow, 222
  mass, 95, 189, 190
  mechanical, 191
  momentum, 95, 96, 99, 101
  of galaxy motion, 188
  of gravity fields, 190, 208
  of motion, 205
  of proton, 79
  of *vacuum*, 200
  photonic, 72, 114
  positive and negative, 191, 192, 204
  regeneration, 71
  rest energy, 96
  sub-physical component, 204
  surface-trapped, 114
  thermal, 105, 190
  total, 95
  total mechanical, 96, 100
  transfer, 224
  unification, 198
energy amplification process, 9, 61, 70, 71, 72, 77, 87, 220
  experimental proof, 80
energy balance in the universe, 150, 204, 207
energy conservation, 76, 188, 202
energy conservation of the Universe, 11, 144, 202, 207
energy triangle
  alternative interpretation, 102
  alternative manipulation, 101
  conventional manipulation, 100
  during mass conversion, 105
  graphic, 96, 97, 191
  mechanical, 96
  relativistic, 96
  schematic, 113
energy-momentum equation, 191
entropy, **317**
escape velocity, 135, **317**
essence medium, 20, 190, 264, 265, **318**
essence oscillators, **318**
essence process, 190
eternal natural order, 240
event horizon, 78, 106, 107, 115, 149, 151, **318**
existence of matter, 150, 262, 265
existence, subquantum level, 293
expansion function, 281
expansion hypothesis
  extrapolated, 47, 216, 275
expansion rate, 278, 279, 280
expansion redshift, 25, **318**
expansion-redshift law, 25

F

facts of reality, 85
Fairall, Anthony P., 232, 236
Feinberg, Gerald, 180
Feldman, Burton (historian), 249
Feynman, Richard, 17, 124, 193
First law of thermodynamics, 144, 153
fluid-flow continuity equation, 266
Folger, Tim, 286
frame dragging, 287
fundamental force, **318**

fundamental force/energy
  particle. *See* photon.
fundamental process of energy,
  205
fusion, **318**

G

galaxy, **318**
  apparent recession, 46, 47
  dark matter mystery, 176
  elliptical, **317**
  ESO444-46 (in Shapley), 232
  kinetic energy, 205
  M87 (in Virgo), 232
  mass-correction multiplier, 176
  NGC4874 (in Coma), 232
  rotation, 249
  rotation curve/graph, **323**
  simulated rotation, 175
  spiral, **324**
  supergiant elliptical, 232
  type *cD*, 232, **314**
  unnecessary *dark matter*, 177
galaxy clusters, 228
  Abell 85, 238
  AGC1656(Coma), 240
  core diameters, 231
  DC1842-63, 239
  distribution, 274
  linear, **320**
  Major node, 228
  Minor node, 228
  network, schematic, 229
  size, 276
  spatial periodicity, 238
  sustainment, 229
  two distinct sizes, 228, 229, 231
gamma-ray bursts, 223
Gamow, George, 28
Geller, Margaret J., 235
general relativity, 52, 253, 261,
  272, 291, 295, **323**
Glashow, Sheldon L., 46
Gödel, Kurt, 180

Gödel's strange universe, 180
Gott, Richard, 244
Grant, Darren, 223
gravitation
  acceleration, 66
  acceleration curve, 33
  acceleration differential, 33, 67
  acceleration expression, 66
  aether-based, 263
  affects radiation, 66
  as negative energy, 199
  causal mechanism, 34, 38
  collapse, 70, 94
  convergent, 272, 280, **314**
  defined, 265
  direct cause, 265
  divergent, 274, 280, **316**
  divergent, schematic, 275, 279
  five manifestations, 259
  force gradient, 32
  general-relativity view, 272
  intensification, 290, 291
  Newtonian, 268, 310
  Newton's problematic premise,
    176
  only three models, 227
  potential differential, 33, 67
  Primary, 263, 267, 268, **322**
  Primary convergent, 266
  secondary, 148, 273
  secondary convergent, 268
  tertiary, 148
  ultimate manifestation, 288
  vorticular effects, 281, 287
  vorticular stress, 289
gravitation processes, **319**
gravitational
  acceleration, 305
  amplification effect, 284, 286
  apparent mass, 103, 285, 311
  apparent mass formula, 311
  energy, **318**
  heating, 95, 114
  lensing, 32, 66, **319**

merger, 143
potential energy, 105, 114
spiral merger, 294
gravitational collapse, 73, 86, 95, 101, 105, 106, 132, 133
  6-solar-mass star, 138, 139, 140
  and density increase, 133
  and magnetic fields, 73
  and rotation, 73, 146
  and surface mass, 122
  denied, 86, 87, 89
  extreme, 132
  final stage, 135
  general-relativity view, 106
  graph, 110
  natural, 108
  of massive star, 102
  of very massive star, 132
  per general relativity theory, 106
  schematic, 107, 110, 134
  thought experiment, 110
  total, 106, 148
  views compared, 149
gravitational redshift, **319**
gravitational spectral shift, 27, 39
gravitons, 120, 121
gravity, **319**
gravity cell (cosmic), **319**
gravity paradox, 149
gravity theory, stress and strain, 295
gravity waves, 13, 292
gravity waves schematic, 293
gravity waves, key points, 294
gravity well of Sun, 41
gravity well schematic, 31, 32
gravity well/sink, 8, 33, 35, 36, 37, 55, 63, 67
Greene, Brian, 261
Griffiths, David, 193, 194
Gursky, Herbert, 107

H

harmony of opposites, 202, 240, 276, 281
Harrison, Edward R., 76, 127, 144, 212, 216, 258, 272, 295
Hawking, Stephen, 155
Helmholtz, Hermann von, 185
hertz (Hz), **319**
Higgs boson / field, **319**
Higgs field, 147
Higgs mechanism, 147
historical evidence, 122
Horgan, John, 17
Hu, Wayne, 43
Hubble constant, **320**
Hubble, Edwin P., 8, 24, 47
Hubble's law, **320**
Hubble's perplexing problem, 46
Hubble's warnings, 25
Huchra, John P., 235
Humason, Milton, 24

I

IceCube Neutrino Observatory, 79, 144, 222
inertia, 179
inflation, **320**
initial conditions, 255
interface, **320**
isotropy, **320**

J

Jayawardhana, Ray, 79
Jeans, James H., 92, 195, 263
Journals
  Advances in Astronomy, 183, 253
  American Journal of Astronomy and Astrophysics, 182
  Frontiers in Astronomy and Space Sciences, 251
  Galaxies MDPI, 254
  International Journal of Astronomy and Astrophysics, 52, 84, 181, 252
  International Journal of Astrophysics and Space Science,

182, 210, 251
Journal of High Energy Physics, Gravitation and Cosmology, 210
Nature, 155, 294
New Scientist, 223
Physics Essays, 52, 54, 55
Physics International, 182
Physics Letters A, 51
Progress in Physics, 183, 211
Scientific American, 52, 170, 179, 180, 237
Space Science International, 211, 251
The Astrophysical Journal, 253

K

Kajita, T., 40
Kaku, Michio, 153
Kant, Immanuel, 230
Kearns, E., 40
Klein, Spencer, 79
Kraan-Korteweg, Renée C., 232

L

Lambda force (Λ), **320**
Landau, Lev, 116, 150
large scale structure. See cosmic structure.
law of Blueshift accrual, 60, 61, 87
  bonus feature, 80
  force-based proof, 66
  no limit, 78
  proof, 63
  solves mystery, 81
law of energy amplification, 60, 61
law of energy conservation of the Universe, 206
  See also cosmic-scale energy conservation.
law of mass extinction by aether deprivation, 10, 148, 154
law of noninteraction mass-to-energy conversion, 115
law of redshift accrual, 60

law of velocity differential propagation, 57, 59, 60, 71
  See also principle of velocity differential propagation.
Levin, Janna, 298
light propagation, 14, 15
  overlooked mechanism, 30
light quantum. See photon.
lightspeed boundary, 10, 68, 74, 101, 106, 108, 111, 137, 142, 149, 174, 223, 288, 323
lightyear, **320**
LIGO, 293
Lorentz gamma factor, 103
Lundmark, Knut, 24

M

Mach, Ernst, 166
  his oversight, 180
  his rotation hypothesis, 167, 179, 287
  his strange universe, 179
  on inertia, 179
Mach's principle, **320**
magnetic channels, 73
magnetic energy columns, 74
magnetic field, 78
  becomes collimated, 73
  collimated, 222, 223
Magueijo, João, 124
Mark, M. B. van der, 196
Maselli, Andrea, 125, 126, 128
mass, **321**
mass as confined photons, 69, 264
mass as *negative energy*, 196
mass energy conversion to motion energy, 101
mass equivalence, 92
mass extinction process, 203, 225, 226
mass extinction/annihilation, 112, 143, 144, 146, 147
mass property acquisition, 12, 147, 263

mass regeneration/formation, 224
mass vanishment, 141, 225
mass varies with speed, 98, 103, 104
mass-energy, **321**
mass-producing collision, 224
mass-to-energy conversion, 9, 69, 70, 83, 104
  aspects, 114
  denial, 83
  implications, 115
  involving collision, 111, 113
  mechanism, 105, 109, 115
  noninteraction, 94, 115, 119
  nuclear fission, 91
  nuclear fusion, 91
  particle-antiparticle annihilation, 91
  schematic, 105
mathematical universe, 247
matter as *negative energy*, 196, 198
matter density limit, **321**
matter extinction, **321**
matter regeneration, 219, 221, 222, 227
matter sustaining process, 155, 205
matter, sustaining its existence, 132, 133, 140, 148, 155
Maxwell, James Clerk, 242
Michelson, Albert A., 15, 16, 17
Michelson-Morley experiment, 14, 15, 255
Miller, Dayton, 15, 16, 46
Mitsuda, Kazuhisa, 231
momentum conservation, 98
momentum energy transfer, 80
Morrison, Philip, 153
Mössbauer effect, 28
myth of universal cosmic expansion, 185
mytho-science views, 212

N

natural eternal order, 241
negative energy processes, 207
Neutrino Observatory, Antarctic, 79
neutrino, schematic, 198
neutrinos, 40, 41, 52, 76, **321**
  ultrahigh energy, 79, 80, 224
neutron, 79, 134
neutron degeneracy pressure, 114
neutron degeneracy state, 111, 138
neutron density, 123
neutron star, 68, 123, 125, 141, 221
neutronium, 106, 111, 138
Newton, Isaac, 164
  his water-bucket experiment, 165, 166, 287
  on rotation as absolute motion, 164, 166
Nobel Physics Award
  1907, 15, 17
  1978, 47
  1993, 290
  2004, 93
  2011, 48, 248
  2019, 48, 176, 249
  2020, 249
nodal galaxy cluster, **321**
nova (event), **321**
Novum Organum, 18
nuclear degeneracy state, 133
nuclear fusion, 91

O

Ockham's razor, 256
Olbers' Paradox, 48
Old Physics, 78, 97, 181, 253, **321**
Ostriker, Jeremiah, 175
Ota, Naomi, 231

P

particle collisions, 79

particle creation, 80
particle physics history, 191
particle-antiparticle annihilation, 91
particles
  acceleration, 93
  acceleration to lightspeed, 92
  antielectron, 91, 197
  antiparticles, **313**
  antiproton, 93
  as *negative energy*, 196
  baryon, **313**
  boson, **313**
  electron, 193, 196, **317**
  fermions, 195, **318**
  hadron, **319**
  ionized, **320**
  lepton, **320**
  meson, **321**
  most fundamental, 207
  neutrino, 197, 198, **321**
  nucleon, **321**
  PI mesons, 93
  pion, 93
  positron, 93, 192, 195, **322**
  propagating in-place, 70, 72
  proton, **322**
  spin ½, 194
Pavlovian conditioning, 19
Pavlovian scientists, 20
Peebles, Philip James Edward, 48, 175, 249
peer review, 7, 51, 154, 250, 251
Penrose, Roger, 7, 249
Penzias, Arno, 47
Perlmutter, Saul, 248
perpetual motion machine, 76
photon, 30, 32, 66, **321**
  as aether excitation, 265
  as *negative energy*, 196
  bombardment, 93
  circularly polarized, 196, 197
  conduction process, **321**
  confinement, 69, 195, 264
  confinement, graphic, 197
  energy balance Table, 203
  energy gain-loss, 203
  fundamental energy particle, 265
  fundamental force particle, 72, 264
  gamma, 91
  momentum transfer, 80, 94
  principle of *loop completion*, 196
  propagating in-place, 221
  propagation, 199, 265
  surface-embedded, 220
  theory of particles, 72, 195, 196, 264
  ultrahigh energy, 80, 223
  unifies *negative energy*, 199
Pioneer-6 anomaly, 41
Platonic solids, **322**
polar portals, 75, 76, 77, 144, 149
Positive energy process, 205, 207, 274
Pound, R. V., 28
primacy of mathematics, 84
primacy of processes, 84
Primary Cause process, 200, 205
principle of centrifugal effect negation, 146, 170
Principle of *mass extinction by aether deprivation*, 148
principle of mass variance, 111
principle of *velocity differential propagation*, 30
  Cartesian-coordinate-system proof, 55
  cause of cosmic redshift, 44
  corollary, 30
  dynamic-aether proof, 34
  force-effect proof, 32
  integral-calculus proof, 31
  overlooked, 48
  prediction, 39
  summary, 81
  validity discussion, 44
process theory of gravity, 120
proton, 79, 93, 104, **322**

publication rejection, 250
pulsar, **322**
pulsar PSR 1913+16, 290

Q

quantum foam, 14
quantum theory, 124

R

radiation pressure, **322**
radioactive decay, 91
Ratcliffe, Hilton, 19
realms of existence, 80
Rebka Jr., G. A., 28
redshift compounding, 38
redshift defined, 38, **322**
redshift factor, 38
redshift in gravity well, 57
redshift interpretation. *See* cosmic redshift interpretation.
redshift misinterpretation, 8
redshift-distance law, 24, **322**
relativistic aberration, 170, 171
relativity
    fixed-mass view, 98
    general theory, 106, 115, 126, 145, **323**
    Lorentz gamma factor, 103
    relativistic mass, 98
    special theory, 97, 98, 120, 152, **323**
relativity theory, 45, **323**
    incomplete, 127
rest mass, 98, 103
revolutionary development, 148
rhombic dodecahedron, 230, **316**
    profile, 230
Riemann, Georg Bernhard, 105, 123
Riess, Adam G., 248
Right Ascension (R.A.), **323**
Rindler, Wolfgang, 123
rotation. *See* centrifugal effect.
rotation and aether drag, 172

rotation drag effect, Table, 284
rotation, absolute vs relative, 164
Rubin, Vera C., 176
Rudolf, Germar, 21

S

Sachs–Wolfe effect, 43
Schmidt, Brian P., 248
Schwarzschild radius, 122, **323**
Schwarzschild singularity, 151
Schwinger, Julian, 15
science, end of, 17
science, goal of, 256
Scientific Age, twilight, 17
scientists, 19, 256
Second law of thermodynamics, 76
secret of the Universe, 265
Seife, Charles, 131
singularity, 106, 107, 108, 115, 149, **323**
singularity paradox, 149
Slipher, Vesto M., 24
Snider, J. L., 28
space, **323**
space, background, **313**
space, definitions of, 20
space, definitions Table, 262
space, DSSU, **323**
space, Einstein's, 261, 262
space, Euclidean, 262, **318**
space, Newtonian, 261, 262
space-filling shapes, 229, 234, 244
space-medium contraction, **323**
space-medium dynamics, **324**
space-medium expansion, **324**
spacetime continuum, 14, 16, 25
spectral shift, 23, 24, 81
spectrum, **324**
speed of light constancy, **324**
spiral galaxies problem, 286
steady state universe, 13, 20, 80, **324**
steady state Universe, 219

stellar collapse, 105
stress and strain
   in DSSU gravity (chart), 296
   in Einstein's gravity (chart), 296
strong nuclear force, **324**
Stückelberg, Ernst, 193
subquantum medium, 147
subquantum phenomenon, 294
supergiant elliptical galaxy, 232
supermassive black hole, **324**
supermassive region, **325**
Superneutron star (SnS). *See* Terminal star.
supernova, **325**
Susskind, Leonard, 145, 216

T

Taurus-A Experiment, 43, 52
Taurus-A radio source, 41
Tegmark, Max, 19, 247, 248
temperature of background universe, 48, 53
Terminal annihilation of matter, 142, 143, **325**
Terminal star, 9, 10, 61, 68, 69, 70, 71, 109, 137, 149
   active terminating state, **325**
   and astrophysical jets, 77
   and conservation law, 76, 143, 144
   and entropy rule, 76
   and infalling particle, 113
   and mass infall, 141
   and *relativistic aberration*, 170
   as energy generator, 74, 144
   defined, 142, 288, **325**
   diameter, 145
   emission beams, 75, 80, 142, 149
   energy escape mechanism, 74, 75
   energy intake, 75
   energy layer, 71, 72, 74
   energy outflow, 77
   escape mechanism, 223
   magnetic beams, 222
   magnetic channels, 75, 222
   merger, 142, 143
   no rotation limit, 171, 172
   polar ejection, 112
   polar magnetic channels, 73
   polar portals, 75, **322**
   radius, 140, 308
   radius calculation, 307
   regenerates energy, 214
   rotation, 73, 146
   rotation, schematic, 289
   surface, 110
   unaffected by self-rotation, 146, 168
   where found, 225
   with rotation, 74, 169, 171, 172
Terminal state, 69, 73, 137, 138, 141, 143, **325**
terminology, 20
Tesla, Nikola, 263
theories, two kinds, 260
theory, **325**
theory subsummation, 298
thermodynamics, First law, 144, 153
thermodynamics, Second law, 76
thermonuclear fusion, **325**
totalitarianism, 21
Totsuka, Y., 40

U

ultimate density, 106, 134, 135, 137, 138, 139, 145, 307, **325**
ultimate electromagnetic barrier, 137
ultimate mass density, 136
unification, 259, 260
unified gravitation cell, **325**
unified gravity theory, 260
unified theory, **326**
universal medium
   stress and strain (Table), 297
   summary of stresses, 296
universe expansion myth, 185, 256

universe genesis myth, 217
Universe's source energy, 9

V

vacuum, 14, 15, **326**
vacuum energy, 16, 261, **326**
validity judgment, 13
Vaucouleurs, Gérard de, 231
velocity differential Blueshifting, 9, 70, 76, 80
velocity differential redshift, 37, 40, 56, 57, 218, **326**
velocity differential redshift evidence, 41
velocity-differential mechanism, 8
velocity-distance law, 24, **326**
Viking Mars-missions anomaly, 42, 52
virtual quantum foam, **326**
Vishniac, Ethan T., 253
voids. *See* cosmic Voids.
Voth, Grant L., 213

W

Walker, Evan Harris, 242
Wang, Ling Jun, 83
wavelength, 57, 65
  See also spectral shift.
weak nuclear force, **326**
white dwarf star, 111, 133, **326**
White, Martin, 43
Whittaker, Edmund, 14
Wilczek, Frank, 93, 224
Williamson, John G., 196, 264
Wilson, Edward O., 303
Wilson, Robert, 47
Wirtz, Carl W., 24
Wright, Edward, 28

Z

zone of divergence (graph), 280
Zwicky, Fritz, 28

Printed in Poland
by Amazon Fulfillment
Poland Sp. z o.o., Wrocław

26248675R00198